Modern Methods in Protein Chemistry
Review Articles Volume 3

# Modern Methods in Protein Chemistry

## Volume 3

## Review Articles

Editor
Harald Tschesche

Walter de Gruyter · Berlin · New York 1988

*Editor*

Harald Tschesche, Dr. rer. nat.
Professor für Biochemie
Lehrstuhl für Biochemie
Fakultät für Chemie
Universität Bielefeld
D-4800 Bielefeld
Federal Republic of Germany

*CIP-Kurztitelaufnahme der Deutschen Bibliothek*

**Modern methods in protein chemistry :**
review articles ... – Berlin ; New York : de Gruyter. Vol. 3 (1988)
ISBN 3-11-011216-7 (Berlin ...) Kunststoff (Pr. nicht mitget.)
ISBN 0-89925-298-2 (New York) Kunststoff


Printing: Gerike GmbH, Berlin. Binding: Lüderitz & Bauer GmbH, Berlin. Printed in Germany.

# Preface

Two books on "Modern Methods in Protein Chemistry" were published in 1983 and 1985 by de Gruyter Publishers. This third volume intends to continue the efforts to survey the present but rapidly developing field of analytical methods available to the protein chemist for isolating, analyzing and characterizing different proteins. It is the aim of this book not to give a complete listing of all efforts made in a particular field, but rather to offer the inexperienced investigator an opportunity to orient himself amongst the literature and to evaluate the chosen method in accordance with his special needs.

The exchange of methodological knowledge between investigators is one of the impulses for scientific progress. This perception was one of the main reasons for establishing the study group for chemical protein analysis within the "Gesellschaft für Biologische Chemie", which kindly supported the last Conference on "Modern Methods in Protein Chemistry" held at the Center of Interdisciplinary Research (ZIF) in Bielefeld, November 6-8, 1986. Some of the papers were given during this meeting and others have been added by later invitation.

Because of the growing importance of genetic engineering techniques for the study and preparation of proteins in the future, the protein chemist has to adopt methods of nucleic acid research and gene cloning techniques. This is the reason for incorporating a chapter dealing with Electrotransformation of Culture Cells.

It is hoped that this book will help to extend methodological knowledge and to facilitate the approach to actual scientific problems.

Bielefeld, January 1988 Harald Tschesche

# CONTENTS

# STRATEGIES FOR THE LOCALIZATION AND SYNTHESIS OF PROTEIN BINDING SITES

M. Zouhair Atassi

Marrs McLean Department of Biochemistry, Baylor College of Medicine, One Baylor Plaza, Houston, Texas 77030, U.S.A.

## Introduction

Definition of the structural features that constitute a given protein binding site is crucial for the understanding of how the protein discharges its function. Furthermore, strategies that would go beyond the level of observation to the level of duplication, by synthetically mimicking protein binding sites, will open up powerful untapped avenues in the exploitation and manipulation of protein activities.

Our initial efforts focussed on the delineation and synthesis of the sites that are involved in the binding of a protein with its antibodies (1,2). Such sites (termed antigenic sites) have served as prototypes for the synthetic mimicking of other protein binding sites. These efforts resulted in the determination of the first complete antigenic structures of several proteins (2-11).

Protein antigenic sites can be categorized into two architectural alternatives (2,3,12). First, a site may occupy a conformationally sensitive surface region of a protein consisting of a continuous segment of the polypeptide chain. Sites exhibiting this architecture have been termed continuous antigenic sites (13) and are illustrated by the antigenic sites of myoglobin (2) and hemoglobin (9-11). Second, an antigenic site may occupy a conformationally dependent surface region of a protein composed of amino acid residues which are distant in sequence but are brought into close spatial proximity by the folding of the polypeptide chain. Thus, the residues involved in the site are not generally directly linked by peptide bonds and such sites are termed discontinuous (13) as exemplified by the antigenic sites of lysozyme (13).

Modern Methods in Protein Chemistry, Vol. 3

The strategies for the localization and synthesis of protein binding sites will be subdivided to describe, in Sections I-III, the approaches that have been devised for the continous antigenic (as well as other protein binding) sites. Section IV will describe the approach devised for the synthetic mimicking of the discontinuous protein-binding sites.

## I. General Chemical Strategy

The complexity of protein molecules has made the determination of their antigenic sites difficult. Early studies from this laboratory, therefore, charted a strategy (1,2) that employed the following five distinct approaches which must be applied in concert to solve the problem: (a) Studying the effects of conformational changes on the immunochemistry of the protein; (b) Examination of the immunochemistry (or other biological activity) and conformational changes of chemical derivatives of the protein; (c) Preparation, isolation and identification of immunochemically reactive overlapping peptide fragments which account for the anti-protein response; (d) Chemical modification of immunochemically reactive peptides at selected residues and study of their immunochemistry and conformation; (e) Verification of the deduced locations of the antigenic sites by synthesis of the indicated regions of the protein. It must be emphasized that each of these approaches has advantages and deficiencies (for details, see refs 2,14,15) which makes each approach incapable of yielding the full antigenic structure of a protein. Therefore, it is necessary to use the results from one approach to confirm and/or correct the results from the others. The antigenic profile is a composite, logical picture of all of the findings.

In addition to the delineation of antigenic sites, the above strategy can be adapted, with appropriate modifications, for the localization and synthesis of other types of protein binding sites.

### 1. Myoglobin

Native sperm whale myoglobin has five major continuous antigenic sites which account for greater than 98% of the antibody response to the protein (2,15-17). The sites, which occupy conformationally-sensitive surface locations, are: (Site 1) residues 16-21, located on the bend between helices A and B. Depending on the antiserum, this antigenic site exhibits a 'shift' or 'displacement' or one residue on either side of the site; (Site 2) residues 56-62, located on the bend between helices D and E; (Site 3) residues 94-99, on the bend between helices F and G; (Site 4) residues 113-119, on the end of the helix G and partly on the bend GH: (Site 5) residues 146-151 ($\pm$ lysine 145 with some antisera), on the end of helix H and partly on the randomly coiled C-terminal pentapeptide. The covalent structures of the five antigenic sites are shown in Figure 1 and their three-dimensional locations in myoglobin are shown in Figure 2.

| Site | Structure and Location |
|---|---|
| | 15 16 21 22 |
| 1 | (Ala)-Lys-Val-Glu-Ala-Asp-Val-(Ala) |
| | 56 62 |
| 2 | Lys-Ala-Ser-Glu-Asp-Leu-Lys |
| | 94 99 |
| 3 | Ala-Thr-Lys-His-Lys-Ile |
| | 113 119 |
| 4 | His-Val-Leu-His-Ser-Arg-His |
| | 145 146 151 |
| 5 | (Lys)-Tyr-Lys-Glu-Leu-Gly-Tyr |

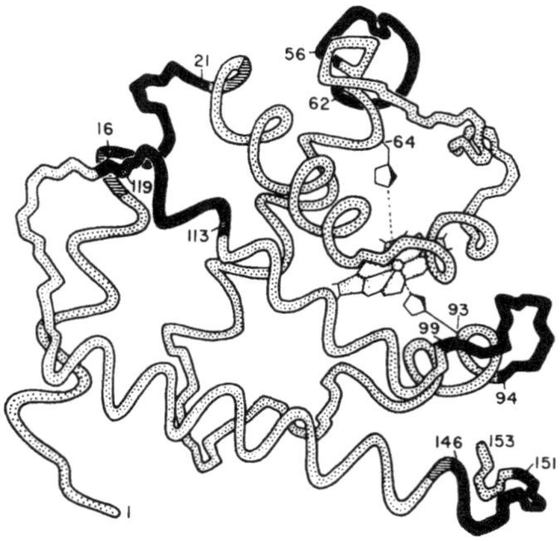


Figure 1: (Left) Covalent structure of the five antigenic sites of sperm-whale myoglobin. Residues in parentheses are part of the antigenic site only with some antisera. For example, the reactive region of site 1 invariably comprises residues 16 to 21 and with some antisera Ala-15 as well (the site would then correspond to residue 15 to 21). With other antisera, Ala-22 is also an essential part of the site (the antigenic region would then correspond to residue 16 to 22). Thus, the site occupies either six or seven residues depending on the antisera examined. A shift in site 2 by two residues to the left was recently found with monoclonal antibody to myoglobin. No such 'shift' or 'displacement' has been yet observed in sites 3 and 4 in those antisera studied so far. In the case of site 5, Lys-145 can be part of the antigenic site only with some antisera. Therefore, this site will comprise six or seven residues, once again depending on the antisera. (From ref. 2).

Figure 2: (Right) A schematic diagram showing the folding of the myoglobin molecule and its antigenic structure. The solid black portions represent segments which have been shown to accurately comprise entire antigenic sites. The striped parts, each corresponding to one amino acid residue only, can be part of the antigenic site with some antisera. The dotted portions represent parts of the molecule which have been shown exhaustively to reside outside the major antigenic sites. (From ref. 2).

The antigenic sites are surprisingly small consisting of 6-7 adjacent residues (19-23 Å in their extended dimension). Each site also possesses discrete boundaries which may be slightly shifted with some antisera. Boundary shifts have been found, so far, with site 1 (18), Site 2 (6,19), Site 3 (20) and Site 5 (21). The sequence and three-dimensional structural features that confer immunogenicity on these regions, however, are not clear at the present time despite attempts by many workers to provide explanations. It should be noted that the immune responses to myoglobin are under H-2 linked Ir gene control (22) with the responses to each site being under separate genetic control (23,17).

The binding of antibody to myoglobin is sensitive to conformational changes. The finding that conformational changes play a significant role, even when triggered in regions outside of the antigenic sites (24,25), and the results obtained with numerous peptide fragments have lead to the conclusion (1,15,24) that the antibody response is directed against the native-dimensional structure of proteins. Although these conclusions were initially derived from studies with early course antisera, they were subsequently shown to apply to antisera obtained up to a year after initial immunization (26). For detailed accounts of the antigenic structure of myoglobin and its implication, references (2,4,6,15) may be consulted.

2. Serum Albumin

Serum albumin has been shown to contain six major antigenic sites which account for the bulk of the antibody response to this protein. These sites have been localized by the general chemical strategy and confirmed by synthesis (5,7,8,27). Sites reside within, but do not necessarily include all of the regions shown in Figure 3. The precise boundaries of the sites, have not yet been determined.

The antigenic sites of albumin include both architectural types. Sites 1,2 and 3 in both bovine and human albumin comprise adjacent residues of the polypeptide chain residing in the third, sixth, and ninth subdomains (disulfide double loops) far from the disulfide bonds and therefore are 'continuous' antigenic sites. On the other hand, sites 4, 5 and 6 can be categorized as 'discontinuous' sites since they comprise the regions on both sides of the disulfide bonds 166-175, 314-359, and 565-566, respectively, in bovine albumin and the bonds 168-177, 315-360 and 566-557, respectively, in human albumin. All of the sites are rich in polar and hydrophilic amino acids which is compatible with an exposed surface location. However, the conformational locations of these sites cannot yet be discussed since the three-dimensional structure of the albumin molecule is not yet known. It is noteworthy that the antigenic sites in the two albumins occupy equivalent structural locations.

The antigenic sites demonstrate a similarity in binding function which is a reflection of the similarity in structure (Figure 3). Sites 1,2 and 3 in a given albumin show this similarity in binding function and exhibit an increase in magnitude of immunochemical cross-reaction with progressively later antisera. Sites 4,5 and 6 also exhibit a functional similarity and their cross-reactivity (antibody binding) increases with progressively later bleedings after initial immunization. As recently shown (28), the cross-reaction at the antibody level is accompanied by extensive cross-reaction at the T-cell level.

```
Site 1:
      137                                     146
BSA:  Tyr-Leu-Tyr-Glu-Ile-Ala-Arg-Arg-His-Pro
      138                                     147
HSA:  Tyr-Leu-Tyr-Glu-Ile-Ala-Arg-Arg-His-Pro

Site 2:
      328                                     337
BSA:  Phe-Leu-Tyr-Glu-Tyr-Ser-Arg-Arg-His-Pro
      329                                     338
HSA:  Phe-Leu-Tyr-Glu-Tyr-Ala-Arg-Arg-His-Pro

Site 3:
      526                                     535
BSA:  Ala-Leu-Val-Glu-Leu-Leu-Lys-His-Lys-Pro
      527                                     536
HSA:  Ala-Leu-Val-Glu-Leu-Val-Lys-His-Lys-Pro

Site 4:
      168                                             179
BSA:  Gln-Ala-Glu-Asp-Lys-Gly-Ala-Cys-Leu-Leu-Pro-Lys
      170                                             181
HSA:  Gln-Ala-Ala-Asp-Lys-Ala-Ala-Cys-Leu-Leu-Pro-Lys

Site 5:
      308                      314 359          362
BSA:  Ala-Glu-Asp-Lys-Asp-Val-Cys Cys-Ala-Lys-Asp
                                S---S
      309                      315 360          363
HSA:  Val-Glu-Ser-Lys-Asp-Val-Cys Cys-Ala-Ala-Asp
                                S---S

Site 6:
      559                      565 556          553
BSA:  Ala-Asp-Asp-Lys-Glu-Ala-Cys Cys-Lys-Asp-Val
                                S---S
      560                      566 557          554
HSA:  Ala-Asp-Asp-Lys-Glu-Thr-Cys Cys-Lys-Glu-Val
                                S---S
```

Figure 3: Structure and location of the six regions of bovine serum albumin (BSA) and human serum albumin (HSA) that we have shown (5,7,8) to carry antigenic sites. It is not implied that the antigenic sites comprise the entire size of the regions shown, but rather that they fall within these regions. (From ref. 8)

## II. Synthetic Overlapping Peptides Encompassing the Entire Protein Chain

The systematic study of overlapping peptide fragments is a key approach (1) which can yield valuable information pertaining to the number, locations, and relative contributions of antigenic sites to the immunoreactivity of the whole protein. This approach has several inherent difficulties (for review see refs. 14,15) that include: the availability of a limited number of suitable cleavage procedures which consequently restricts the spectrum of overlapping peptides that can be generated; the risk of inadvertent scission within an antigenic site which would render these regions inactive or only slightly reactive; the potential with chemical cleavage procedures for modifications of internal residues which again may yield inactive or weakly active fragments; possible contamination of the fragments with other fragments or with parent protein. In addition, the distribution of possible cleavage sites within a protein may not permit the generation of peptides with the desired overlaps.

A novel and comprehensive synthetic approach which circumvents the aforementioned disadvantages of fragmentation and enables the delineation of continuous sites of molecular recognition on a protein in a reasonable period of time was introduced by this laboratory (9). In this approach, a series of consecutive overlapping peptides which encompass the entire polypeptide chain of a protein are synthesized (Figure 4). The synthetic peptides are uniform in size and each peptide overlaps its adjacent neighbors by a fixed number of residues. The size of the peptides and the number of residues in the overlap regions are selected in order to optimize the synthetic load and to obtain reasonable resolution of individual sites. In examination of T-cell activity, the size and overlap should take into consideration the effects of peptide size (29-31) on *in vivo* T-cell responses. Thus, a systematic screening of an entire protein chain for continuous regions of immunochemical or other binding activities is possible. Using this approach, the full profiles of antigenic (i.e. antibody binding), T-cell recognition and other binding sites have been localized on several proteins.

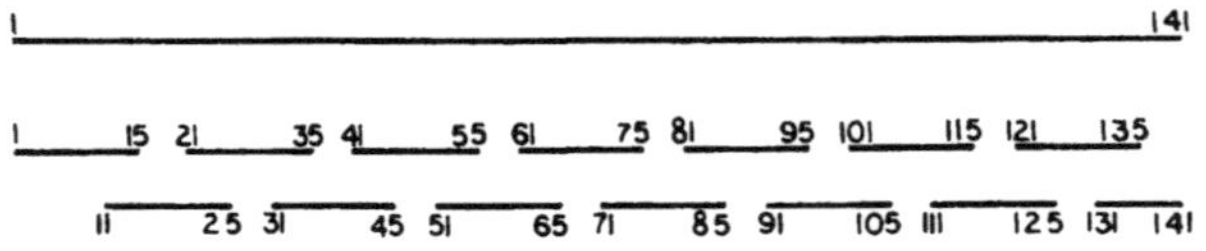


Figure 4: A schematic diagram showing the principle of the comprehensive synthetic overlapping peptide strategy, using its application to the α chain of human hemoglobin as an example. The strategy relied on the synthesis of the entire molecule in 15-residue peptides (except for the C-terminal region 131-141), each overlapping its two adjacent neighbors by 5 residues on both sides. In application to other proteins, the size of the overlapping peptides is adjusted according to the overall length of the protein chain in order to optimize the synthetic load. (From ref. 9)

A. Localization of Antigenic Sites

1. Hemoglobin

The full profile of continuous antigenic sites on human hemoglobin, was determined (9-11) by a comprehensive synthetic strategy, employing 28 overlapping peptides encompassing the entire α chain (α1-15, α11-25 etc.) and β chain (β1-15, β11-25 etc.).

Each of the hemoglobin chains has five antigenic sites (Figures 5 and 6) which are recognized by antibodies raised in several different species goat, rabbit, mouse). Similar sites are recognized by all three species (9-11). The precise boundaries of sites 1,2, and 5 of the alpha chain have been delineated (32-34) while the other sites have been broadly assigned as shown in Figures 5 and 6 and reside within, but not necessarily include all of, the regions indicated. It should be noted that the boundaries were intentionally enlarged to increase the probability of correct assignment (10,11). Further studies with synthetic peptides within the assigned regions, as was previously done with the sites of myoglobin (18,21,35,36) and of the hemoglobin α chain (33,34) will be required to precisely delineate the remaining sites.

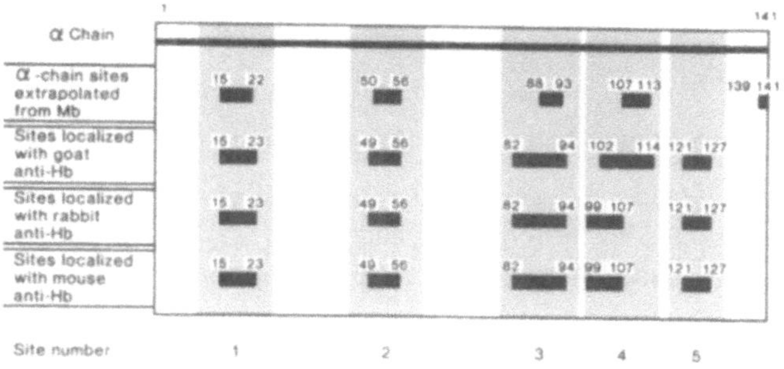


Figure 5: Schematic diagram of the alpha chain of hemoglobin, alpha chain regions that have been extrapolated from the five antigenic sites of sperm-whale myoglobin and the antigenic sites of the alpha chain that were localized with goat, rabbit and mouse antisera to hemoglobin. The general areas of the sites are highlighted by shading. The numbers refer to the locations of the residues in the alpha chain. The asterisk (*) indicates that, owing to differences in the sizes of the polypeptide chains of myoglobin and the alpha subunit, the antigenic site (145-151) of myoglobin does not really have full structural counterpart in the alpha chain. Note that sites 3 and 4 have not yet been narrowed down to their precise boundaries. The regions shown imply that these two sites fall within, but do not necessarily include all of the indicated area (From ref. 10 and the precise boundary delineation of sites 2 and 5 is from refs. 33 and 34).

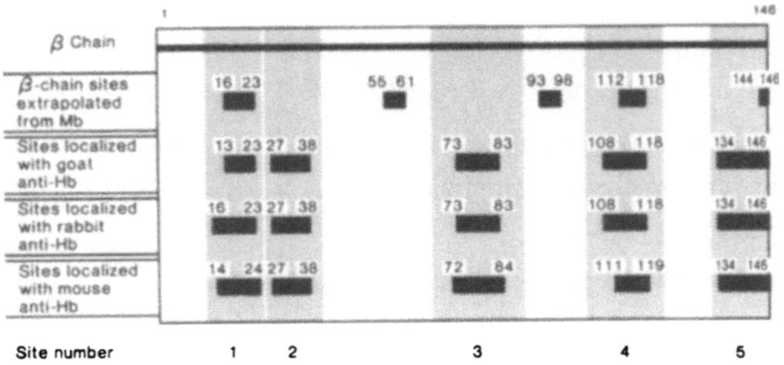


Figure 6: Schematic diagram showing the beta chain of hemoglobin, its regions that have been extrapolated from the five antigenic sites of sperm-whale myoglobin and its antigenic sites that have been localized with goat, rabbit and mouse antisera to hemoglobin. The general areas of the sites are highlighted by shading for clarity. The numbers refer to the locations of the residues in the beta chain. It is not implied that the antigenic sites comprise the entire size of the regions shown, but rather that they fall within these regions. (From ref. 11).

Several of the antigenic sites of hemoglobin (four alpha chain sites and three beta sites) (Figures 5 and 6] reside in regions of the molecule which are conformationally similar to those regions occupied by the myoglobin antigenic sites. The correlation between hemoglobin and myoglobin is in agreement with the concept of 'structurally inherent antigenic sites' (37), namely that antigencity is conferred on certain surface regions of a protein by virtue of their three-dimensional locations. The individual alpha and beta subunits of hemoglobin exhibit an overall similarity in tertiary structure to that of myoglobin. However, the tetrameric subunit structure of hemoglobin endows it with unique conformational properties that contrast to the monomeric structure of myoglobin (38). Therefore, as was previously stressed (37), hemoglobin would also have antigenic sites that are unique to its structure.

Inspection of the three-dimensional structure on hemoglobin (38,39) reveals that several of the antigenic sites occupy regions of the alpha and beta chains which are involved in subunit interactions. Three antigenic sites of the alpha chain contain residues which participate in $alpha_1$-$beta_1$ (sites 4 and 5) or $alpha_1$-$beta_2$ (site 3) contacts. Of the beta chain antigenic sites, almost all of the residues of site 2 are involved in $beta_1$-$alpha_1$ or $beta_2$-$alpha_1$ contacts, while most of site 4 participates in $beta_1$-$alpha_1$ contacts. Thus, these site residues are potentially obstructed in the hemoglobin dimer or tetramer. This implies that the immune recognition of hemoglobin occurs, for the most part, at the level of individual subunits (10,11). The immune response to each of the hemoglobin subunits is under separate genetic control (40–43). Therefore, it is highly probable that, on immunization, hemoglobin is recognized predominately as individual subunits and to a much smaller degree in higher states of subunit association.

2. Ragweed Allergen Ra3

The antigenic and allergenic (IgE binding) sites of Ra3 were recently localized (44,45). Quantitative immunoadsorbent titrations of mouse, rabbit and human IgG and human IgE anti-Ra3 antibodies with overlapping synthetic peptides which encompass the entire Ra3 molecule enabled, for the first time, the localization of the full profile of the IgG and IgE responses (Figure 7) to an allergen (44,45). These antisera localized similar regions indicating, as expected (for review see Atassi, 1984), that the locations of the antigenic sites are independent of the immunized species and that they are inherent in their structural locations (37). Further, the results showed for the first time that IgG and IgE antibodies recognize the same sites on an allergen.

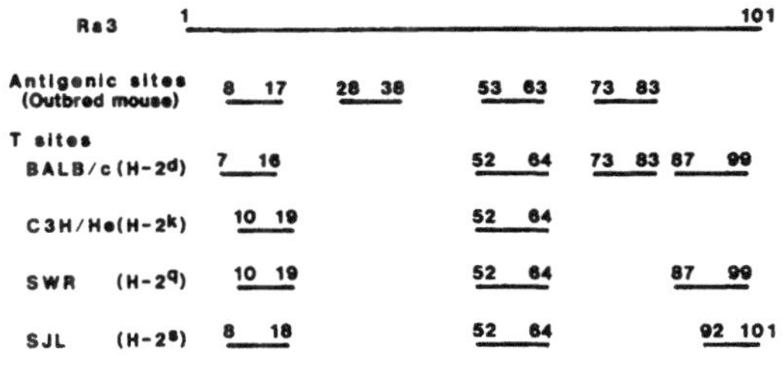


Figure 7: Schematic diagram showing the full profile of the regions of the Ra3 that carry the continuous sites of T cell recognition (T sites) in four mouse strains. The T cell recognition sites are also compared with the sites of outbred mouse antibody (B cell) recognition (i.e. antigenic sites) which were localized in the preceding article (12). It is not implied that the entire regions shown comprise the T sites, rather that the sites reside within these regions. Indeed, in the localization by the overlapping peptide strategy, the T sites are localized within intentionally larger regions than their expected size (From ref. 58).

Anti-peptide antibodies, generated to the most reactive peptides by immunization with free synthetic peptides, bound to both peptide and native protein in comparable amounts suggesting the possibility that these synthetic regions, which are also the target of IgE recognition may be used in the future as the basis for design of desensitization strategies.

3. α Chain of Acetylcholine Receptor

Eighteen synthetic overlapping peptides which spanned the entire extracellular part (residues 1-210) of the α chain of the Torpedo californica acetylcholine receptor (AChR) were used to systematically screen for the profiles of the continuous regions that are recognized by antibodies against free, or membrane-sequestered AChR. Eight antigenic sites were localized (46) on the α chain by antibodies from rabbit and mouse. The sites reside within, but do not necessarily include all of, the regions shown in Figure 8.

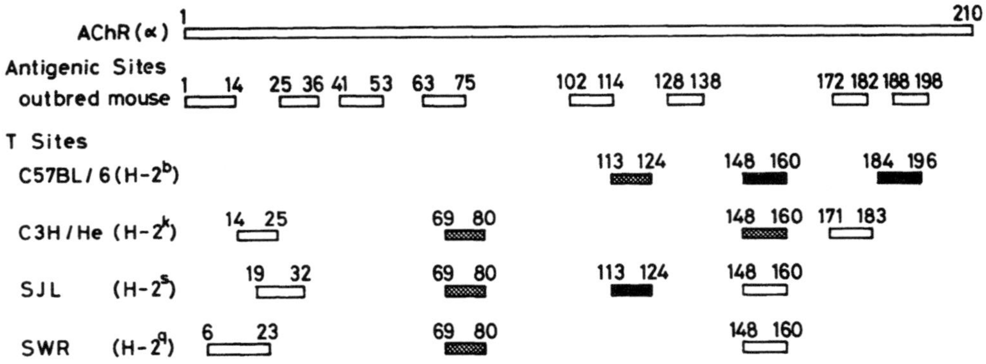


Figure 8: Schematic diagram showing the full profile of the regions of the AChR α subunit that carry the continuous antigenic (i.e. antibody-recognition sites and the sites of T cell recognition (T sites) in four mouse strains. The T cell recognition sites are compared with the sites of antibody (B cell) recognition in the outbred mouse (which were also similarly recognized by other host species (46). The T site notations are as follows: filled bars, regions stimulating high T cell responses (Δ cpm > 20,000); stippled bars, regions stimulating intermediate responses (Δ cpm, 10,000-20,000); open bars, regions stimulating low responses (Δ cpm, 5000-10,000). It is not implied that the entire regions shown comprise the T sites, rather that the sites reside within these regions. Indeed, the T sites are localized within intentionally larger regions than their expected size. (From ref. 59).

The antigenic sites of AChR are similarly recognized by rabbit and mouse antibodies (46). This is consistent with the evidence (2-11,44,45; see ref. 6 for review) that the regions of protein antigenic sites are not dependent on the immunized species, but rather are determined by their conformational locations (32,37). The immunodominance of a given site (i.e. the amounts of anti-site antibodies) varied with the antiserum (46). This is consistent with the antibody response to other protein sites (2-11,44,45,47). In the recognition of a multi-determinant complex protein antigen, the responses to each site are under separate Ir-gene control (23,24). It has been postulated (48,49) that AChR contains a 'main immunogenic region' (MIR) to which an overwhelming fraction of anti-AChR antibodies are directed. However, no single MIR was found (46), but rather a fluctuating dominance of several sites.

4. Human DR2β Chain

Nineteen consecutive synthetic 15-residue peptides (except for peptide 181-198) that spanned the entire extracellular part (residues 1-198) of the HLA DR2β chain and overlapped by 5-residues were examined for antibody binding activities with human alloantisera and with antisera raised in other species (50,51). This established the full antigenic profile of the entire extracellular part of the β chain of HLA DR2 (50,51). Thus, the continuous antigenic sites on the extracellular part of the DR2β subunit reside within seven regions (Figure 9) with an eighth site being recognized somewhat weakly with some antisera. As was found for the other protein antigens, the immunodominance of a given site varied with the antisera and boundary shifts were present. The submolecular regions recognized on the extracellular part of the DR2β chain were independent of the host species from which the antisera were obtained.

B. T-Cell Recognition Sites

Definition of the submolecular features involved in antibody recognition of proteins has rendered a wealth of information on the molecular basis of B-cell recognition and factors affecting this recognition. In contrast, other than fragmentary information, very little has been known about T-cell recognition of proteins. The molecular localization of the full T-cell recognition profile of a complex multi-determinant protein antigen has only been accomplished relatively recently (52-55).

Very recently, by applying the comprehensive synthetic overlapping peptide approach, we have been able for the first time, to localize the full profile of the sites of T-cell recognition (T sites) for several protein antigens. For each protein, a series of consecutive overlapping peptides encompassing the entire respective protein chain were synthesized and examined systematically for their capacity to stimulate in vitro proliferation of protein-primed lymph node cells or protein-specific long term T-cell cultures.

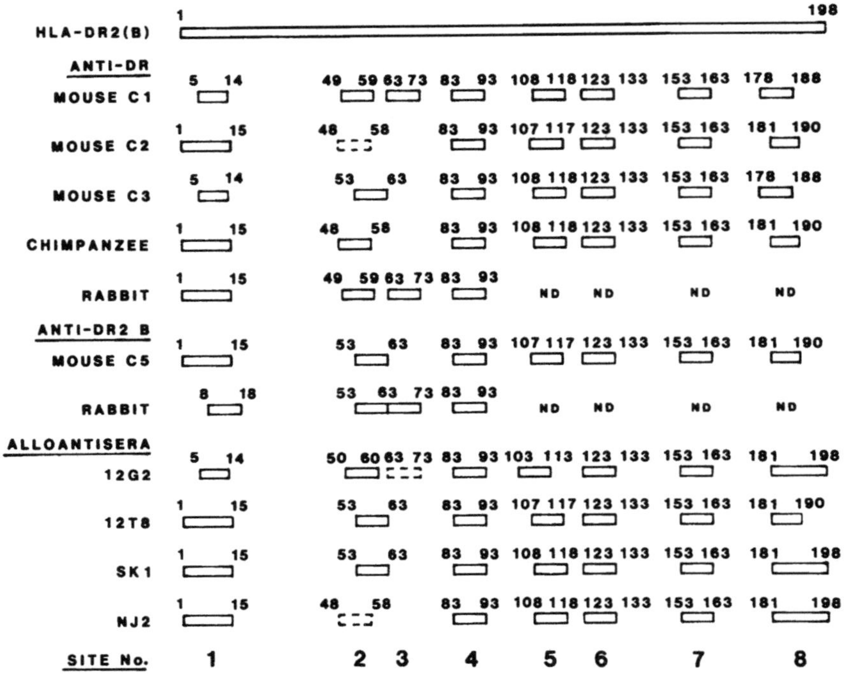


Figure 9: Schematic diagram showing the antigenic regions on the extracellular part of DR2β subunit that have been localized with human alloantisera and various xenoantisera against the DR2 molecule or the DR2β chain. Regions shown by a broken line are weak antigenic sites with these antisera. (ND) implies that the reactivity of the indicated region with the antiserum was not determined. It is not implied that the antigenic sites comprise the entire size of the regions shown, but rather that they fall within these regions. (From ref. 51).

1. Myoglobin

Six regions within the polypeptide chain of myoglobin (Figure 10) evoked proliferative responses in protein-primed lymph node cells obtained from three mouse strains (DBA/2, Balb/c and SJL). In localizing the sites, we employed the procedures previously described for the delineation of the continuous antigenic sites of hemoglobin (9,10) and considered the activities residing in the overlap regions of adjacent peptides. Shifts in the location of the T sites from strain to strain were observed and were not unexpected. Similar shifts in the boundaries of the sites that bind antibody in myoglobin (2,6,18,21) and in other proteins (10,11) have been discussed. Further, as discussed below, subtle shifts in the T sites of other proteins were also apparent.

Of the six T sites in myoglobin, five were found to coincide with the known (12) antigenic (antibody binding) sites while one site was exclusively T-cell specific. The latter T site was located within region 68-80 in the H-2$^d$ haplotypes and region 71-82 in the H-2$^s$ haplotype.

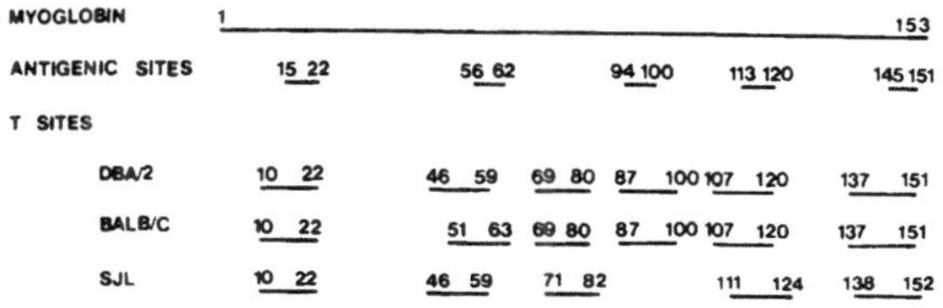


Figure 10: Schematic diagram showing the full profile of the regions of sperm-whale myoglobin that carry the continuous sites of T-cell recognition (T-sites) in three mouse strains. It is not implied that the entire regions shown comprise the T site, rather that the sites reside within these regions. Indeed, in the localization by the overlapping peptide strategy, the T sites are localized within intentionally larger regions than their expected size. (From refs. 52 and 53).

2. Lysozyme

The application of the overlapping peptide strategy to lysozyme has enabled a systematic localization of the full profile of continuous regions of T-cell recognition within the polypeptide chain of lysozyme (54,55). Examination of the proliferative responses of lymph node cells from several strains of lysozyme-primed mice, representing the haplotypes a,d,k,q and s, revealed the presence of four continuous sites of recognition (Figures 11). The boundaries of T sites 1 and 4 exhibited shifts in the mouse strains so far examined. The shifts in the boundaries, once again, are reminiscent of those previously observed in protein sites recognized by antibody (2,10,11,18,21).

In studies of the submolecular T-cell recognition of myoglobin, the sites recognized by antibody were also recognized by T cells (23,29,30, 52,53). It was difficult to examine this phenomenon in lysozyme using lysozyme-primed lymph node cells. By enriching for lysozyme-reactive T cells through long-term culture, it was found that each of the three surface-simulation sites provoked a response from lysozyme-specific T cells of B10.BR. Of the three sites, site 1 was the most immunogenic in this strain

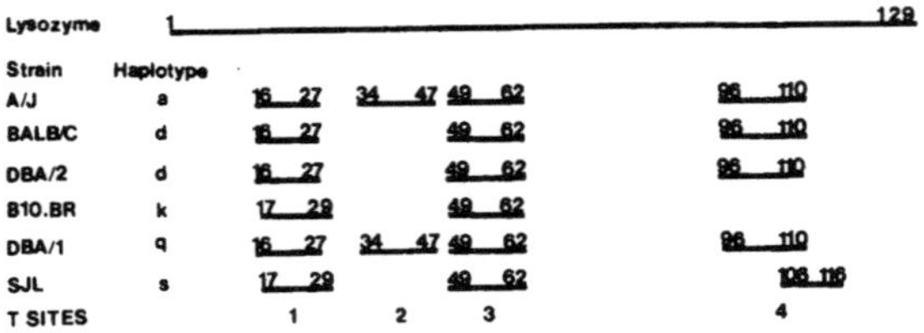


Figure 11: Schematic diagram showing the regions on hen egg lysozyme which carry the continuous sites of T-cell recognition (T-sites) with several mouse strains (54,55). The sites were localized by the overlapping peptide strategy. Note that each T site shown falls within but may not necessarily include all the region indicated. In addition to the four T sites shown in this diagram, long term cultures of lysozyme-primed T cells were found (77) to mount a proliferative response to the three surface-simulation synthetic sites shown in Figure 3. (From refs. 54 and 55).

In summary, the discontinuous sites of lysozyme that are recognized by antibodies are also recognized by T cells, clearly showing that at least some T-cell recognition is dependent on native conformation. Lysozyme, however, has in addition four T sites that are recognized only by T cells and to which no antibody responses are detectable.

3. Hemoglobin

Using the comprehensive synthetic strategy, the full profile of T cell recognition for human adult hemoglobin beta chain has been determined. In two mouse haplotypes which were earlier identified (40) as high or intermediate responders to the protein, five continuous T sites (Figure 12) were localized (56) which fall within the regions 12-24, 60-69, 73-83, 112-124 and 133-146.

Examination of the three-dimensional structure of hemoglobin (38,39) reveals that the T site residing within the region 112-124 constitutes almost entirely residues involved in the $beta_1$-$alpha_1$ contacts (i.e. residues 112, 115, 116, 119, 122, 123 and 124). Further, the last T site, residing within the region 133-146, carries residues 145 and 146 that are normally involved in the $beta_2$-$alpha_1$ contacts. Since such residues would be buried in the tetramer, they would not be expected to participate in immune recognition by T cells unless this recognition takes place at the subunit level (56). Similar conclusions were made from the locations of the antigenic sites, using anti-hemoglobin antibodies, of the alpha chain (10) and the beta chain (11).

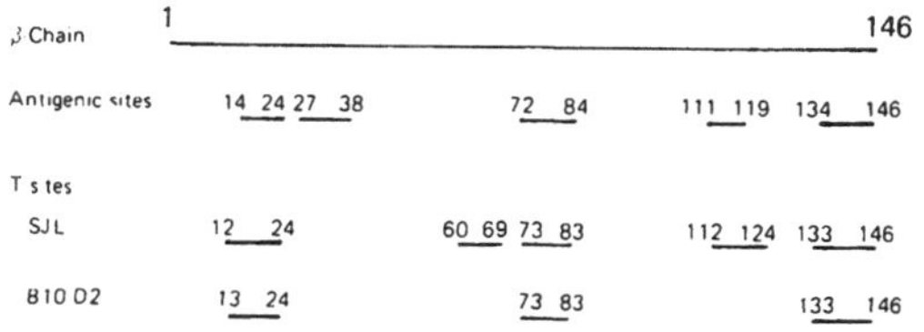


Figure 12: Schematic diagram showing the full profile of the regions of the β-chain of HbA that carry the continuous sites of T-cell recognition (T sites) in two mouse strains. The T-cell recognition sites are also compared with the sites of mouse antibody (B-cell) recognition (i.e. antigenic sites). It is not implied that the entire regions shown comprise the T site, rather that the sites reside within these regions. Indeed, in the localization by the overlapping peptide strategy, the T sites are localized within intentionally larger regions than their expected size. (From ref. 54)

It is significant to point out that four of the T sites on the beta chain (56) coincides with the antigenic (antibody binding) sites that are recognized by mouse antibodies (11). There is, however, at least one T site which is exclusively T-cell specific and is not recognized by B cells. This site resides within region 60-69 for the $H\text{-}2^s$ haplotype. This is similar to the findings with both hen egg lysozyme (54,55) and myoglobin (52,53). The roles of the various T sites (i.e. those that are specific for T cells and those that are recognized by both T and B cells) in the immune recognition of proteins are yet to be delineated. In addition, the shift in the locations of the T sites from strain to strain is also not unexpected.

The levels of the responses to the T sites by the $H\text{-}2^d$ and $H\text{-}2^s$ haplotypes were different. Earlier studies had shown that the immune responses to hemoglobin and its subunits is under genetic control (40). The most recent studies of T-cell recognition of the beta chain of hemoglobin are consistent with previous findings which showed that the immune response to the individual antigenic sites in a protein are each under separate Ir gene control (17,23,52,53,57).

4. Ragweed Allergen Ra3

From recent work employing the comprehensive overlapping peptide strategy it is now evident that Ra3 contains several T sites (58). Five T sites were localized [Figure 7] in four high responder (to Ra3) mouse strains [Balb/c ($H\text{-}2^d$); C3H/He ($H\text{-}2^k$); SWR ($H\text{-}2^q$) and SJL ($H\text{-}2^s$)]. As summarized in Figure 7, slight shifts in the locations of the T sites from strain to strain were observed within regions 7-19 and 87-101. Three of the T sites of Ra3 (regions 8-17, 52-64 and 73-83) coincide with the antigenic sites that are recognized by antibodies from outbred mice. However, there is at least one T site which is exclusively T-cell specific and is not recognized by B cells. This site resides within region 87-99 for the H-$2^d$ and H-$2^q$ haplotypes and within region 92-101 in the H-$2^s$ haplotype.

Each of the four mouse haplotypes responded well to the T site residing within residues 53-63. However, they differed in their responses to the other T sites which reflected genetic control operating at the antigenic site level.

5. Acetylcholine Receptor α Chain

The synthetic overlapping peptide strategy has been employed to localize the T-cell recognition sites of the α chain of AChR from *Torpedo californica* in four mouse strains (59) representing different haplotypes (H-$2^b$, H-$2^k$, H-$2^q$ and H-$2^s$). Two of the strains (H-$2^q$ and H-$2^s$) were high responders to AChR, one strain (H-$2^b$) was a moderate responder, while the fourth strain (H-$2^k$) was essentially a non-responder. A broad localization of the major continuous T sites recognized by the four strains is given in Figure 8. The boundary frame shifts observed for the T sites from strain to strain are similar to boundary shifts of T sites on the other aforementioned proteins.

The responses to the T sites were subject to genetic control operating at the antigenic site level (59) (i.e. each site is under separate genetic control). It is significant to note that AChR-primed T cells of C3H/He (H-$2^k$) mice gave significant responses to at least three different peptides even though they exhibited virtually no *in vitro* response to intact AChR. The lack of responsiveness of a given strain may be due to Ir-gene restriction. On the other hand, regulating inter-site cellular influences (help or suppression) play an important role in the expression of the response to a protein (4,60). It would appear, therefore, that in the H-$2^k$ haplotype, the cellular responses of some site(s) have the effect of turning off the responses to other sites. These site-directed cell-cell interactions are now being studied.

These findings permitted comparison, at the submolecular level, of the sites of T- and B-cell recognition in AChR (59). Three of the T sites on the α chain coincide with the antigenic (antibody binding) sites. There are, however, at least two sites (and possibly three) which are exclusively T-cell specific and are not recognized by B cells. These sites reside within region 113-124, 148-160 and possibly 14-25 (which shifts in $H-2^s$ and $H-2^q$). This is similar to the aforementioned findings with several other proteins, each of which contains sites that are recognized by both B and T cells as well as sites that are recognized exclusively by T cells. The functions of the T sites that are specific for T cells may be concerned with intersite regulatory influences (60) mentioned above.

C. Other Protein Binding Sites

The comprehensive synthetic approach, although first devised and applied (9) for the localization of continuous protein regions that are recognized by the immune system, was also immediately recognized as a powerful tool for the localization of binding regions involved in other protein functions. Two examples of such applications are described here.

1. <u>Subunit Interacting Surfaces in Oligomeric Proteins</u>

A synthetic approach has been introduced (61) for localization of subunit interacting surfaces in oligomeric proteins. It consists of studying the binding activity of consecutive uniform overlapping peptides encompassing an entire subunit to the other, radiolabelled subunit. This permits the establishment of the full profile of peptides that bind the other intact subunit. The approach has been demonstrated (61) with hemoglobin, using the β-chain peptides and has enabled the localization on the β-chain of the submolecular regions responsible for its binding to the α-chain in solution. The results are summarized in Figure 13. There was good agreement between the binding surfaces found in solution (61) and those expected from the crystal structure (38,39). There were also, however, some significant differences in the levels of binding found in solution and those expected from the crystal. Peptide 21-35 possessed much higher binding activity than would be expected from its contribution to subunit association in the crystal. Conversely, other regions expected to possess considerable binding capacity for α-chain either showed low (peptides 111-125 and 121-135) or almost no binding (peptides 91-105 and 101-115) capacity. On the other hand, two interacting surfaces (within peptides

11-25 and 71-85) that make a contribution in solution do not appear to play a role in the crystal. It was concluded (61) that the regions of subunit association in solution are close to, but not identical with, those in the crystal. The approach should serve as an effective method for localization of subunit interacting surfaces of unknown proteins, even those that can be isolated only in traces.

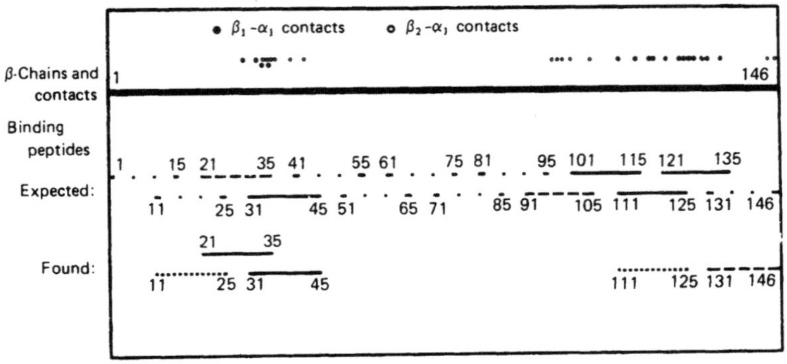


Figure 13: Schematic diagram summarizing the positions of the subunit contact residues on the β-chain and comparing the expected and found binding activities of the overlapping β-peptides. $\beta_1$-$\alpha_1$ contacts; 0, $\beta_2$-$\alpha_1$ contacts; both being placed at appropriate positions to scale along the polypeptide chain. High binding activity; ---, moderate binding; ....., low binding activity; -. . - no binding. (From ref. 61)

## 2. Hemoglobin Binding with Haptoglobin

The comprehensive synthetic approach was employed (62,63) to identify the haptoglobin-binding sites on the α and β chains of human hemoglobin. This approach consisted of the synthesis of a series of consecutive overlapping peptides that, together, systematically represent the entire protein chain. Twenty-eight synthetic peptides (α1-15, α11-25 etc.; β1-15, β11-25 etc.) were examined for their ability to bind human haptoglobin by quantitative solid-phase radiometric titrations of $^{125}I$-labelled haptoglobin. Of the 14 α-chain peptides, only peptide α121-135 bound haptoglobin and accounted quantitatively for the haptoglobin binding by the α chain (62). Of the 14 β-chain peptides only peptides β11-25 and β131-146 bound haptoglobin significantly. Peptide β21-35 exhibited a small binding activity as a consequence of the overlap with peptide β11-25. Figure 14 shows the two haptoglobin binding sites on the β chain. On this basis and by examination of the three-dimensional structure of hemoglobin, it was concluded that the α-chain of hemoglobin has one haptoglobin-binding site within the region α121-135 (62) and that the β-chain has two binding sites for haptoglobin that reside within, but do not necessarily encompass all of, the regions β11-25 and β131-146 (63).

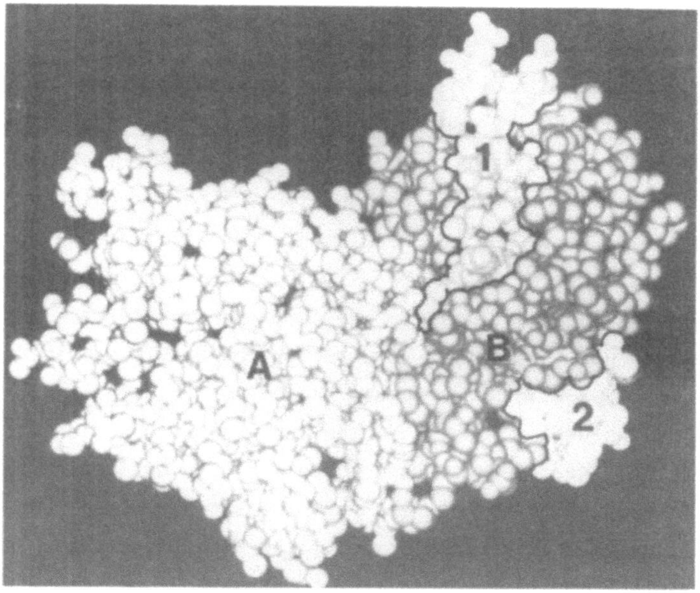


Figure 14: Computer-generated space-filling model of the αβ-dimer of Hb showing the relative locations of regions β11-25 and β131-146. Key: A, α-chain; B, β-chain; 1, region 131-146 of the β-chain; 2, region 11-25 of the β-chain. (From ref. 63)

III. Comprehensive Synthetic Surface Scanning

Although the systematic screening of the polypeptide chain for continuous regions of immunochemical or other binding activity has been quite effective, it is nevertheless costly and labor-intensive, especially when applied to large molecules such as hemoglobin where the alpha and beta chains comprise 287 residues. Because of the systematic overlaps, the total synthetic load was actually 417 residues. The determination, however, of several protein antigenic structures has shown that the antigenic sites of proteins occupy, without exception, surface regions of the molecule (2-4,6,9-11). Any approach that focuses the screening effort to the surface of the protein should shorten the time required for the localization and synthesis of immune recognition (and perhaps other binding) sites. Thus, from the known three-dimensional structure of a protein, all of the exposed, continuous regions of the molecule can be identified, synthesized and examined for immunological or other binding activity. Such an approach was introduced and applied (47) to the localization of continuous sites of T- and B-cell recognition on the hemagglutinin of influenza A virus. It should be emphasized that comprehensive synthetic surface screening can be applied at this stage only for the localization of the 'continuous' antigenic sites. Discontinuous sites cannot realistically be scanned by this method.

1. Antibody and T-Cell Recognition of Influenza A Hemagglutinin

This approach has been applied (47,64,65) to the localization of continuous sites of T- and B-cell recognition on the hemagglutinin of influenza A X-31 virus and has resulted in the localization of twelve continuous antigenic sites on the HA molecule (Figure 15). Of the twelve antigenic sites localized (47) by the comprehensive surface scanning, at least six (and possibly seven) were recognized by virus-primed T cells from $H\text{-}2^d$, $H\text{-}2^b$, $H\text{-}2^k$ and $H\text{-}2^s$ mouse haplotypes (Figure 16).

Although virus-primed T cells of the four haplotypes did not stimulate detectable proliferative responses to some of the antigenic (antibody binding) sites, this does not mean that the sites will not be recognized by primed T cells of other haplotypes. The responses to the HA molecule are consistent with the findings from other proteins which show that the responses to individual sites (23,55,57) are regulated by Ir-gene control, with each site being under separated genetic control.

## IV. Surface-Simulation Synthesis

It is known that the 'contact' residues in a protein binding site are not necessarily always in direct peptide bond linkage but may very often be spatially adjacent in the molecule so as to form the binding surface. In several proteins and enzymes, many of the residues involved in ligand or substrate binding have been identified by X-ray crystallography. However, this knowledge could not be translated into a useful experimental approach by which a site could be mimicked synthetically. A new chemical approach was needed to synthetically mimic a given protein activity. This was to come about from our studies on the antigenic structure of lysozyme.

We rationalized (12) that if the requirements of amino acid arrangement and shape of the binding site could be duplicated chemically on a matrix other than the parent protein, then it should be feasible to duplicate the binding activity in a synthetic substance of an appropriate design. This rationale led us to devise (12) a novel concept by which the spatially adjacent residues constituting a protein binding site are directly linked via peptide bonds with appropriate spacing and directionality. Thus, a peptide is generated which does not exist in the native protein but which attempts to mimic the construction of a specific surface area which forms the binding site (12). Soon, thereafter, we proposed (66) the term "surface-simulation" synthesis to describe this concept.

HA 1 Peptides

| | | |
|---|---|---|
| Peptide 1 | 23–36 | G.T.L.V.K.T.I.T.D.D.Q.I.E.V |
| Peptide 2 | 124–134 | G.F.T.W.T.G.V.T.Q.N.G |
| Peptide 3 | 138–152 | A.G.K.R.G.P.G.S.G.F.F.S.R.L.N |
| Peptide 4 | 154–167 | L.T.K.S.G.S.T.Y.P.V.L.N.V.T |
| Peptide 5 | 175–188 | D.K.L.Y.I.W.G.V.H.H.P.S.T.N |
| Peptide 6 | 183–199 | H.H.P.S.T.N.Q.E.Q.T.S.L.Y.V.Q.A.S |
| Peptide 7 | 201–218 | R.V.T.V.S.T.R.R.S.Q.Q.T.I.I.P.N.I.G |
| Peptide 8 | 272–288 | A.P.I.D.T.G.I.S.E.G.I.T.P.N.G.S.I |
| Peptide 9 | 300–315 | I.T. .G.A.G.P.K.Y.V.K.Q.N.T.L.K |

HA 2 Peptides

| | | |
|---|---|---|
| Peptide 10 | 1–11 | G.L.F.G.A.I.A.G.F.I.E |
| Peptide 11 | 56–68 | I.E.K.T.N.E.K.F.H.Q.I.E.K |
| Peptide 12 | 68–84 | K.E.F.S.E.V.E.G.R.I.Q.D.L.E.K.Y.V |

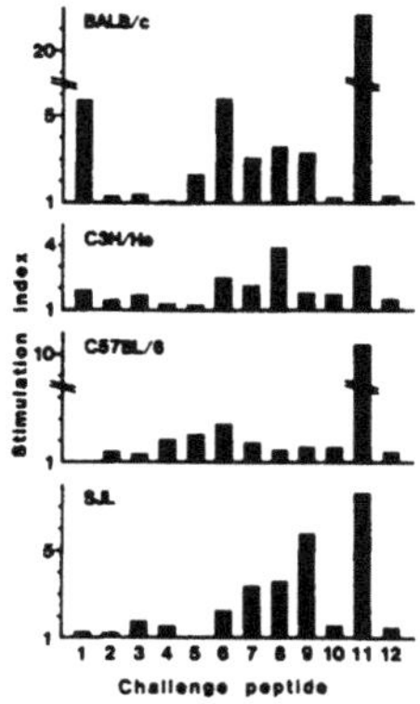


Figure 15 (Left) Covalent structures of the surface peptides of influenza virus hemagglutinin which bind anti-viral antibodies. In the synthetic peptides, the residues Cys-277, Cys-281 and Cys-305 were replaced by glycine. It is not implied that the antigenic sites constitute the full size of the peptides shown, but rather that the sites reside within the boundaries of these peptides. (From ref. 47)

Figure 16: (Right) The peptide T-cell recognition profile in four mouse strains that had been immunized with an optimum dose of X31 virus. The results, which are expressed in stimulation index (cpm of stimulated cells/ cpm of unstimulated cells), summarize the values of maximum responses at the optimum *in vitro* challenge dose for each peptide (From ref. 65)

Before any synthetic studies can be done, it is of course necessary to know the identity of the 'contact' residues of the site. Usually these are determined either through chemical methods (for reviews, see refs. 14, 5,67) and/or from X-ray crystallography of the protein complex with the appropriate ligand. The $C^{\alpha}$-to$C^{\alpha}$ distances separating the 'contact' residues of the site can be calculated from the X-ray co-ordinates.

It has been found that, generally, only one synthetic direction is appropriate for binding (68-70). The requirement for directionality is needed for appropriate side-chain orientation (68,69). Studies with computer graphics have shown (Atassi and Barry, unpublished results) that, for a site mimicked by surface-simulation synthesis, only one synthetic direction, can assume a conformation in which the side-chain orientation superimposes with that in the native protein. In the reverse synthetic direction, the side-chain orientation will not fit.

The application of surface-simulation synthesis (12,66) is not confined to definition of protein antigenic sites but can also be applied to protein binding sites that discharge other functions. The following sections give examples of its application and demonstrate its versatility.

A. The Antigenic Sites of Lysozyme

Through the systematic application of the general chemical strategy outlined in Section I (see refs. 1-3), we found that hen-egg lysozyme has three antigenic sites, each constituting spatially adjacent residues (that are otherwise distant in sequence) and describing a discrete area on the surface topography of the protein (for review, see ref. 3). Although we had proposed the existence of this type of protein antigenic sites quite early (71), we subsequently found (2,16,17) that they did not exist in myoglobin and it was in lysozyme that we identified (12) the first such sites. The need for the chemical verification of such sites led us to introduce (12,66) the concept of 'surface-simulation' synthesis which enabled us to precisely delineate and chemically synthesize the three antigenic sites of lysozyme (for review, see ref. 3).

The identities of the lysozyme antigenic sites are shown in Figure 17 and their locations in the three-diensional structure of lysozyme can be seen in Figure 18. The antigenic sites are briefly described below:

Site 1: This antigenic site is constructed (70) by the side-chains of the spatially adjacent five surface residues: Arg-125, Arg-5, Glu-7,

Arg-14 and Lys-13. The extended dimension of the site from Arg-125 to Lys-13 ($C^{\alpha}$-to-$C^{\alpha}$) is 30 Å. These residues, which describe a region of the surface topography of the protein, bind with antibody as if they are in direct peptide linkage (12). In fact, the immunochemical reactivity of this site is fully satisfied by the surface-simulation synthesis peptide Arg-Gly-Gly-Arg-Gly-Glu-Gly-Gly-Arg-Lys, which does not exist in native lysozyme, but mimics the arrangement of the residues in this surface region. The surface-simulation synthetic site exhibits a directional preference (Arg-125 to Lys-13), which appears to be independent of the species of the immunized animal, at least with the species so far tested (70). The site has a restricted conformational freedom clearly evidenced by its sensitivity to variation of the spacing between its constituent residues in surface-simulation synthesis (70). The intactness of the disulfide bond 6-127 in native lysozyme is critical for the intergrity of this site (72,73).

Site 2: This antigenic site consists (12,68) of the spatially adjacent surface residues: Trp-62, Lys-97, Lys-96, Asn-93, Thr-89 and Asp-87. The extended dimension of the site from Trp-62 to Asp-87 ($C^{\alpha}$-to-$C^{\alpha}$ distance) is 27.3 Å. As with site 1, site 2 also describes part of the surface topography of the protein molecule. The residues forming the site behave functionally as if they were directly linked by peptide bonds (12). Thus, the surface-simulation synthetic peptide Phe-Gly-Lys-Lys-Asn-Thr-Asp, which does not exist in lysozyme but simulates a surface region of it, carries full reactivity of the site (12,68). With the antisera so far studied, the antigenic site exhibits a preferred directionality in surface-simulation synthesis (Trp-62 to Asp-87). This site is subject to conformational restrictions indicated by the requirement for appropriate spacing between the constituent residues (68). The intactness of the disulfide bonds 64-80 and 76-94 is essential to bring together the various constituent residues of the site (73,74). This antigenic site overlaps with the enzymic active site because they both share Trp-62 (12,68,74).

Site 3: Like sites 1 and 2, this site is also constructed by spatially adjacent surface residues. The antigenic site comprises (69) the five residues: Lys-116, Asn-113, Arg-114, Phe-34 and Lys-33, and, in its extended form, has a $C^{\alpha}$-to-$C^{\alpha}$ dimension (from Lys-116 to Lys-33) of 21 Å. These residues describe a region on the surface of the molecule and act functionally towards the antibody as if they are in direct peptide bond linkage. In fact, the surface-simulation synthetic peptide having the structure Lys-Asn-Arg-Gly-Phe-Lys (which does not exist in lysozyme) carries the full immunochemical reactivity of the site (69). With the antisera so far studied the synthetic antigenic site exhibited a preferred directionality (Lys-116 to Lys-35), since the reverse surface-simulation synthetic sequence was immunochemically inefficient (69). the intactness of the disulfide bond 30-115 is critical for the integrity of this antigenic site in lysozyme (72). Antigenic site 3 overlaps with the hexasaccharide substrate binding site at the carbonyl group of Phe-34 and the side-chain of Arg-114 (69).

It is important to note that the three surface-simulation synthetic peptides had the capacity to bind the bulk (97-99 ) of the antibody response against native lysozyme (3,75). These three surface-simulation synthetic peptides were coupled to succinylated bovine serum albumin, and the conjugates were injected into rabbits (76). Antibodies against each peptide reacted, as expected, only with that peptide, and also specifically with lysozyme (76). This finding is extremely important in view of the fact that a surface-simulation synthetic peptide does not exist in the native protein but attempts to mimic a surface region of it. This was the first demonstration that antibodies against surface-simulation synthetic peptides will infact bind to the native protein.

In a more recent study, T-cell recognition of the surface-simulation synthetic antigenic sites has been explored (77) in mouse strain B10.BR, which is a high responder to lysozyme (78). The discontinuous antigenic sites of lysozyme also had the capacity to stimulate proliferation of lysozyme-specific, long term T-cell cultures (77). This constituted the first clear demonstration that, contrary to a long-held impression, T-cell recognition is not restricted only to sequence features, but also can be directed against protein discontinuous surface areas of high conformational dependency.

Thus, the surface-simulation synthetic peptides (of the correct spacing and directionality) posses the ability to bind antibodies raised against native lysozyme and to be recognized by T cells that are specific for the native protein. Furthermore, antibodies raised against the peptides bind to native protein. These findings clearly indicate that these peptides, even though they do not exist in lysozyme, must be reasonable mimicks of the respective discontinuous surface regions.

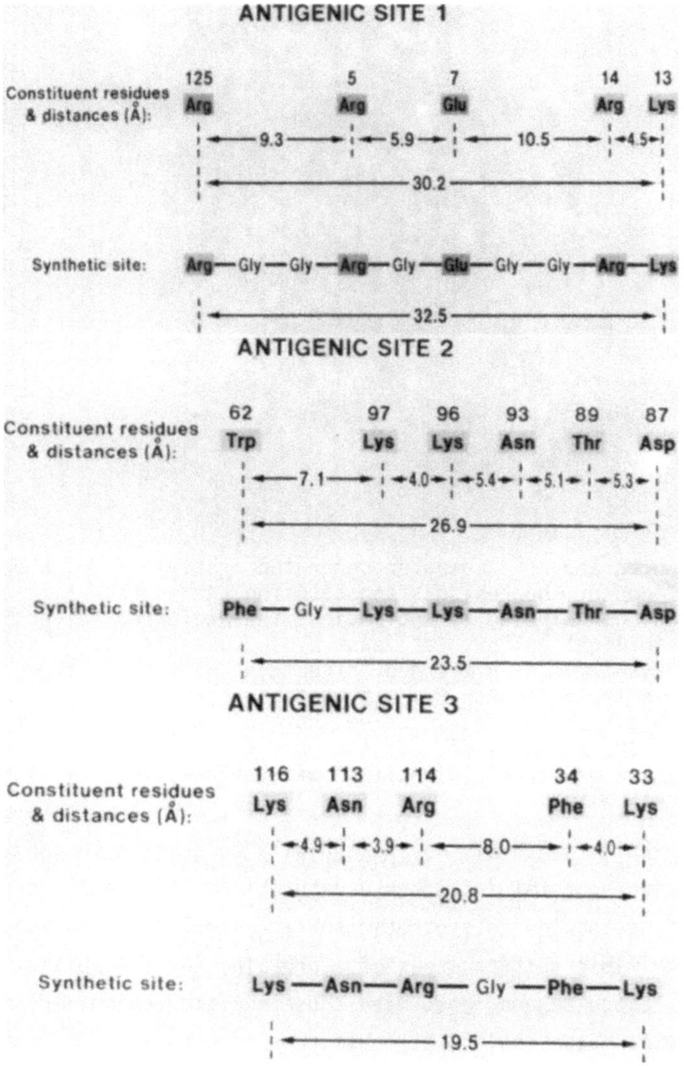


Figure 17: The three antigenic sites of lysozyme. The diagram shows the spatially adjacent residues constituting each antigenic site and their numerical position within the primary structure. The distances (in nm) separating the consecutive residues and the overall dimension of each site (in its extended form) are given, together with the dimension of each surface-simulation synthetic site. The latter assumes an ideal distance of 0.362 nm between alpha carbons. The precise boundary, conformational and directional definitions of the sites are described in the text. The three sites account quantitatively for the entire (96 to 100%) of the antigenic reactivity of lysozyme. (From ref. 3)

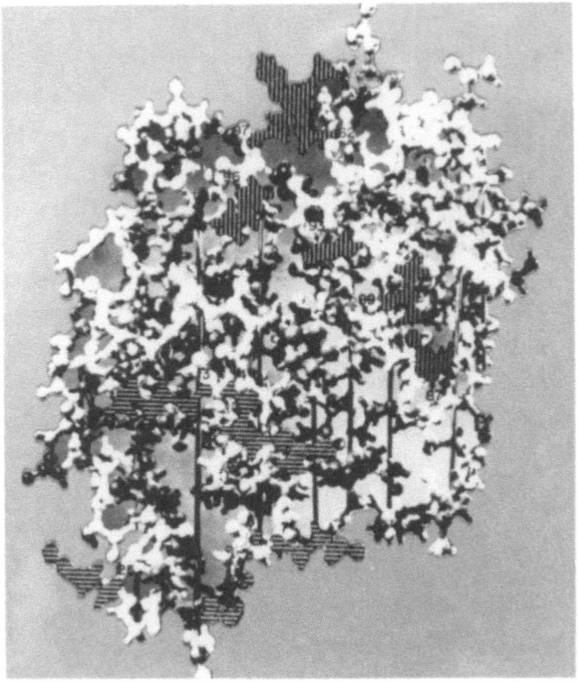

Figure 18: Photograph of a computer-generated lysozyme model showing the relative positions of the residues constituting the surface-simulation antigenic sites. In binding to antibodies, the preferred 'direction' of site 1, shown in light yellow, (at least by surface-simulation synthesis) is Arg-125 to Lys-13. The preferred 'direction' of site 2, shown in blue, with antisera is Trp-62 to Asp-87. It must be noted that only Trp-62 should be visible from this perspective; however, with the aid of computer graphics all of the residues of this site are shown. Site 3, shown in orange, exhibits a preferred 'direction' of synthesis from Lys-116 to Lys-33 (From ref. 3)

Surface-simulation synthesis may potentially be applied to monitor protein conformation or refolding after denaturation. This depends on the preparation of antibodies to synthetic surface-simulation peptides that are designed to mimic surface areas of a protein, and the ability of these antibodies to recognize and react with those surface regions in the native protein molecule. Such antibodies could then be used as conformational probes to double-check the three-dimensional structure of a protein in solution (3). This approach would afford a direct region-by-region scanning of the proximity of several residues on the surface. Thus, for example, the three surface-simulation synthetic sites of lysozyme report simultaneously on the spatial interrelationships of 16 surface residues in the native protein. The approach would therefore afford much more valuble and precise information than that derived from chemical (e.g. availability of side-chains to chemical modification), spectral or other physical measurements in solution. Of course, in principle, antibodies could be made against several surface-simulation peptides representing surface areas that jointly encompass the entire surface of a protein molecule (76). Such antibodies would provide powerful conformational probes for monitoring the three-dimensional structure of a protein in solution. The ability to make antibodies to surface areas of a protein molecule that are preselected and mimicked by surface-simulation synthesis presents a versatile and powerful tool in exploiting therapeutic applications for helper, suppressive, tolerogenic and allergenic sites on proteins and invasive agents (76).

B. Surface-Simulation Synthesis of Antibody Combining Sites

From the foregoing short treatment, it is quite obvious that surface-simulation synthesis constituted a conceptual breakthrough without which the precise definition of the entire antigenic structure of lysozyme would not have been possible. However, this concept can be used to synthetically mimic any type of protein binding sites involving the interaction of a protein with other proteins or with non-protein molecules (3, 12,66,75). One of the most fascinating applications is the surface-simulation synthesis of antibody-combining sites. The mimickry of the antibody-combining site, at least in terms of binding function, by surface-simulation concept constitutes a major contribution in immunology.

1. <u>Surface-Simulation Synthesis of the Immunoglobin New Combining Site to the γ-Hydroxyl Derivative of Vitamin $K_1$</u>

We have examined whether an antibody-combining site can be reconstructed by surface-simulation synthesis. For our first attempt (79) to test the feasibility of this idea, we chose the myeloma protein IgG New, which binds a hydroxyl derivative of vitamin $K_1$ (Vit. $K_1$ OH) (80,81). The combining site is shown in Figure 19. The site residues (Ile-100 H, Ala-101 H, Asn-30 L, Tyr-90 L, Ser-93 L, Leu-94 L, Arg-95 L, Trp-47 H, Tyr-50 H, Tyr-33 H) were directly linked by peptide bonds, with appropriate intervening spacers (79) (Figure 20). Also, a control peptide was synthesized having exactly the same amino acids but which were in a different random sequence. The surface-simulation peptide showed remarkable binding activity towards Vit. $K_1$ OH while the control peptide exhibited no binding activity. Inhibition studies confirmed the specificity of the binding between Vit. $K_1$ OH and the surface-simulation peptide. The results clearly showed that a complex binding site, can be successfully mimicked by surface-simulation synthesis (79).

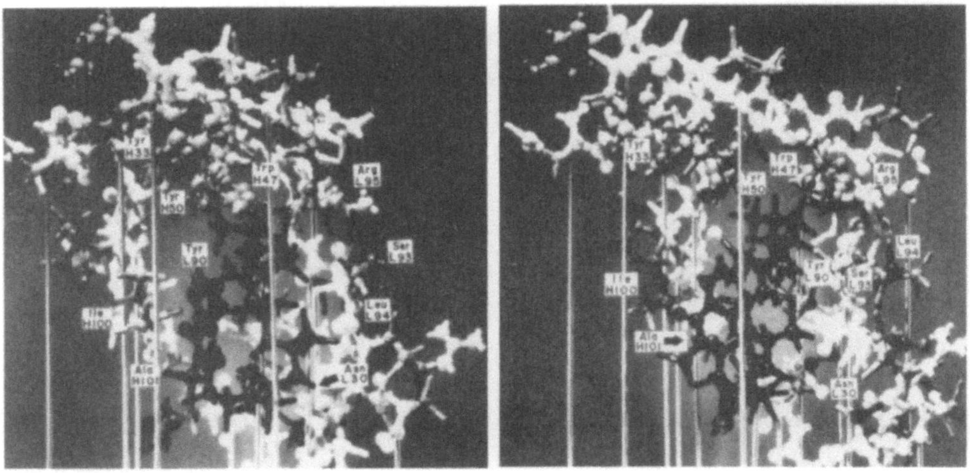

Figure 19: Two views of a model of the IgG New combining site with bound Vit. $K_1OH$ in place. Hypervariable light chain residues 27 to 31 and 90 to 95 and hypervariable heavy chain residues 31 to 33, 47 tO 58, and 97 to 103 are shown. The labelled residues were linked, with appropriate spacing, by surface-simulation synthesis (see Fig. 20) into a single peptide with Ile-100 H as its $NH_2$ terminus (From ref. 79)

2. Surface-Simulation Synthesis of the Phosphorylcholine Combining Site of the Myeloma Protein M-603

We have also applied (82) the concept of surface-simulation synthesis to the combining site of the myeloma protein M-603. The X-ray coordinates of the binding site are known to a 3.1 Å resolution (83,84). We have mimicked (82) the phosphorylcholine (PC) binding properties of M-603 in the surface-simulation peptide: Ser-Tyr-Gly-Gly-Arg-Tyr-Gly-Glu-Try-Val (Figure 21). In contrast, a control peptide which had the same amino acid composition but a different sequence (see Figure 21) showed little or no binding activity. The specificity of binding to PC-Sepharose by the surface-simulation peptide was further demonstrated by the inability of the control peptide to significantly inhibit the binding of M-603, while the surface-simulation peptide inhibited by more than 90% in the concentration range examined (82).

These findings suggest that surface-simulation synthesis can be effectively employed to synthetically mimic antibody combining sites and may in the future be a valuable tool with which to examine idiotypic specificities of antibodies and to manipulate the immune response to clinically important antigens. These avenues of investigation are being pursued in this laboratory.

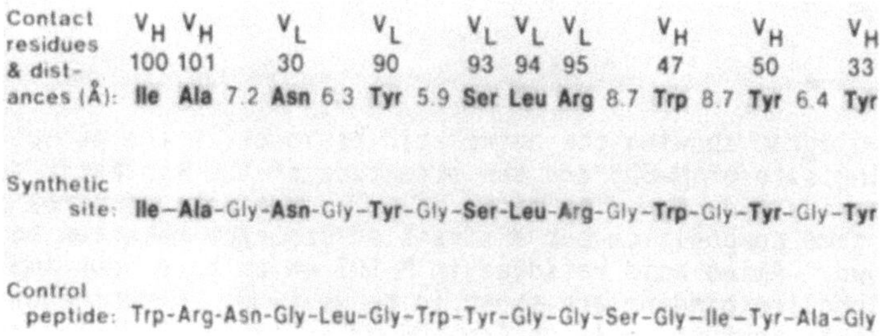


Figure 20: A diagram showing the amino residues which make close contact with Vit. $K_1OH$ in the hypervariable region of IgG New and the structures of the two surface-simulation peptides (A and B) designed to mimic the binding site. The diagram also gives the distances (in $C^{\alpha}$-to-$C^{\alpha}$) between the constituent residues of the combining site. The surface-simulation synthetic peptides A and B differed only in the absence (peptide A) or presence (peptide B) of a spacer between Tyr-90L and Ser-93L. The control peptide has an identical amino acid composition to peptide B, except that its sequence is randomized. It should be emphasized that the diagram is not intended to imply that the site is an extended surface. It is in fact a cavity (See Fig. 19). (From ref. 79).

3. The Possible Surfce-Simulation Synthesis of Antibody-Combining Sites to Lysozyme Antigenic Sites

It is evident that the proper application of surface-simulation synthesis to a protein requires knowledge of the residues constituting a binding site as well as their conformational spacing and directional requirements. We reasoned (85) that in all likelihood, the antibody-combining site will be expected to comprise residues that are complementary to those in the corresponding antigenic sites. Furthermore, the directionality of the antigenic site and the spacings between its constituent residues and the corresponding parameters of the antibody-combining sites must be equivalent or comparable in order for appropriate binding to take place. Thus, by the precise knowledge of all the parameters of an antigenic site, it should be possible to create a reasonable design of its complementary antibody-combining site. This has been done (85) for two antigenic sites in native lysozyme.

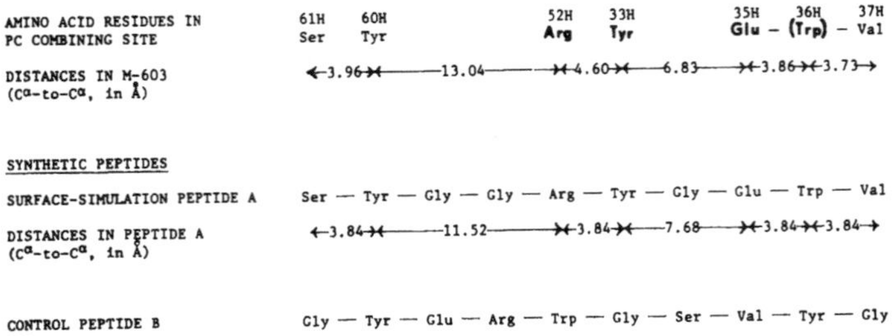


Figure 21: A diagram showing the amino acid residues in the phosphorylcholine binding site of M-603 and the structure of the synthetic surface-simulation peptide A designed to mimic it. The synthetic control peptide B having the same composition but a scrambled sequence relative to peptide A is also shown. Amino acid residues in M-603 which have been implicated in phosphorylcholine binding are shown in heavy type. Parentheses around Trp-35 H indicate that it replaces Trp-104 aH. The $C^{\alpha}$–$C^{\alpha}$ distances in M-603 were calculated from the $C^{\alpha}$ atomic coordinates. The distances in peptide A were calculated assuming a $C^{\alpha}$–$C^{\alpha}$ distance of each peptide bond to be 3.84 Å. (From ref. 82)

Two peptides CS-2 and CS-3 (Figure 22) were designed (85) on the basis of complementarity to lysozyme antigenic sites 2 and 3, respectively, in ionic, hydrophobic, hydrophilic and side-chain length of the constituent amino acids. Appropriate binding and inhibition studies with lysozyme and control proteins (85) indicated that the antibody-combining sites against antigenic sites 2 and 3 of native lysozyme were successfully mimicked synthetically, at least in terms of binding function. Since each of these two peptides (CS-2 and CS-3) mimics an antibody-combining site in terms of its binding function to a lysozyme antigenic site, antibodies against such a peptide reacted with anti-lysozyme antisera.

Antibodies were raised (86) against one of these peptides (CS-3) which is complementary to antigenic site 3 of lysozyme. These anti-peptide antibodies were found to react with anti-lysozyme antibodies. Thus, by precise knowledge of a protein antigenic site, a complementary peptide can potentially be designed (85,86) to mimic the arrangement of residues in the combining-sites of antibodies against that antigenic site. The success of the design can be tested by immunochemical studies on the complementary peptide and finally by the ability of its antibodies to react with anti-protein antisera. The achievement of this opens up many hiterto untapped avenues in immunology, in particular those pertaining to regulation and manipulation of the immune response by anti-combining site antibodies.

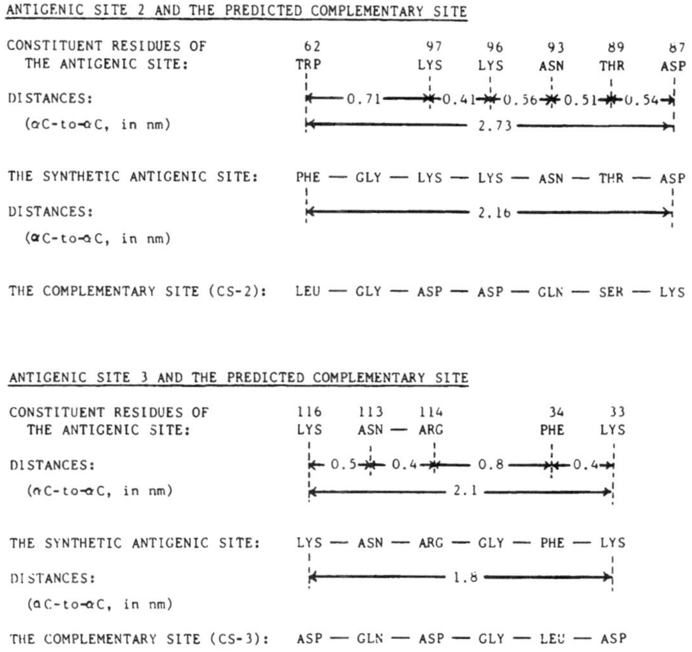


Figure 22: A diagram showing antigenic sites 2 and 3 of hen egg-white lysozyme and their predicted complementary antibody-combining sites. The spatially adjacent surface residues constituting each antigenic site and the numerical positions of these residues in the primary structure of lysozyme are shown. The distances (in nm) separating the consecutive residues and the overall dimension of each site are given, together with the dimension of the respective surface-simulation synthetic antigenic site. Below each antigenic site is given the structure of the respective complementary surface-simulation peptide which was predicted to mimic the antibody-combining site directed against the antigenic site. (From ref. 85).

C. Surface Simulation Synthesis of the Substrate-Binding Site of Trypsin

Recent studies (87) were aimed at mimicking the substrate-binding site of bovine trypsin into a surface-simulation peptide that expresses the binding activity of the enzyme molecule with substrates and inhibitors. Trypsin was chosen as the model for these studies because its X-ray structure is known in fine detail (88,89). The structure of iPrP-F-inhibited trypsin, initially determined by Stroud *et al.*, (88,90), was subsequently refined to

1.5 Å resolution (91). Also, the binding of trypsin to soya-bean trypsin inhibitor afforded the making of a convenient assay by which the surface-simulation peptides could be screened for binding activity. Figure 23 shows two perspectives of the substrate-binding site of bovine trypsin.

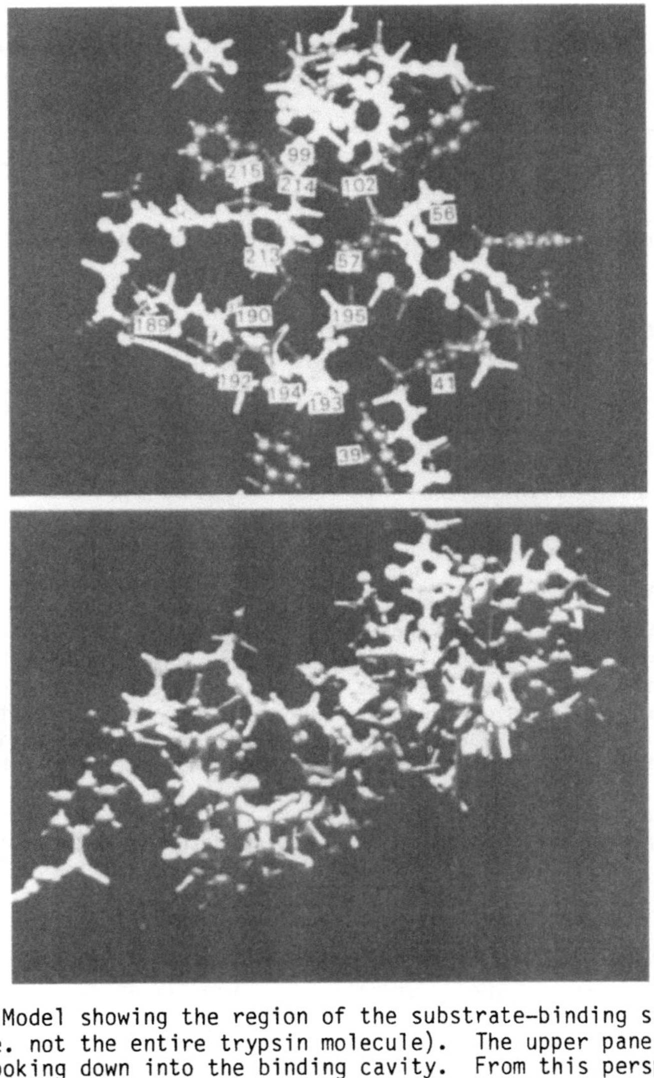


Figure 23: Model showing the region of the substrate-binding site of trypsin (i.e. not the entire trypsin molecule). The upper panel gives a top view, looking down into the binding cavity. From this perspective, the following contact residues can be seen: Tyr-39, Phe-41, His-57, Ala-56, Asp-102, Leu-99, Trp-215, Ser-214, Val-213, Ser-195, Asp-194, Gly-193, Gln-192, Ser-190 and Asp-189. The lower panel gives a side view of the cavity, taken from the left side of the upper panel. (From ref. 87)

A peptide (TP) was designed (87) by surface-simulation synthesis to mimic the binding site of trypsin. It should be emphasized that this peptide does not exist in trypsin, but rather mimicks the topographic arrangement of the residues in its substrate-binding cavity. The element of the design, which is based on spatial and distance considerations, are shown in Table 1. A control peptide (CTP) was also prepared (87) that incorporated all the contact and spacing residues but in random fashion. The control peptide was crucial in order to confirm, should binding activity be found in the TP peptides, that such binding is indeed specific and not caused by some unusual charge or hydrophobic effects.

The radiolabelled TP peptide was able to bind to benzamidine and trypsin inhibitor adsorbents, whereas the control CTP peptide exhibited no binding activity (87). The TP peptide conjugate to succinylated lysozyme was able to bind soya-bean trypsin inhibitor and to $\alpha_{-1}$-anti-trypsin (87). There was no binding activity in the CTP peptide conjugates, thus confirming the correctness of the peptide design (87).

We have also investigated (87) the action of other inhibitors known to react at the substrate-binding site. The action of $PhMeSO_2F$ and iPrP-F on acetylcholine-sterase, chymotrypsin, trypsin and other serine proteinases (92-95) results in the formation of a covalent bond via the active serine residue of the enzyme. The TP peptide reacted with both iPrP-F and $PhMeSO_2F$, the reaction in each case taking place at a serine residue in the peptide. Again the lack of reaction with the CTP peptide confirmed the specificity of reaction with the TP peptide. We have not yet determined which of the three serine residues on the peptide participates in the reaction with $PhMeSO_2F$.

The TP peptide, either free or conjugated, exhibited no enzymic activity with protein substrates (87). Also, the soluble TP peptide-Suclysozyme conjugate had no hydrolytic activity on $N^{\alpha}$-p-tosyl-L-arginine methyl ester, a substrate commonly used to monitor trypsin enzymic activity. Thus, it was evident that the TP peptide, generated by surface-simulation, expressed the binding activity of trypsin with substrates and inhibitors but not its catalytic activity (87).

Catalysis is subject to very stringent conformational and distance requirements and, since the distances and conformational relationships of the residues in a surface-simulation synthetic peptide do not precisely duplicate the corresponding parameters in the intact molecule, it is not surprising that only binding but not catalysis is duplicated (87). Whether catalysis can be obtained by improvement in design and employment of other synthetic strategies is yet to be determined.

Table 1

**Design of Surface-Simulation Peptide Mimicking Substrate Binding Site of Trypsin**[a,b]

| Contact residues and distances (A) | Surface-simulation synthetic peptide (TP) (B) | Control synthetic peptide (CTP) (C) |
|---|---|---|
| Asp 189 | Asp (N terminus) | Trp (N terminus) |
| Ser 190 | Ser | His |
| 0.49 nm | Gly | Gly |
| Gln 192 | Gln | Asp |
| Gly 193 | Gly | Gly |
| Asp 194 | Asp | Asp |
| Ser 195 | Ser | Val |
| 0.85 nm | Gly | Ala |
| | Gly | Gly |
| Val 213 | Val | Gln |
| Ser 214 | Ser | Phe |
| Trp 215 | Trp | Ser |
| 0.85 nm | Gly | Gly |
| | Gly | Gly |
| Leu 99 | Leu | Asp |
| 0.55 nm | Gly | His |
| Asp 102 | Asp | Gly |
| 0.53 nm | Gly | Ser |
| Ala 55 | Ala | Gly |
| Ala 56 | Ala | Ser |
| His 57 | His | Gly |
| 0.50 nm | Gly | Ala |
| Cys 42 | Gly | Gly |
| Phe 41 | Phe | Leu |
| His 40 | His | Gly |
| Tyr 39 | Tyr (C terminus) | Tyr (C terminus) |

a From ref. 87

b Column A identifies the contact residues in the binding cavity of the enzyme and the distances ($C^{\alpha}$ to $C^{\alpha}$, in nanometers) separating appropriate residues. Column B indicates the surface-simulation synthetic peptide (TP) designed to mimic the binding site. Column C indicates control peptide (CTP) that has all the essential residues of the binding site except that they are randomized. The peptides are written with the N terminus on the top and the C terminus on the bottom.

## Conclusions

These studies have charted strategies not only for determining protein antigenic sites but also for delineating and synthesizing other types of protein binding sites (3,4,12,75). Furthermore, the concept of 'surface-simulation' synthesis, which we introduced and developed during our determination of the antigenic structure of lysozyme (12,66-70), has afforded a dimension of unlimited versatility in the synthetic mimicking of any type of protein binding sites (for a recent review of surface-simulation synthesis, see ref. 96). Already, the strategy and concepts thus developed during our determination of the antigenic structures of Mb and lysozyme have also enabled us to synthetically mimic sites related to other protein activities such as combining sites of antibodies (79,82,85, 86) and substrate binding sites of enzymes (87).

The strategies which we have devised should be extremely useful in developing immunological therapeutic approaches (based on use of free small synthetic peptides as immunogens) to antigens of clinical important (e.g. allergens, viruses, toxins, tissue antigens, etc.). Furthermore, antibodies against surface-simulation peptides would provide powerful probes for the conformation of the parent protein. Another approach (4) of yet untapped clinical potential would rely on the identification of the antibody combining site of selected monoclonal antibodies to the antigen. The antibody combining site can then be mimicked by the concept of surface-simulation synthesis. The synthetic peptides, suitably designed, can in principle be used to permanently neutralize the invasive antigen (e.g. a virus). These chemical strategies, have generated active peptides the exploitation of which in biochemical, therapeutic and diagnostic applications is now being realized by many investigators.

Acknowledgements

The work that resulted in the findings outlined in this article was supported by grants AI-21386, AI-21336 from the National Institutes of Health, AHA-81-1012 from the American Heart Association and Q994 from the Welch Foundation. The author holds the Robert A. Welch Chair of Chemistry, which is endowed by the Welch Foundation, and this support is gratefully acknowledged.

References


1. Atassi, M.Z. (1972) In: Specific Receptors of Antibodies, Antigens and Cells, 3rd Int. Convocation on Immunol. p. 118, Karger, Basel.
2. Atassi, M.Z. (1975) Immunochemistry 12, 423.
3. Atassi, M.Z. (1978) Immunochemistry 15, 909.
4. Atassi, M.Z. (1980) Mol. Cell. Biochem. 32, 21.
5. Atassi, M.Z. Biochem. Biophys. Acta 704, 552.
6. Atassi, M.Z. (1984) Europ. J. Biochem. 145, 1.
7. Sakata, S. and Atassi, M.Z. (1980) Biochim. Biophys. Acta 625, 159.
8. Sakata, S. and Atassi, M.Z. (1980) Molecular Immunol. 17, 139.
9. Kazim, A.L. and Atassi, M.Z. (1980) Biochem. J. 191, 261.
10. Kazim, A.L. and Atassi, M.Z. (1982) Biochem. J. 203, 201.
11. Yoshioka, N. and Atassi, M.Z. (1986) Biochem. J. 234, 457.
12. Atassi, M.Z., Lee, C.L. and Pai, R.C. (1976) Biochim. Biophys. Acta 427, 745.
13. Atassi, M.Z. and Smith, J.A. (1978) Immunochemistry 15, 606.
14. Atassi, M.Z. (1977) In: Immunochemistry of Proteins (Atassi, M.Z., Ed.) 1, 1, Plenum Press, New York.
15. Atassi, M.Z. (1977) In: Immunochemistry of Proteins (Atassi, M.Z., Ed.) 2, 77, Plenum Press, New York.
16. Twining, S.S. and Atassi, M.Z. (1979) J. Immunol. Methods 30, 139.
17. Twining, S.S., David, C.S. and Atassi, M.Z. (1981) Molecular Immunol. 18, 447.
18. Koketsu, J. and Atassi, M.Z. (1974) Immunochemistry 11, 1.
19. Schmitz, H.E., Atassi, H. and Atassi, M.Z. (1983) Immunol. Comm. 12, 161.
20. Leach, S.J. (1983) Biopolymers 22, 425.
21. Koketsu, J. and Atassi, M.Z. (1973) Biochim. Biophys. Acta 328, 289.
22. Okuda, K., Christadoss, P., Twining, S.S., Atassi, M.Z. and David, C.S. (1978) J. Immunol. 121, 866.

23. Okuda, K., Twining, S.S., David, C.S. and Atassi, M.Z. (1979) J. Immunol. 123, 182.

24. Atassi, M.Z. (1967) Biochem. J. 103, 29.

25. Andres, S.F. and Atassi, M.Z. (1970) Biochemistry 9, 2268.

26. Twining, S.S., Lehmann, H. and Atassi, M.Z. (1981) Molecular Immunol. 18, 473.

27. Atassi, M.Z., Sakata, S. and Kazim, A.L. (1979) Biochem. J. 179, 327.

28. Atassi, M.Z., Long, P.M., Beisel, K., Sakata, S., Peters, T., Jr. and David, C.S. (1982) Molecular Immunol. 19, 313.

29. Young, C.R. and Atassi, M.Z. (1983) J. Immunogen. 10, 161.

30. Yoshioka, M., Bixler, G.S. and Atassi, M.Z. (1983) Molecular Immunol. 20, 1133.

31. Bixler, G.S., Bean, M. and Atassi, M.Z. (1986) Biochem. J. 240, 139.

32. Kazim, A.L. and Atassi, M.Z. (1978) Immunochemistry 15, 67.

33. McCormick, D.J. and Atassi, M.Z. (1983) Fed. Proc. 41, Abstr. 2411.

34. McCormick, D.J. and Atassi, M.Z. (1983) Fed. Proc. 41, Abstr. 2412.

35. Koketsu, J. and Atassi, M.Z. (1974) Biochim. Biophys. Acta 342, 21.

36. Atassi, M.Z. and Pai, R.C. (1975) Immunochemistry 12, 735.

37. Atassi, M.Z. and Kazim, A.L. (1978) In: Immunobiology of Proteins and Peptides (Atassi, M.Z. and Stavitsky, A.B., Eds.) 1, 19, Plenum Press, New York.

38. Perutz, M.F., Muirhead, H., Cox, J.M. and Goaman, L.C.G. (1968) Nature, London 219, 131.

39. Fermi, G. (1975) J. Mol. Biol. 97, 237.

40. Krco, C.J., Kazim, A.L., Atassi, M.Z. and David. C.S. (1981) J. Immunogen. 8, 315.

41. Krco, C.J., Kazim, A.L., Atassi, M.Z. and David, C.S. (1981) J. Immunogen. 8, 395.

42. Krco, C.J., Kazim, A.L., Atassi, M.Z., Melvold, R. and David, C.S. (1981) J. Immunogen. 8, 471.

43. Krco, C.J., Abramson, E.J., Yoshioka, N., Atassi, M.Z. and David, C.S. (1984) J. Immunogen. 11, 11.

44. Atassi, M.Z. and Atassi, H. (1985) FEBS Letters 188, 96.

45. Atassi, M.Z. and Atassi, H. (1986) Europ. J. Immunol. 16, 229.

46. Mulac-Jericevic, B., Kurisaki, J. and Atassi, M.Z. (1987) Proc. Natl. Acad. Sci. USA. (in press)

47. Atassi, M.Z. and Kurisaki, J. (1984) Immunol. Comm. 13, 539.

48. Tzartos, S. and Lindstrom, J.M. (1980) Proc. Natl. Acad. Sci. USA 77, 755.

49. Tzartos, S., Langeberg, L., Hochschwender, S., and Lindstrom, J. (1983) FEBS. Lett. 158, 116.

50. Ulrich, R.G., Atassi, H., Lutz, P., Cresswell, P. and Atassi, M.Z. (1987) Europ. J. Immunol. (in press)

51. Atassi, H., Ulrich, R.G. and Atassi, M.Z. (1987) Europ. J. Immunol. (in press)

52. Bixler, G.S. and Atassi, M.Z. (1983) Immunol. Comm. 12, 593.

53. Bixler, G.S. and Atassi, M.Z. (1984) J. Immunogen. 11, 339.

54. Bixler, G.S., Yoshida, T. and Atassi, M.Z. (1984) Expl. Clin. Immunogen. 1, 99.

55. Bixler, G.S., Yoshida, T. and Atassi, M.Z. (1984) J. Immunogen. 11, 327.

56. Yoshioka, M., Yoshioka, N. and Atassi, M.Z. (1986) Biochem. J. 234, 449.

57. David, C.S. and Atassi, M.Z. (1982) In: Immunobiology of Proteins and Peptides (Atassi, M.Z., Ed.) 2, 97, Plenum Press, New York.

58. Bixler, G.S., Yoshida, T. and Atassi, M.Z. (1984) J. Immunogen. 11, 327.

59. Yokoi, T., Mulac-Jericevic, B. and Atassi, M.Z. (1987) Adv. Exptl. Med. Biol. (in press)

60. Atassi, M.Z., Yokota, S., Twining, S.S., Lehmann, H. and David, C.S. (1981) Mol. Immunol. 18, 945.

61. Yoshioka, N. and Atassi, M.Z. (1986) Biochem. J. 234, 457.

62. Kazim, A.L. and Atassi, M.Z. (1981) Biochem. J. 197, 507.

63. Yoshioka, N. and Atassi, M.Z. (1986) Biochem. J. 234, 453.

64. Atassi, M.Z. (1985) In: High Technology Route to Virus Vaccines, (Dreesman, G.R., Bronson, J.G. and Kennedy, R.C., Eds.), p. 17, A.S.M. Publications, Washington, D.C.

65. Atassi, M.Z. and Kurisaki, J. (1986) Proteides of the Biological Fluids 34, 157.

66. Lee, C.L. and Atassi, M.Z. (1976) Biochem. J. 159, 89.

67. Atassi, M.Z. and Habeeb, A.F.S.A. (1977) In: Immunochemistry of Proteins (Atassi, M.Z., Ed.) 2, 177, Plenum Press, New York.

68. Lee, C.L. and Atassi, M.Z. (1977) Biochim. Biophys. Acta 495, 354.

69. Lee, C.L. and Atassi, M.Z. (1977) Biochem. J. 167, 571.

70. Atassi, M.Z. and Lee, C.L. (1978) Biochem. J. 171, 419.

71. Atassi, M.Z. and Saplin, B.J. (1968) Biochemistry 7, 688.

72. Atassi, M.Z., Habeeb, A.F.S.A. and Ando, K. (1973) Biochim. Biophys. Acta 303, 203.

73. Atassi, M.Z., Koketsu, H. and Habeeb, A.F.S.A. (1976) Biochim. Biophys. Acta 420, 358.

74. Lee, C.L. and Atassi, M.Z. (1975) Biochim. Biophys. Acta 405, 464.

75. Atassi, M.Z. (1979) Crit. Revs. Biochem. 6, 371.

76. Atassi, M.Z. and Sakata, S. (1982) Biochem. J. 201, 669.

77. Bixler, G.S. and Atassi, M.Z. (1984) J. Immunogen. 11, 245.

78. Okuda, K., Sakata, S., Atassi, M.Z. and David, C.S. (1979) J. Immunogen. 6, 447.

80. Amzel, L.M., Poljak, R.M., Saul, F., Varga, J.M. and Richard, F.F. (1974) Proc. Natl. Acad. Sci. USA. 71, 1427.

81. Saul, F.A., Amzel, L.M., Poljak, R.J. (1978) J. Biol. Chem. 253, 585.

82. Kazim, A.L. and Atassi, M.Z. (1980) Biochem. J. 187, 661.

83. Padlan, E.A., Davies, D.R., Rudikoff, S. and Porter, M. (1976) Immunochemistry 13, 945.

84. Segal, M.F., Padlan, E., Cohen, G.H., Rudikoff, S., Potter, M., and Davies, D.R. (1974) Proc. Natl. Acad. Sci. USA 71, 4298.

85. Atassi, M.Z. and Zablocki, W. (1977) J. Biol. Chem. 252, 8784.

86. Atassi, M.Z. and Sakata, S. (1982) Biochem. J. 201, 669.

87. Atassi, M.Z. (1985) Biochem. J. 226, 477.

88. Stroud, R.M., Kay, L.M. and Dickerson, R.E. (1974) J. Mol. Biol. 83, 185.

89. Huber, R. and Bode, W. (1978) Acc. Chem. Res. 11, 114.

90. Stroud, R.M., Kay, L.M. and Dickerson, R.E. (1971) Cold Spring Harbor Symp. Quant. Biol. 36, 125.

91. Chambers, J.L. and Stroud, R.M. (1977) Acta Crystallogr., Sect. B B33, 1924.

92. Fahrney, D.E. and Gold, A.M. (1963) J. Am. Chem. Soc. 85, 997.

93. Singer, S.J. (1967) Adv. Protein Chem. 22, 1.

94. Turini, P., Kurooka, S., Steer, M., Corbascio, A.N. and Singer, T.P. (1969) J. Pharmacol. Exp. Ther. 167, 98.

95. Morgan, P.H., Robinson, N.C., Walsh, K.A. and Neurath, H. (1972) Proc. Natl. Acad. Sci. USA 69, 3312.

96. Atassi, M.Z. (1986) In: Protein Engineering (Inouye, M. and Sarma, P., Eds.) p. 125, Acad. Press, New York, N.Y.

# CONTINUOUS FLOW PEPTIDE SYNTHESIS

R. Frank, H.Gausepohl

European Molecular Biology Laboratory, Heidelberg

## Introduction

The introduction of solid phase peptide synthesis by R.B. Merrifield in 1962 (1) induced tremendous activity in peptide synthesis. A large number of biologically active peptides have been synthesized using reaction conditions close to Merrifield's original version. The repetitive nature of the operations in the solid phase concept soon led to the development of several automated instruments (2, 3), for the assembly of peptide chains. However, a number of mechanical problems and chemical inadequacies reduced the practical value of the instruments and prompted many chemists to return to manual methods. A new era in peptide synthesis research was initiated during the last decade, when synthetic peptides became increasingly important experimental tools in molecular biology and immunology. Consequently the development of new improved synthesis strategies and the construction of new instruments for automated peptide synthesis has been stimulated (4, 5).

In this paper we describe the construction and operation of a new automated solid phase peptide synthesizer employing a continuous flow reaction system. The mechanical design is based on our experience with the construction of a line of biochemical instruments (oligonucleotide synthesizer and protein sequencer) (6) and on a critical evaluation of the mechanical parameters affecting the performance and efficiency of an automatic instrument. Departing from Merrifields proposals an orthogonal synthesis strategy is employed which can be carried out under reasonably mild reaction conditions.

Modern Methods in Protein Chemistry, Vol. 3

## Synthesis strategy

Transient N-α-protection

In Merrifield´s original procedure one fundamental weakness is the dependence on differential acidolysis of N-α-protection on one hand and side chain protection and peptide anchoring on the other hand. The experience of the first two decades has shown that differential acidolysis is never absolutely selective (7) and can cause serious problems due to the accumulative effect of side reactions. Therefore a number of modifications have been developed to increase the difference in acidolytic susceptibility of N-α-protection and side chain protection and peptide anchoring, respectively (8, 9).
One solution to this problem is the application of an orthogonal protection scheme. In this strategy the transient protecting groups of the N-α-amino function and the semi permanent blocking groups of side chains are cleaved by completely different chemicals introducing chemical selectivity into the protection scheme. Since base treatment is routinely used in classical solid phase procedures to deprotonate α-amino functions after acidolytic deprotection, several orthogonal methods have been proposed, which employ base labile N-α protection (10, 11). Because of its rapid and quantitative cleavage by nonhydrolytic base treatment (in aprotic solvents) the 9-fluorenylmethyloxycarbonyl-group (Fmoc) (12) appeared to be best suited for this strategy when used in combination with tertiary butyl based side chain protection which is resistant against nucleophiles. Using a solution of 20% piperidine in dimethylformamide (DMF) the Fmoc-group is cleaved off very fast and the final dibenzofulvene-piperidine adduct (13) is formed within seconds. This strategy introduces the desired chemical selectivity and simplifies the synthesis procedure by omitting the deprotonation step. Furthermore, the high UV-absorption of the Fmoc-group and its piperidine adduct can be used to monitor the concentration of Fmoc-derivatives during the coupling and deprotection procedures. The principal problems of optical density measurements in highly concentrated solutions of fluorophores such as the Fmoc-group and the potential of quantitative evaluations are discussed below.

Side chain protection and anchoring

Using a base labile N-α-protecting group the side chain functions of trifunctional amino acids can be blocked by more acid labile groups than the benzyl derived blocking groups originally introduced by Merrifield. The most commonly employed groups are based on the tertiary butyl residue such as t-butyl ester for glutamic and aspartic acid and t-butyl ether for serine and threonine. A complete listing of the side chain protection groups used with other trifunctional amino acids is given in Materials.

The Fmoc N-α–protection and acid labile side chain protection is advantageously combined with p-alkoxybenzyl type anchoring (14) of the peptide chain to the solid support. This type of anchoring is base stable and linkers can be constructed with varying degrees of acid lability (15). Thus, the semipermanent t-butyl side chain blocking groups and the peptide resin link can easily be cleaved under relatively mild acidic conditions in 50% trifluoroacetic acid (TFA) in dichloromethane (DCM).The ease of the final deprotection and cleavage procedure represents one of the major advantages of the orthogonal synthesis strategy. Harmful side reactions by treatment of the final product with strong and hazardous acids such as liquid hydrogen fluoride (16) are avoided and no complicated apparatus is required. During longer syntheses valuable information on the synthesis quality can be obtained and possible side reactions can be localised by taking resin samples from each synthesis step, cleaving the peptide and analysing the product purity by high performance liquid chromatography (HPLC). However, for certain combinations of side chain protection groups and sensitive amino acids in a given peptide the final deprotection strategy must be carefully selected. Effective scavengers have to be used in the deprotection and cleavage reaction, since the tertiary butyl carbonium ions which are produced as intermediates are extremely reactive alkylating agents. The proper selection of scavengers is easily and rapidly evaluated using 2-5 mg resin samples for test cleavages and analysis of the product by HPLC.

Peptide bond formation

In the chain elongation step the liberated amino group of the polymer-bound peptide is acylated by another N-α-protected amino acid building block bearing an activated carboxyl group. To date dicyclohexylcarbodiimide (17) is the most utilized reagent for activation of N-α-protected amino acids because of its high reactivity and convenience of application. However, this reagent is not ideal for *in situ* activation in an automated instrument. Several diimide-dependent side reactions (for review see 18) and the formation of insoluble acylurea derivatives reduce the utility of the method. To overcome these problems and to simplify mechanisation, preformed symmetrical anhydrides (19) have been introduced as acylating species (in automated synthesizers), thus avoiding contact of the peptide-amino function with the activating agent. Furthermore, activated esters have been used to avoid the need for manual preactivation in each step (20). Recently an instrument has been described (21) which is capable of unattended formation of symmetrical anhydrides and removal of urea precipitates.

While searching for a universal acylation procedure, which is easy to incorporate in an automated process we have reinvestigated the benzotriazole-1-yl-oxy-tris-(dimethylamino)-phosphonium-hexafluorophosphate (BOP) reagent introduced in 1975 by Castro *et al.* (22). This reagent facilitates a fast and efficient acylation reaction without formation of insoluble precipitates. It represents a stable N-hydroxybenzotriazole (HOBt) derivative which is not hygroscopic and soluble in dimethylformamide even at high concentrations. The acylating species is the highly reactive HOBt-ester (23), which is aminolysed at a rate about $10^3$-fold faster than similar active esters.

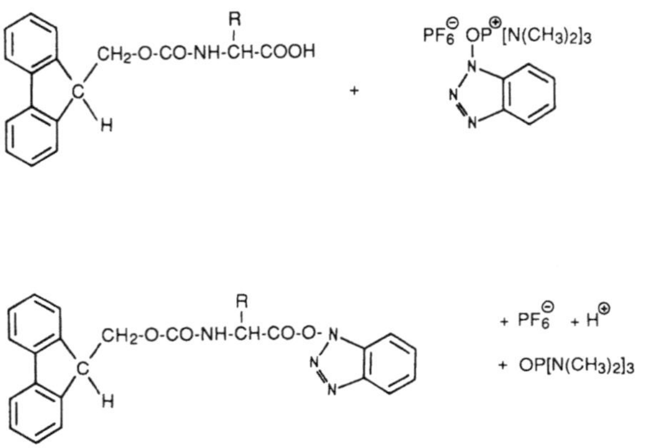


*Fig. 1: Activation of Fmoc-amino acids with benzotriazole-1-yl-oxy-tris-(dimethylamino)-phosphonium hexafluorophospate*, BOP.

The same intermediate is formed when dicyclohexylcarbodiimide is used in combination with HOBt as catalyst (24). This active ester intermediate is known to suppress racemisation of amino acids and to prevent side reactions on the amide groups of asparagine and glutamine, especially when used in polar solvents such as dimethylformamide. Nevertheless, BOP has been used only rarely in liquid phase peptide synthesis (25) and no example of its application in solid phase peptide synthesis has been reported.

The experiments reported below suggest that BOP may be a generally suitable activating reagent when used in DMF solution and in the presence of a sterically hindered tertiary base as a catalyst. For activation the N-$\alpha$-protected amino acid is mixed with one molar equivalent of BOP and base catalyst in DMF. The mixture is then immediately transferred to the reactor containing the resin-bound peptide chain. The reaction is extremely fast and in several cases already after five minutes complete acylation was detected by the ninhydrin assay (26).

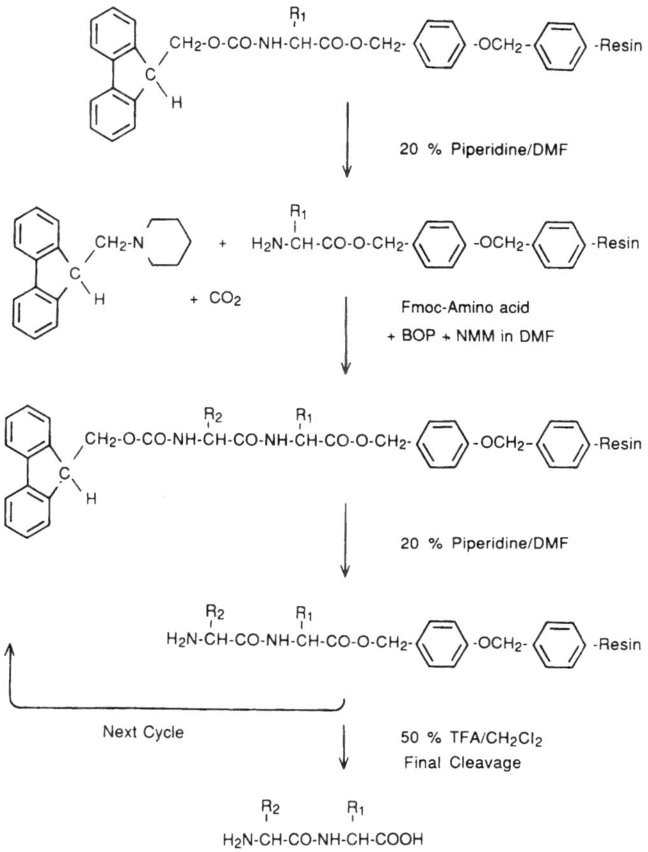


*Fig. 2: Synthesis cycle scheme for Fmoc-solid phase peptide synthesis using BOP activation and p-benzyloxybenzyl ester anchoring.*

The solid support

To date the most effective and best understood support for solid phase peptide synthesis is a copolymer of styrene and 1% divinylbenzene (27), consisting of mechanically stable and chemically inert beads of 50 μm average size. The polymer network is highly mobile and solvated and the attached peptide chains behave like macromolecules in concentrated solution showing normal reaction kinetics . The physico-chemical properties of the polymer seem to enhance the solvation of the growing peptide chain. This helps to overcome the principal problem of "solution" synthesis where the poor solubility of protected peptide intermediates is often the limiting factor. Since high loadings of peptide on the resin may be achieved (0.5 to 0.8 mmol/g) and the reactant concentrations can be kept at their upper limit, the second order coupling reaction is performed under optimal concentration conditions.

## The instrument

### The continuous flow principle

In the original Merrifield concept the synthesis process is carried out in a discontinuous batchwise manner using a shaker to mix solid and liquid phases. As early as 1971 the advantages of flow through reactors were recognized and several systems have been described (18, 28) that use centrifugal and column reactors. Especially when gelatineous polymeric supports with high liquid retention are used, the continuous flow principle offers several advantages over the conventional mode. Removal of reagents from the resin bed and exchange of solvents are performed faster and more efficiently. Furthermore, recirculation of reagents and solvents in the reaction system permits on-line photometric monitoring of the reaction progress and provides a direct control of the instruments most important functions. This qualitative information is obtained in real time and is extremely useful for the control of wash steps, proper reagent delivery and coupling rates.

However, several groups have reported serious problems with gelatinous supports in pumped flow systems (29), resulting from a compression of the soft resin beads and subsequent developement of high back pressures. Besides the technical problems encountered with high pressure flow systems (HPLC pumps are sensitive against air bubbles and precipitates) the compression of the resin beads leads to closure of the interstitial channels and reduces reactant accessibility. This makes the diffusion controlled reagent exchange with the polymer matrix more difficult. There have been several attempts to overcome the compression problem in continuous flow systems by use of highly crosslinked macroporous resins or even surface functionalized silica (30) and glass (31). To our knowledge none of these systems has proven to be fully suited to preparative peptide synthesis. Recently the preparation and successful application of a composite resin has been described, where dimethylacrylamide is polymerized within a rigid Kieselguhr matrix (32). Although this support shows excellent flow and solvation characteristics, its applicability to longer sequences (above 30 residues) has not yet been shown.

However, careful construction of the continuous flow reaction system, and operation of the reactor as a loosely packed fixed bed reactor at low pressure allows for routine use of the

known 1% crosslinked polystyrene resin. Compared to the commercially available dimethylacrylamide composite supports the polystyrene resins are cheaper, offer a five to ten times higher peptide loading and higher reactant concentrations can be achieved due to the lower internal dead volume of the resin bed.
Therefore we constructed a low pressure flow-system using a large diameter column reactor with variable bed size and a valveless reciprocating pump which is completely insensitive to air bubbles in the flow path. With an appropriate set of valves the system can easily be switched from the flow-through mode to the recirculation mode.

The electronic control system

The instrument is controlled by an eight bit microprocessor (Z80) equipped with 32 input/output ports. The main program is written in Basic and assembler language is used for real time software. Editing and execution of single operations, cycles and synthesis runs is supported by a set of menues. In the automatic run mode manual intervention down to the valve level is possible by either using the manual switchboard or by using a set of functions e.g. "hold","stop" or "next step", wich are defined on soft keys on the terminal keyboard.

Liquid delivery system

The major problem in instrumentation of chemical processes is the construction of a reliable delivery system for reagents and solvents. The valves in such a system have to be chemically inert, resistant towards corrosion and should allow for the construction of a dead volume free liquid distribution system to avoid any crosscontamination of reagents. Therefore we have chosen the so called zig-zag valves, pneumatically activated diaphragm valves originally developed by the Wittmann-Liebold group in Berlin (33). In these valves a protective air pad is kept between the pneumatic driver valves (solenoids) and the teflon diaphragms which are in direct contact with agressive chemicals. To simplify manufacture and to obtain optimum reliability the original design has been modified by replacing the zig-zag bore in the central valve body by a straight grove which serves as a collect line for different inlet ports (Fig.3). The straight grove and the inlet ports are covered by a PFA (perfluoralkoxy) membrane which is pressed down by compressed air or nitrogen to separate inlet ports and grove. To open a port the membrane is lifted at that position by a smooth vacuum, allowing the liquid to flow from the port to the collect grove. As shown in Figure 3, the ports can be combined to dead-volume free manifolds of any length.

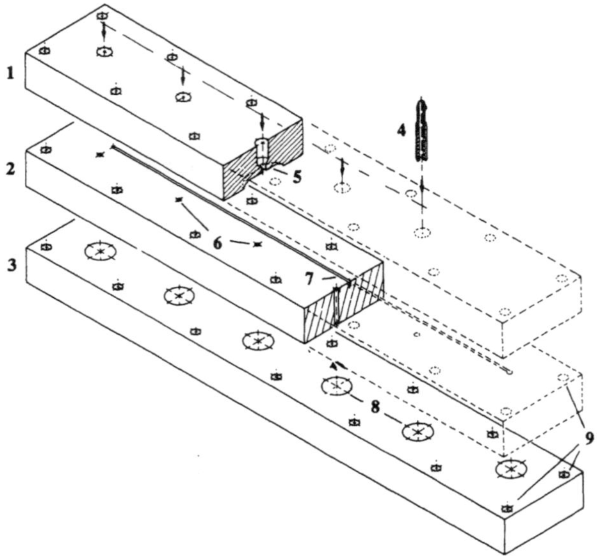


*Fig. 3: Explosion view of the modified valve manifold used. The pneumatic connection (4) is in the top part (1), solvent lines are connected to the bottom (3). The ports (6) are opened to the collect grove (7) by lifting the PFA membrane (not shown) between the top part and the valveblock body (2) into the small groves (5) cut into the top part. Top and bottom parts are made from aluminum, the valveblock body from Kel-F.*

The reaction loop

The flow scheme of the instrument is outlined in Fig.4. The shaded rectangles represent the valve manifolds distributing liquids and/or gas. The containers for reagents and solvents are kept under nitrogen pressure and flow rates are either pressure controlled or determined by the flow rate adjusted at the recirculating pump or the metering syringe.

N-α-protected amino acids are stored in a fraction collector turntable and dissolved immediatly before use. Precise metering and mixing of amino acids, activator and catalyst is performed in a motor-driven syringe from where the activated mixture is transferred into the reaction loop. The loop comprises the reaction column, the recirculating pump, a small volume compensation vial, an UV-flow cell and a combination of valves to fill and empty the loop. The volume compensation vial allows for the free choice of the volume to be recirculated in the reaction loop and serves as a mixing vial and bubble trap. The reaction column (Fig. 5) consists of a 10 ml Hamilton syringe fitted with glass sinter frits at the bottom and in the modified plunger. The syringe body is equipped with an injection port through which resin samples can be withdrawn.

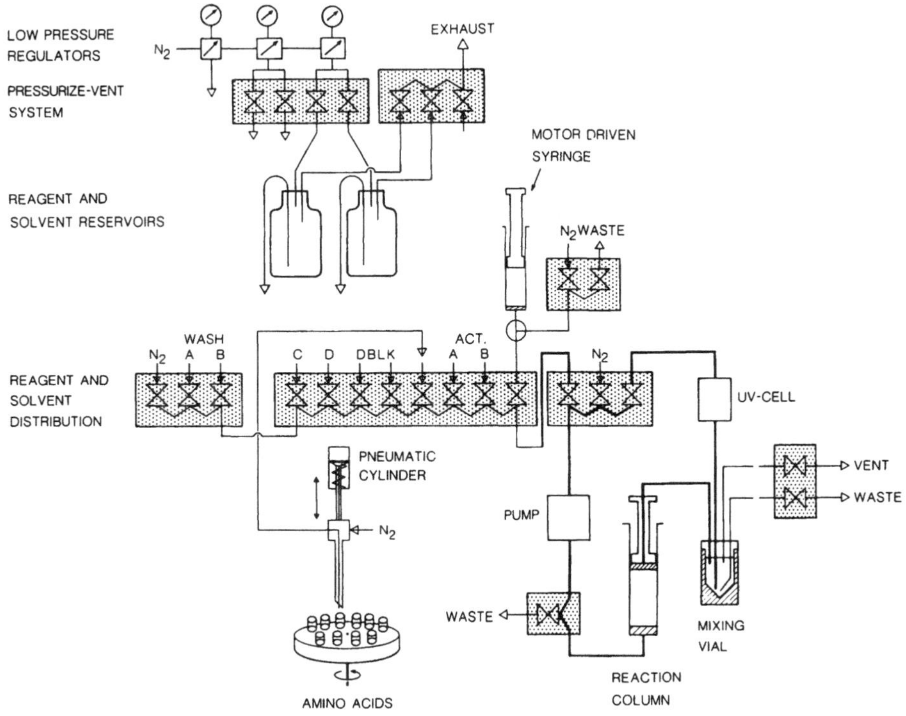


*Fig. 4: Flow scheme of the peptide synthesizer. The bold line indicates the liquid flow in the reaction loop under recycling conditions. The shaded rectangles represent the valve manifolds.*

In the recirculating mode liquids are pumped through the reaction loop and column by a valveless piston pump. This pump was chosen because of its insensitivity towards air bubbles in the flow path and its chemical inertness. The backpressure created by the reaction column usually ranges between 0.1 and 0.5 bar. Due to the fact that only one solvent (DMF) is used during the synthesis cycle, swelling and shrinking of the resin are reduced to a minimum. This also prevents shrunken beads from entering the frits at the ends of the column. Furthermore, the aprotic dipolar solvent DMF easily penetrates the polystyrene matrix and is a good solvent for protected peptide chains. These effects provide an optimum resin-peptide interaction and help to prevent the formation of peptide aggregates unaccessible to reagents.

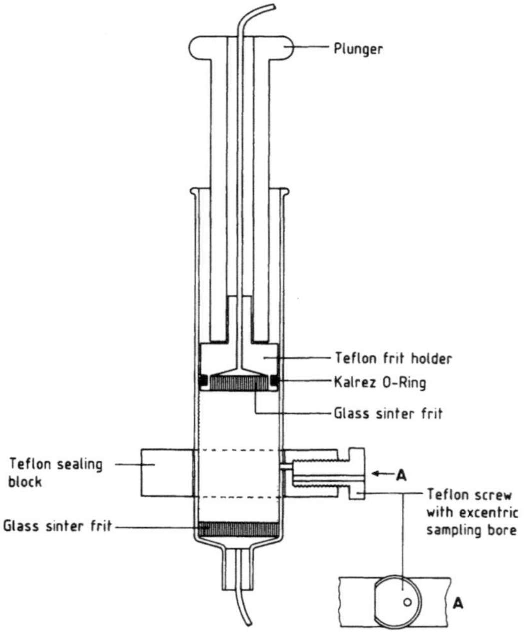


*Fig. 5: Cross section of the reactor column with sampling port. To take a sample the teflon screw is turned a half turn, a syringe needle inserted and 20 μl of the resin removed.*

Operation

A program for operation of the synthesizer depicted in Figure 4 is shown in Table 1. For clarity minor operations such as line purges have been omitted.The UV-recorder operates during the entire cycle and the signal is stored on a chart recorder or an electronic data acquisition system. The first synthesis cycle is started with a single wash of the resin carrying the first N-α-protected amino acid. Following this procedure the repetitive synthesis cycle is entered by performing two deprotection steps under recycling conditions. In the first deprotection step a plateau is reached in UV-absorption which can be quantitatively compared with the one in following cycles. To ensure complete deprotection a short second base treatment is used. After deprotection the reaction column and the mixing vial are emptied by applying nitrogen pressure to the loop and opening the waste valve between reactor and pump and the mixing vial waste valve. This operation is called"empty loop" in the program scheme (Tab.1) and is repeated before each reagent or solvent change.

*Table 1: Peptide Synthesizer Program.*
*Only major functions are listed. Functions listed in italic are performed parallel to the ones above.*

| Step | Operation | Delay (sec) |
|---|---|---|
| 01 | DMF to loop | 50 |
| 02 | recycle (pump on) | 140 |
| 03 | empty loop | 40 |
| 04 | piperidine to loop | 50 |
| 05 | recycle | 600 |
| 06 | empty loop | 20 |
| 07 | piperidine to loop | 50 |
| 08 | recycle | 200 |
| 09 | empty loop | 40 |
| 10 | DMF to vial | 15 |
| 11 | DMF to loop/flow through | 140 |
| 12 | empty loop | 40 |
| 13 | DMF to vial | 15 |
| 14 | DMF to loop/flow through | 140 |
| 15 | empty loop | 40 |
| 16 | DMF to loop | 50 |
| 17 | recycle | 140 |
| *18* | *fraction collector step* | |
| *19* | *catalyst to syringe* | *35* |
| *20* | *amino acid to syringe* | *70* |
| *21* | *BOP to syringe* | *35* |
| 22 | empty loop | 40 |
| 23 | inject amino acid | 140 |
| 24 | recycle (coupling) | 2600 |
| *25* | *wash syringe* | *40* |
| *26* | *wash syringe* | *40* |
| 27 | empty loop | 40 |
| 28 | DMF to vial | 15 |
| 29 | DMF to loop/flow through | 120 |
| 30 | empty loop | 40 |
| 31 | DMF to vial | 15 |
| 32 | DMF to loop/flow through | 120 |
| 33 | empty loop | 40 |
| 34 | DMF to loop | 50 |
| 35 | recycle | 140 |
| 36 | empty loop | 40 |
| 37 | start next cycle | |

In the first step of each washing procedure the system is switched to the flow through mode by opening the waste valve on the vial. The UV-detector line and the vial are washed by delivering DMF directly to the vial . Subsequently, DMF is pumped through the resin bed and the vial to waste. The loop is emptied and this operation repeated once.

The DMF delivered in the next step is recirculated to measure residual UV-absorption and to assess the completeness of the washing procedure. During this recirculation time the predissolved amino acid is drawn from the fraction collector into the metering syringe and is activated by adding BOP-solution and base catalyst. The solution is then introduced across the UV-detector cell into the loop, thus providing a reference value for the UV-absorption. Recirculation then starts for a coupling time of usually 45 minutes. Samples of resin can be taken at any time by just setting the computer into "stop" mode and resuming operation when the sample has been taken.
Coupling times can easily be prolonged if neccessary by setting the computer to "hold" mode during recycling. The cycle is then finished by a series of wash procedures as described above. A summary of reagent composition and solvent consumption is given in Table 2.

*Table: 2. Synthesis Cycle Scheme.*
*Synthesis cycle scheme used for the synthesis of 0.25 mmoles of peptide in a reactor of 14 mm diameter. Resin loading ranges from 0.3 - 0.7 mmoles/g corresponding to a volume of resin swollen in DMF of 1.8 - 3.5 ml, respectively. Two equivalents of Fmoc-amino acid and two equivalents of BOP are used.*

| Function | Volume (ml) | Time (min) |
|---|---|---|
| DMF wash, recycle | 5 | 5 |
| Deprotection, 20 % Piperidine/DMF | 5 | 10 |
| Deprotection, 20 % Piperidine/DMF | 5 | 4 |
| DMF wash through | 9 | 5 |
| DMF wash through | 9 | 5 |
| DMF wash recycle | 5 | 4 |
| Activation of amino acid: | | |
| 0.5 mmol amino acid/2.0 ml DMF | | |
| + 0.5 mmol BOP /1.2 ml DMF | | |
| + 0.875mmol NMM/1.2 ml DMF | | |
| Add activated amino acid | 5 | 2 |
| Coupling, recycle | | 45 |
| DMF wash through | 7 | 4 |
| DMF wash through | 7 | 4 |
| DMF wash, recycle | 5 | 4 |

UV-monitoring

Besides the high speed and efficiency of solvent and reagent utilisation the ability to make optical density measurements of the reactant stream represents one of the most important advantages of the continuous flow system. Direct photometric monitoring of the recirculating reagent and wash solutions allows for a qualitative and - with some restrictions - quantitative evaluation of reaction progress and of the completeness of the washing procedures. In general a very useful real time control of the correct operation of the instrument is provided. The spectrometric record obtained during synthesis cycles 5 to 7 of peptide II is given in Figure 6. The record starts with the addition of the deprotecting reagent (20% piperidine in DMF) to the N-α-protected peptide resin. The speed of the deprotection reaction and of formation of the dibenzofulvene-piperidine adduct is reflected in the sharp peak at time zero. Since any solvent has been blown out of the reaction loop before adding the deprotection reagent, dilution is reduced to the volume of solvent retained by the polymer matrix of the support. Therefore a uniform concentration of reagents throughout the reaction loop is already achieved after 5 minutes.

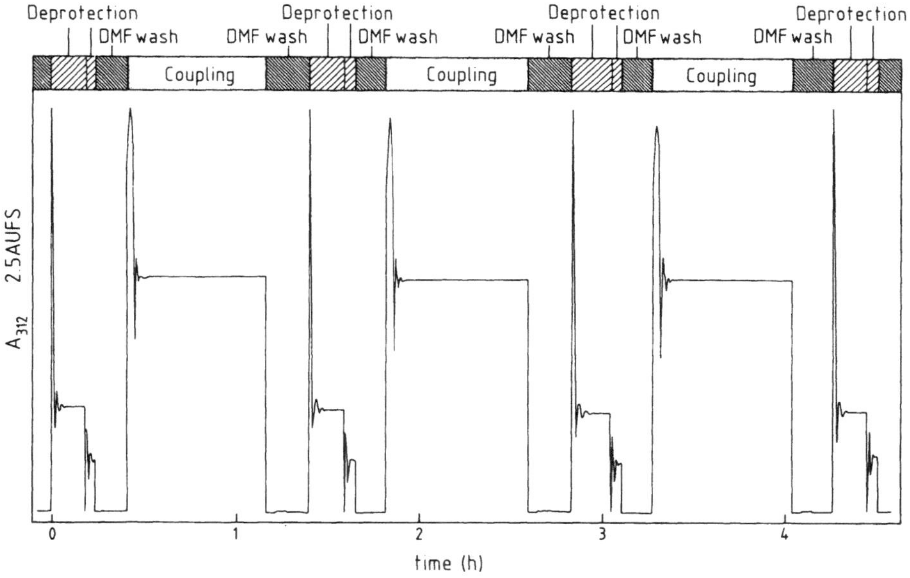


*Fig. 6:.Spectrophotometric record of synthesis cycles 5 to 7 of peptide II. UV-absorption was measured at 312 nm using a flow cell of 0.1 mm path length. The shaded bar at the top indicates the operations corresponding to the UV-profile.*

The plateau height of the UV-absorption during deprotection is directly proportional to the amount of Fmoc-groups cleaved from the peptide on the resin. Based on the known volume of reagent in the reaction loop and using a calibration curve a quantitative value can be calculated. This value can now be related to that of the preceeding and following cycles providing an important control of the efficiency of the synthesis.

A similar picture is produced during the coupling steps where the first peak represents the UV-absorption of the activated Fmoc-amino acid which is delivered through the UV-cell into the mixing vial. Due to the very fast coupling reaction the concentration of the Fmoc-amino acid is decreasing rapidly when the system is switched to the recirculation mode and after a few oscillations a stabile plateau is reached indicating that the reaction is almost complete. The amount of Fmoc-derivative coupled to the peptide resin may now be calculated by subtracting the Fmoc concentration still present in the mixture from the known amount of Fomc-amino acid introduced into the reaction loop. Complete acylation should therefore result in a 50% decrease in Fmoc-amino acid concentration since usually a two-fold excess over the peptide amino functions is used.

However, several mechanical and spectrometric problems reduce the accuracy attainable by this method. Since all concentration measurements are dependent on the precise metering of solvents and reagents even minor variations in the volumes introduced into the reaction loop create an uncertainity. A further source of inaccuracy in volume determination is solvent retention in the polymer matrix which changes with the size and the composition of the peptide loading. The removal of resin samples for analytical purposes is an additional manipulation which introduces errors. Taking together all the factors mentioned above, the step to step reproducibility of the concentration measurement in the deprotection step is charged with a minimal error of two percent. Nevertheless our experience has shown that the value obtained is a good indication that the synthesis is proceeding well and a decrease in coupling yield which is higher than four percent is detected reliably. Athough this information is obtained at a time where remedial operations are no longer possible a direct localisation of difficult coupling reactions is given.

Due to the high reactant concentration and to the chemistry involved, the optical density measurement during the coupling reaction is even more complicated. In addition to the inaccuracy in volume determination discussed above the deviation from Lambert-Beer's law, fluorescence phenomena, detector drift and the release of new UV-absorbing molecules produce a very complex basis for quantitative calculations. Using a 0.1 mm flow cell and a wavelength setting of 312 nm it was possible to obtain a calibration curve for the UV-absorption of activated Fmoc-amino acids. However, there is no linear correlation between absorption and concentration. Due to the increasing contribution of the fluorescence of the fluorenylmethl group the curve flattens at higher concentrations and changes in concentration become less significant. A further source of inaccuracy is the production of UV-absorbing HOBt molecules during the coupling reaction and by a slow decomposition of amino acid active ester intermediates (see Fig.1) Therefore the increase in UV-absorption due to the growing concentration of free HOBt molecules counteracts the decrease caused by the uptake of Fmoc-amino acids by the peptide resin to some extent. Whilst in the example shown in Figure 6 a good agreement with the ninhydrin test was obtained, indicating that the formation of the absorption plateau corresponds with complete acylation, during some very slow coupling reactions after 60 minutes a plateau was reached although three to five percent of free amino termini were still present.

Summarising all errors introduced by mechanical, spectrophtometric and chemical reasons it is obvious that the precision attainable at this stage of development is still below the level required for quantitative determination of the coupling efficiency. However important semi-quantitative information is obtained, allowing for identification of slow acylations which require double couplings. Together with the information produced during the deprotection and washing procedures a valuable set of analytical data and a reliable real time control of the correct operation of the instrument is provided.

## Experimental

### Synthesis of peptide I.

H-Lys-Lys-Ser-Ala-Glu-Asp-Tyr-Glu-Tyr-Pro-Ser-Met-Gly-OH

Fmoc-Glycine resin (242 mg = 0.15 mmoles glycine) was placed in the reactor and allowed to swell in DMF for 15 min. All coupling and deprotection cycles were performed according to the scheme outlined in Table 1, except for variations in coupling time. Ninhydrin tests gave none or only a faint blue colour with the first sample taken after 30 min. of coupling. Two equivalents (0.3 mmoles) of protected amino acids were used and the coupling allowed to proceed for the times indicated below:

Fmoc-Met-OH (60 min), Fmoc-Ser(tBu)-OH (45 min), Fmoc-Pro-OH (40 min), Fmoc-Tyr(tBu)-OH (30 min), Fmoc-Glu(OtBu)-OH (30 min), Fmoc-Tyr(tBu)-OH (60 min), Fmoc-Asp(OtBu)-OH (60 min), Fmoc-Glu(OtBu)-OH (35 min), Fmoc-Ala-OH (35 min), Fmoc-Ser(tBu)-OH (35 min), Fmoc-Lys(Boc)-OH (35 min), Fmoc-Lys(Boc)-OH (35 min).

For activation the Fmoc-amino acids dissolved in 2 ml DMF were mixed with solutions of BOP (420 mM in DMF) and N-methyl morpholine (730 mM in DMF) at molar ratios of 1.0 to 1.0 to 1.5 respectively. Deprotection was performed using 20 % Piperidine in dry DMF. After the last cycle a final deprotection was performed and the resin washed with DMF. The resin was then transferred to a sinter frit funnel, washed extensively with methanol and dried. Cleavage from the resin and deprotection of side chains was performed in 5 ml TFA/DCM (1:1) with 5 % thioanisole for 1 h. Cleavage was 90 % effectice as shown by amino acid analysis of the resin. The peptide was precipitated with ether and washed three times. The precipitate was dissolved in water, remaining ether evaporated, the solution filtered through a 0.45 µm filter and lyophilized, yielding 171 mg of raw product (76% of the theoretical yield). The analytical HPLC-diagram is shown in Fugure 8. The peptide was then dissolved in 2 ml 10% acetic acid and gel filtrated on Sephadex G-25 (2.8x30 cm) in 1 % acetic acid. The final product was sequenced by automatic gas phase sequencing. Results are shown in Figure 7. The peptide was successfully used as a substrate for a tyrosine sulfotransferase.

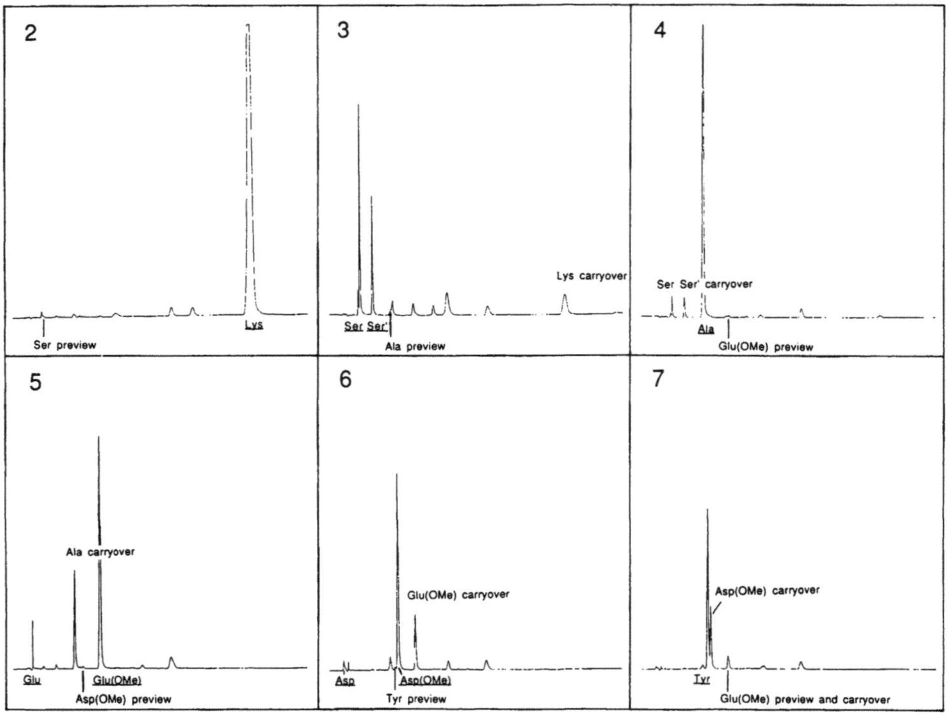


*Fig. 7. Steps 2 to 7 of Edman degradation of Peptide I, final product after gel filtration. The preview values as percentage of the main product found in the corresponding steps are: Step 2: 0.2% Ser; step 3: 0.8% Ala, step 4: 0.4% Glu, step 5: 1.1% Asp, step 6: 1% Tyr The values are not corrected for step yield and overlap.*

Synthesis of peptide II.
H-Pro-Ala-Cys-Lys-Ile-Pro-Asp-Asp-Leu-Gly-Phe-Pro-Glu-Glu-OH

The peptide was synthesized on 510 mg of resin loaded with 0.49 mmoles of Fmoc-Glu(OtBut) (0.25 mmoles total) per gram resin . 0.5 mmoles of Fmoc-amino acid dissolved in 2 ml DMF and activated by addition of one molar equivalent of BOP in DMF and 1.75 equivalents of NMM were used in each coupling. Coupling times were 45 minutes throughout, except for a 2 x 45 min double coupling of isoleucine. Fmoc-Glu(OtBut)-OH, Fmoc-Asp(OtBut)-OH, Fmoc-Lys(Boc)-OH and Fmoc-Cys(Trt)-OH were used.

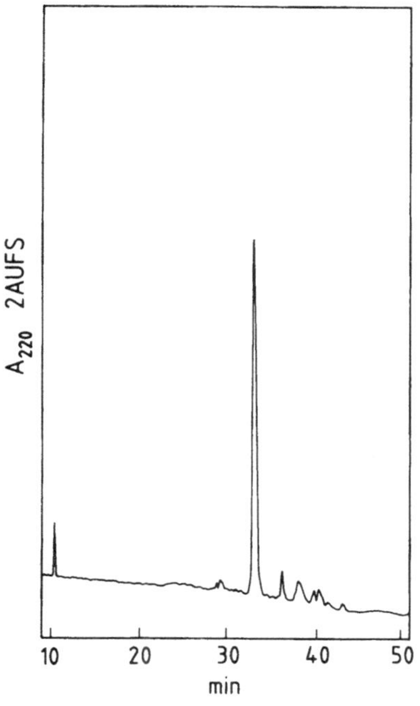


*Fig. 8. Analytical HPLC-diagram of: peptide I, KKSAEDYEYPSMG, raw product before gel filtration.*
*Column: Lichrosorb RP 18, 4x250 mm*

*Eluent A: 10 mM $(NH_4)HCO_3$, pH 8;*
*Eluent B: 80% $CH_3CN$, 20% A.*
*Gradient: 5 to 100% B in 45 min.*
*Detection at 220 nm.*

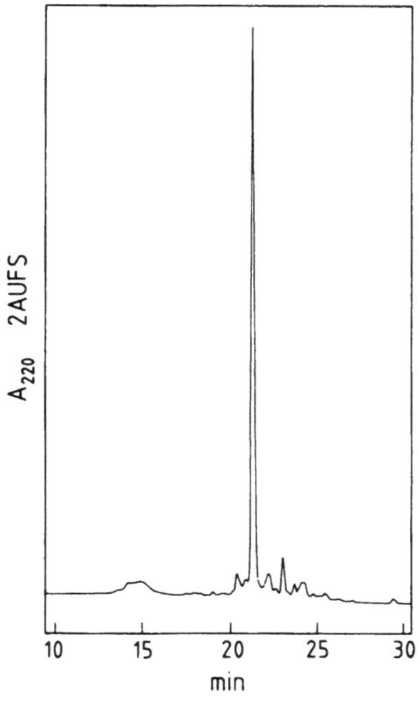


*Fig. 9. Analytical HPLC-diagram of peptide II, PACKIPDDLGFPEE, raw product.*
*Column: Lichrosorb RP 18, 4x250 mm.*
*Eluent A: 40 mM triethylammonium phosphate, pH 2.5*
*Eluent B: 80% $CH_3CN$, 20%A.*
*Gradient: 5 to 100% B in 45 min.*

The peptide was cleaved from the resin and deprotected with 90 % TFA, 10 % ethanethiol in one hour at room temperature, precipitated in ether and washed. The precipitate was dissolved in water, filtered through a 0.45 μm filter and lyophilized.The product was analyzed by HPLC (Fig. 9), assayed for free thiol groups and used for immunisation without further purification. 290 mg crude product (78% of theoretical yield) were obtained. In this case small scale experiments had indicated, that detritylation was only 50 % complete when the peptide was cleaved with TFA/DCM or TFA/10 % phenol. In TFA/10% ethanethiol cleavage of trityl is rapid and back addition is prevented effectively by the competing scavenger ethanethiol.

## Materials

### The synthesizer

The peptide synthesizer was constructed as outlined above and shown in Figure 4. Liquids were pumped by a valveless piston pump (model RHOCKC, Fluid Metering Inc., New York 11771), set at a flow rate of 3 ml/min. For the reactor a 10 ml glass syringe (#1010 TLL, Hamilton, Switzerland) was fitted with a glass sinter frit (DURAN P3, 16-40 um pore size, Schott) at the bottom and a second frit in the modified plunger to facilitate flow through the reactor (Fig. 5). This reactor was used for 0.1 -1.0 g resin at a swollen volume of 0.5-6 ml. Resin samples were withdrawn by removing the plunger or using the sampling bore in the reactor wall. All tubing in the synthesizer was made from 0.5, 0.8 and 1.0 mm i.d. teflon tube using either pull through or Altex flange type fittings. UV-absorption was measured at 312 nm using a SPD-2A (Shimadzu, Japan) variable wavelength detector fitted with a 0.1 mm flow cell (Labomatik, Sinsheim, FRG).

A 5 ml conical vial (Pierce, USA) with the tubing pulled through a teflon stopper in the cap was incorporated into the loop to provide for a variable volume and a bubble trap. The loop volume apart from reactor and vial was 1.0 ml. Reagent flow in the synthesizer was controlled by pneumatically activated membrane valves with the pneumatic valves (PK-2-SA, FESTO, Esslingen) mounted on a standard electronic circuit board. The zig-zag bore in the original Wittmann-Liebold design was replaced by a 0.6 x 0.8 mm groove and the fluid inlets were straight 0.8 mm bores (Fig. 3). The valveblock body was made from Kel-F (DuPont, USA) and the membranes from PFA (Type 1000 LP, o.25mm, Pampus, Willich). Apart from the PFA membrane no other sealing devices are required.

Fmoc-amino acids were stored in a fraction collector (BÜCHI 660, Switzerland) and taken up by a pneumatically actuated injector. Fmoc-amino acids, BOP and N-methyl morpholine solutions were mixed in a motor driven metering syringe (Hamilton, Switzerland). All reagents except for amino acids were stored under 0.5 bar nitrogen pressure. All functions of the synthesizer were controlled by a Z-80 8 bit microprocessor of 64 k RAM, equipped with a floppy disk drive, a Televideo terminal and a printer.

Chemicals

1% crosslinked polystyrene resin with p-benzyloxybenzyl linker and with the first Fmoc-amino acid already attached was purchased from Novabiochem, Switzerland. The same company supplied N-α-Fmoc protected amino acids and benzotriazole-1-yl-oxy-tris-(dimethylamino)-phosphoniumhexaflourophosphate, BOP. The side chain protected derivatives generally used are: Fmoc-Arg(Mtr)-OH, Fmoc-Asp(OtBu)-OH, Fmoc-Cys(Acm)-OH, Fmoc-Cys(tButhio)-OH, Fmoc-Cys(Trt)-OH, Fmoc-Glu(OtBu)-OH, Fmoc-His(Bum)-OH, Fmoc-Lys(Boc)-OH, Fmoc-Ser(tBu)-OH, Fmoc-Thr(tBu)-OH and Fmoc-Tyr(tBu)-OH. Piperidine and N-methyl morpholine (NMM) were p.a. grade from Merck. Dimethylformamide (DMF) (Chromatography grade) was purchased from PromoChem and dried over 4 Å molecular sieve prior to use. Dichloromethane p.a. (DCM), thioanisole p.a., triflouroacetic acid p.a. (TFA) and ethanethiol synthesis grade were from Merck.

References

1. Merrifield, R. B.. 1962. Fed. Proc. 21, 412.

2. Merrifield, R. B., J. M. Stewart, M. Jernberg.1966. Anal. Chem. 38, 1905.

3. Brunfeld, K., J. Halstrom, P. Roepstorff.1969. Acta Chem. Scand. 23, 2830.

4. Clark-Lewis, I., S. Kent. 1985. In: Synthetic Peptides in Biology and Medicine (K. Alitalo, P. Partanen, A. Vaheri, eds.). Elsevier Science Publishers B. V.. p. 27.

5. Dryland, A., R. C. Sheppard. 1986. Chem. Soc. Perkin Trans. I. 125.

6. Frank, R., M. Trosin. 1985. In: Modern Methods in Protein Chemistry, Vol. 2 (H. Tschesche, ed.). Walter de Gruyter, Berlin New York. p. 287

7. Schnabel, E., H. Klostermeyer, H. Berndt. 1971. J. Liebigs Ann. Chem. 749, 90.

8. Sieber, P., B. Iselin. 1968. Helv. Chim. Acta 51, 622.

9. Birr, C.. 1972. Liebigs Ann. Chem. 763, 162.

10. Kunz, H.. 1976. Chem. Ber. 109, 2670.

11. Carpino, L.A., G.Y. Han. 1970. J. Am. Chem. Soc. 92, 5748.

12. Carpino, L.A., G.Y. Han. 1972. J. Org. Chem. 37, 3404

13. Carpino, L.A., G.Y. Han. 1973. J. Org. Chem. 38, 4218.

14. Wang, S.S.. 1973. J. Am. Chem. Soc. 95, 1328.

15. Albericio, F., G. Barany. 1985. Int. J. Pept. Prot. Res. 26, 92.

16. Sakakibara, S., Y. Shimomishi, Y. Kishida, M. Okada, H. Sugihara. 1967. Bull. Chem. Soc. Jap. 40, 2164.

17. Sheehan, J.C., P.G. Hess. 1955. J. Am. Chem. Soc. 77, 1067.

18. Birr, C.. 1978. In: Aspects of the Merrifield Peptide Synthesis. Springer-Verlag, Berlin New York. p. 47.

19. Flor, F., C. Birr, T. Wieland. 1973. Liebigs Ann. Chem. 1601.

20. Atherton, E., R.C. Sheppard. 1985. J. Chem. Soc Chem. Commun. 165

21. Kent, S.B.H., L.E. Hood, H. Beilan, S. Meister, T. Geiser. 1984. In: Peptides 1984 (Ragnarson U. ed.) Almquist and Wiksell Int. Stockholm. p. 185.

22. Castro, B., J.R. Domroy, G. Evin, C. Selve. 1975. Tetrahedron Lett. 14, 1219.

23. Castro, B., J.R. Domroy, G. Evin, C. Selve. 1977. J. Chem. Res. 182.

24. König, W., R. Geiger. 1970. Chem. Ber. 103, 788.

25. Nguyen, D. L., R. Seyer, A. Heitz, B. Castro. 1985. J. Chem. Soc. Perkin Trans. I. 1025.

26. Sarin, V.K., S.B. Kent, J.P. Tam, R.B. Merrifield. 1981. Analytical Biochem. 117, 147.

27. Merrifield, R.B., V. Littau. 1968. In: Peptides 1968 (Bricas, E. ed. )North Holland Publ. Co. Amsterdam. p. 179.

28. Erickson, B.W., T.J. Lukas, M.B. Prystowsky. 1980. In: Polypeptide Hormones (Beer, R.F. Jr., Basset, E.G. eds) Raven Press, New York. p. 121.

29. Erickson, B.W., M.B. Prystowsky. 1979. In: Biological / Biomedical Applications of Liquid Chromatograghy (Hawk, G.L. ed. )Dekker, New York. p. 293.

30. Scott, R.P.W., K.K. Chan, P. Kucera, S. Zolty. 1971. J. Chromatogr. Sci. 9,577.

31. Bayer, E., E. Jung, I. Halasz, I. Sebastian. 1970. Tetrahedron Lett. 4503.

32. Atherton, E., E. Brown, R.C. Sheppard, A. Rosevear. 1981. J. Chem. Soc. Chem. Commun. 1151.

33. Wittman-Liebold, B.. 1983. In: Modern Methods in Protein Chemistry, Vol 2, (Tschesche, H. ed. ) Walter de Gruyter, Berlin.New York, p. 229.

# SPECIFICITIES OF AUTOANTIBODIES TO SMALL NUCLEAR RIBONUCLEOPROTEINS (U snRNPs) IN SERA FROM PATIENTS WITH CONNECTIVE TISSUE DISEASES

Reinhard Lührmann and Rolf Reuter

Max-Planck-Institut für molekulare Genetik
Otto-Warburg-Laboratorium
Ihnestrasse 73, D-1000 Berlin 33 (FRG).

## I. Introduction

A common feature of rheumatic disorders such as systemic lupus erythematosus (SLE) and overlap syndromes is the occurrence of autoantibodies directed against various nuclear components (see ref. 1 for a review). A number of antinuclear antibodies have been demonstrated which bind naked nucleic acids such as DNA or ribosomal RNA, chromatin components such as histones, or else, small ribonucleoprotein complexes. Prototypic autoantibodies of the latter group are the socalled anti-Sm and anti-RNP antibodies. The two antibody systems and the corresponding autoantigens have received much interest in the past because of their close association with SLE (Sm) and mixed connective tissue disease (RNP), respectively, and their value as diagnostic markers (2,3). For detection of these antibodies, most laboratories have previously employed crude cellular salt extracts as a source of antigens and conventional immunological assays such as immunodiffusion and counter-immunoelectrophoresis. With these tests, anti-RNP and anti-Sm antibodies have been operationally defined by differential ribonuclease sensitivity and

Modern Methods in Protein Chemistry, Vol. 3

lines of identity or nonidentity with reference sera on immunodiffusion.

It was not before 1979 that the identity of the Sm and RNP antigens was demonstrated at the molecular level. Lerner and Steitz showed that immune precipitates obtained after reaction of fresh cell extracts with anti-Sm sera contained all major nucleoplasmic snRNAs U1, U2, U4, U5 and U6 (4). Anti-RNP sera precipitated U1 RNA only. Six to seven polypeptides were also present in either immune precipitate which indicated the RNP status of these antigens. The antigenic determinants reacting with both antibody systems were located on the proteins associated with the snRNAs. Since these observations numerous laboratories have investigated the pattern of autoantibodies produced by patients against snRNP proteins for various reasons. (i) Autoantibodies of defined specificity towards individual snRNP proteins can be invaluable tools for the study of structure-function relationships of snRNPs. (ii) Correlation of defined anti-snRNP autoantibody specificities with certain connective tissue diseases might improve the diagnostic value of such antibodies, and (iii), last but not least, the knowledge of the autoantibody profile against snRNPs in the sera is a prerequisite for a better understanding of the mechanisms responsible for the spontaneous development of these autoantibodies in SLE patients. The aetiology of the antinuclear autoantibody production in these patients is largely unknown. Immunoregulatory disorders, polyclonal B-cell activation and specific antigen drive have been suggested as being among the factors contributing to the spontaneous development of these antibodies (5-8).

We here describe the various immunological procedures that we and others have used for the analysis of the snRNP-reactive autoantibody populations in sera of patients with connective tissue diseases. Specificity of the antibodies against snRNPs and their components has been studied at different structural levels. (i) Reactivity with individual snRNPs as a whole was investigated by immunoprecipitation of the particles from fresh radiolabelled cellular extracts. (ii) Specificity of the antibodies for individual snRNP polypeptides was studied by immunoblotting methods using total proteins from purified snRNPs as antigenic material. This included also the study of immunological relationships between snRNP polypeptides by reblotting techniques. (iii) Reactivity of autoantibodies with distinct epitopes on an individual snRNP polypeptide was studied by competition between patient sera and monoclonal antibodies raised against snRNP proteins. This technique also allowed the definition of immunodominant regions on the snRNP proteins against which patients primarily produce antibody specificities.

These studies not only revealed the great diversity and distribution on the various snRNP polypeptides of antigenic determinants recognized by naturally occurring autoantibodies but also provided insight into possible mechanisms of the autoantibody production and suggested strongly that the anti-Sm/RNP autoimmune response is driven by the endogeneous snRNP particles.

## II. Protein composition of U snRNPs

Before the reactivity of autoantibodies with snRNP polypeptides is discussed, the molecular composition of the U snRNPs will briefly be described.

The U snRNPs are a group of ribonucleoprotein particles which are found in the nucleus of all eucaryotic organisms investigated so far, including mammals, amphibians, insects and even unicellular organisms (for reviews see ref. 9-11). The names of the snRNPs derive from the particular snRNA that they contain. The five major nucleoplasmic snRNAs U1, U2, U4, U5 and U6 are organized in four discrete ribonucleoprotein complexes. The snRNAs U1, U2 and U5 exist as monomeric RNP particles, whereas the snRNAs U4 and U6 are organized together in a single RNP complex, the snRNP U4/U6 (12,13). In the higher eucaryotes these snRNPs are present in an abundance of $10^5$-$10^6$ copies per cell. The snRNA molecules themselves are characterized by a high content in uridylic acid residues - that is why they are called U snRNAs - and by the presence of many modified nucleosides such as pseudouridylic acid residues and methylated nucleosides. A hallmark of the snRNAs is their 5'-terminal cap structure which possesses a 2,2,7-trimethylguanosine ($m_3G$) at its 5' end. The only exception is U6 RNA, which is also capped but does not contain an $m_3G$ residue. The chain lengths of the snRNAs varies between 189 (for U2) and 106 nucleotides (for U6). In the cell the snRNPs U1, U2, U5 and U4/U6 appear to play an essential role in the splicing of nuclear pre-mRNAs (for reviews see ref. 14-16).

Various procedures have been developed during recent years in order to purify the individual snRNP particles and to identify their protein constituents, some using strictly biochemical approaches, others using immunological methods or a mixture of both (see ref. 11 for a review). A very quick and efficient procedure for the one-step purification of all snRNPs U1, U2, U5 and U4/U6 from nuclear extracts uses antibodies against the snRNAs' $m_3G$ cap structure by immune affinity chromatography (17). Anti-$m_3G$ antibodies were first raised in rabbits (18) and it could be demonstrated that they react with the $m_3G$ cap structure of all snRNAs as part of the intact snRNPs (19). (Anti-$m_3G$)-bound snRNPs are desorbed from the affinity column by elution with excess nucleoside $m_3G$, which allows the isolation of native snRNP particles. Pure U1 snRNPs can be obtained by subjecting the mixture of isolated snRNPs U1 to U6 to an ion exchange chromatography step on DEAE-Sepharose (20). Individual snRNPs U2, U5 and U4/U6 can be isolated by serial passage of the particles over affinity columns constructed either with monoclonal antibodies against selected snRNP proteins or with antibodies against 6-methyladenosine ($m^6A$) (21). One $m^6A$-residue each is present in the snRNAs U2, U4 and U6 and these nucleosides are accessible for anti-$m^6A$ antibodies in the intact snRNPs U2 and U4/U6, respectively (22). An alternative way for the purification of individual snRNPs other than U1 RNP uses equilibrium density gradient centrifugation in CsCl and ion exchange chromatography on DEAE-Sepharose columns (23).

A typical pattern of the proteins contained in individual snRNPs from HeLa cells as isolated by immune affinity chromatography with nucleoside-specific antibodies is shown in Fig. 1. The nomenclature for the proteins is that introduced by Lerner and Steitz (4). Seven polypeptides denoted B' (29k), B (28k), D (16k), D' (15.5k), E (12k), F (11k) and G (9k) are present in each of the snRNPs U1, U2, U5 and U4/U6. In addition to these common polypeptides, some of the snRNPs contain additional proteins. U1 RNP contains three unique proteins denoted 70k, A (34k) and C (22k). U2 RNP contains two specific polypeptides of molecular weights 33k (A') and 28.5k (B"). (Protein B" is frequently not resolved from protein B on one-dimensional SDS-polyacrylamide gels). U5 RNP may also contain one or two specific polypeptides (23-25); the assignment of these candidate proteins to the U5 RNP particle, however, has not yet been proved by immune precipitation of U5 RNPs with antibodies specific for the respective polypeptides.

## III. Classification of autoimmune sera by immunoprecipitation of snRNPs

A powerful means of defining the specificities of antinuclear antibodies that react with protein-RNA complexes is the technique of immunoadsorption of $^{32}P$-labeled cell lysates on solid phase staphylococcal protein A and subsequent poly-

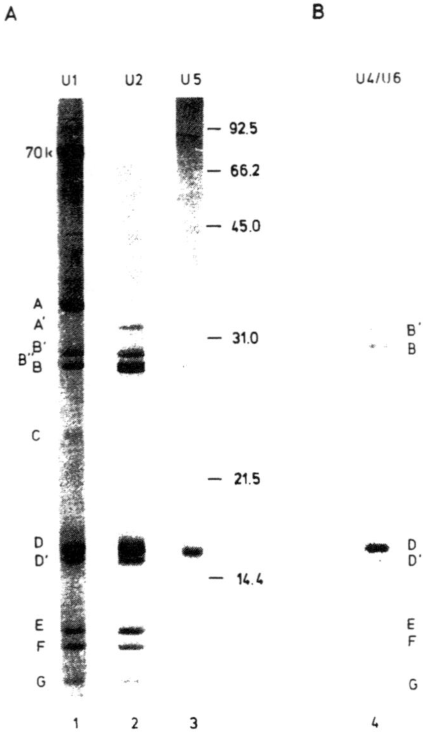


Fig. 1
Polypeptide composition of snRNPs U1, U2, U5 and U4/U6. The individual snRNPs were purified as described previously (21). The photograph shows the Coomassie-stained proteins from isolated snRNPs U1, U2, U5 and U4/U6 (lanes 1 to 4) after fractionation on 15 % SDS-polyacrylamide gels.

acrylamide gel electrophoresis of the extracted RNA (4,26). A schematic drawing of the principle of this procedure is shown in Fig. 2. We have used a $^{32}$P-labeled HeLa nuclear extract as a source of antigen for the screening of 54 sera from patients with various rheumatic diseases for the presence of autoantibodies capable of precipitating U snRNPs. In a typical experiment, 10 $\mu$l of a serum, pre-adsorbed to 25 $\mu$l of swollen Protein A-Sepharose, is allowed to react with a cell extract equivalent to approximately 1 x $10^6$ HeLa cells.

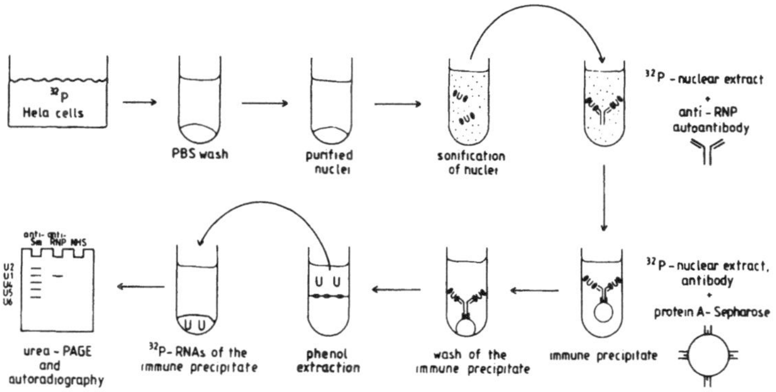


Fig. 2
Principle of radioimmunoprecipitation for the analysis of ribonucleoprotein.

The antigen-antibody complex immobilized to Protein A-Sepharose is then extensively washed, resuspended in a 2 % SDS solution; the RNA contained in the immune precipitates is recovered after phenol extraction and separated by electrophoresis in a 10 % denaturing polyacrylamide gel. The gel profiles of $^{32}$P-labeled snRNAs found in the immunoprecipitates with sera from representative patients are shown in Fig. 3A. Sera that precipitate only U1 RNP (lanes 6 to 8) are classified as anti-RNP sera. Sera that precipitate all snRNAs U1, U2, U4, U5 and U6 are termed anti-Sm. Pure anti-RNP sera are in the minority. Most of the 34 sera which were positive for snRNPs displayed anti-Sm specificity (see also Table 1). One serum does not conform to either of the above classifications insofar that it precipitates exclusively the snRNPs U1 and U2 and none of the others (Fig. 3A, lane 11). By definition this serum does not contain anti-Sm activities, but antibodies that react with U2-specific polypeptides in addition to antibodies which recognize proteins confined to the U1 RNP particle.

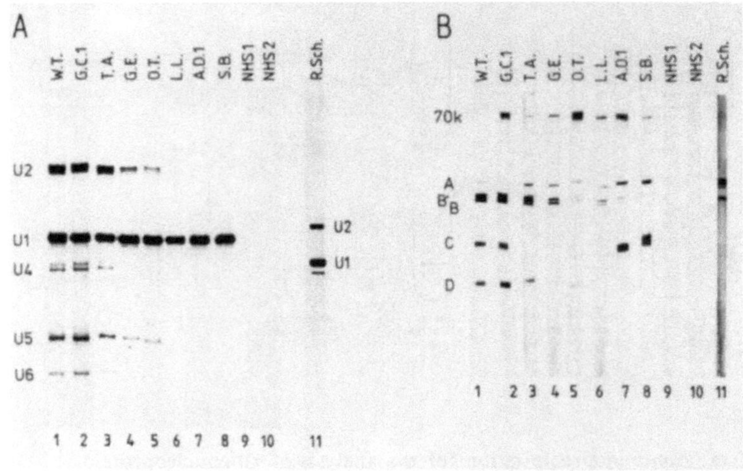


Fig. 3
Characterization of snRNP-specific antibodies by (A) radioimmunoprecipitation and (B) immunoblotting. Part A. Nuclear extracts from HeLa cells, radioactively labeled with $^{32}P$-phosphate, and prepared as described previously (19), served as a source of antigens for the immunoprecipitation assay (see Fig. 2) with the 9 stated patient sera (lanes 1-8,11) or with normal human sera (lanes 9,10). The Figure shows an autoradiogram of the radioactively labeled RNA that was contained in the immune precipitates after polyacrylamide gel electrophoresis in the presence of 7M urea. Part B. Proteins of anti-$m_3G$ affinity purified snRNPs U1 to U6 served as antigens for the immunoblot in which 1:40 dilutions of the stated patient sera (lanes 1-8,11) or two normal human sera (lanes 9,10) were used for the detection of the antigenic proteins.

## IV. Antigenicity of snRNP polypeptides towards autoantibodies as determined by immunoblotting

The immunoprecipitation assay described above allows the detection of antibodies directed towards polypeptides which are intimately associated with a stable ribonucleoprotein complex such as the snRNP particles. Given the complex protein composition of the various snRNP species, however, it cannot reveal which polypeptide(s) in the complex are the antigens against which the antibodies contained in a particular serum react; antibodies against any of the protein constituents of a stable RNP complex will have the same effect, namely precipitation of the RNP particle. A convenient means to identify the polypeptides that bear antigenic determinants reacting with snRNP-specific autoantibodies in autoimmune sera is the immunoblot assay. We have blotted the 34 sera which contained anti-snRNP antibodies, using total proteins of the mixture of anti-$m_3G$ affinity-purified snRNPs U1, U2, U5 and U4/U6 from HeLa cells as antigenic material. For this purpose the snRNP polypeptides were fractionated preparatively on 15 % SDS-polyacrylamide gels and transferred electrophoretically to nitrocellulose sheets (27). Detection of autoantibodies bound to particular protein bands on nitrocellulose strips was accomplished by incubation with goat anti-human Ig coupled to alkaline phosphatase.

Fig. 3B shows the immunoblot patterns of protein reactivity of the same representative sera for which the corresponding profiles of immunoprecipitated snRNAs is displayed in Fig. 3A. Sera that exclusively precipitate U1 RNAs, i.e. anti-RNP sera, exhibit strong staining of the U1 RNP-specific proteins 70k, A and C (Fig. 3B, lanes 6-8). Thus these proteins must bear the RNP-antigenic determinants (17, 28-35). Sera which contain high titres of anti-Sm antibodies, i.e. which strongly precipitate all snRNAs U1 to U6, exhibit strong reactivity with the proteins B', B and D (Fig. 3, lanes 1-5). One Sm serum reacts only with polypeptides B' and B, but not with D (Fig. 3B, lane 4). Thus the main snRNP polypeptides bearing Sm antigenic determinants are B', B and D which are common to all snRNPs (17, 29-35).

Of the large number of patient sera whose reactivity towards snRNP proteins in immunoblot is shown in Table 1, only three can be classified as pure anti-Sm and six as pure anti-RNP sera. Most contain both anti-RNP and anti-Sm antibodies, that is, antibodies against U1-specific proteins and against proteins common to all snRNPs. Of particular interest are sera such as A.D.1 which behave as an anti-RNP serum in the immunoprecipitation assay, but exhibit reactivity with

Table 1: Autoantibody Specificities to Individual snRNP Proteins in Sera from Patients with Connective Tissue Diseases.

| | snRNP Proteins | | | | | | | |
|---|---|---|---|---|---|---|---|---|
| Patients | 70k | A | A' | B/B' | B" | C | D | F |
| A.D.1 | * | * | | * | | * | | |
| A.G. | * | * | | * | | * | * | |
| A.R. | * | * | | * | | * | | |
| B.A. | * | * | | * | | | * | |
| B.H. | * | * | | * | | * | | |
| B.K. | * | * | | | | | | |
| C.K. | * | * | | | | | | |
| Ch.K. | | * | | * | | | * | |
| E.K. | * | | | * | | | * | |
| E.R. | * | * | | * | | * | * | |
| G.C.1 | * | | | * | | * | * | |
| G.C.2 | * | * | | * | | * | | |
| G.E. | * | * | | * | | | | |
| G.G. | * | * | | * | | * | | |
| G.P. | | | | * | | | * | |
| G.W. | | | | * | | | | |
| H.D. | * | | | | | | | |
| J.F. | * | | | * | | * | | |
| K.A. | | * | | * | | | * | |
| K.L. | * | * | | * | | * | | |
| K.Sch. | | * | | * | | | | |
| L.E. | * | * | | * | | * | * | |
| L.L. | * | * | | * | | * | * | |
| L.R. | * | | | * | | | | |
| M.M. | | * | | * | | | * | * |
| O.T. | * | * | | * | | | * | |
| R.K. | * | * | | * | | * | * | |
| R.M.1 | | * | | * | | | | |
| R.M.2 | | * | | * | | * | | |
| R.Sch. | | * | * | | * | | | |
| S.B. | * | * | | | | * | | |
| T.A. | * | * | | * | | | * | |
| T.L. | * | * | | | | * | | |
| W.T. | | * | | * | | * | * | |

Autoantibodies directed to the indicated snRNP proteins were determined by immunoblotting as described in Fig. 3.

proteins B' and B in addition to their reaction with U1-specific polypeptides by immunoblot. This phenomenon is very probably to be attributed to the presence in such sera of very low amounts of B'/B-specific anti-Sm antibodies, that can be detected in immunoblot but not in immunoprecipitation assays (see also ref. 33). The co-existence of anti-RNP and anti-Sm antibodies in a serum thus appears to be a general phenomenon. The existence of such mixtures was first detected some years ago by Mattioli and Reichlin who investigated the auto-antisera by immunodiffusion against an extract of nuclear antigen (36).

Given the distribution of RNP-antigenic determinants on three distinct polypeptides (70k, A and C) and the fact that these antibodies are often found together in autoimmune sera with anti-Sm antibodies, a question arises concerning the previously established association of anti-RNP antibodies with the syndrome mixed connective tissue disease (MCTD). In a cross-sectional, longitudinal study, Pettersson et al. have recently provided experimental evidence that high-titered antibodies against the U1-specific 70k protein are consistently found in sera from MCTD patients, suggesting that anti-70k specificities may be best correlated with this disease (35). Anti-RNP antibodies in SLE patient sera which exhibit high titres of anti-Sm antibodies appear to be primarily directed against the proteins A and C.

While antibodies against the proteins 70k, A and C as well as against the major Sm-antigenic proteins B', B and D preponderate in snRNP-positive sera of patients with connective tissue diseases, antibody specificities against other snRNP polypeptides can also be found, though at a much lower frequency. One serum (Fig. 3B, lane 11) contains a new class of anti-snRNP antibodies, described recently, and called anti-(U1/U2)RNP, that react specifically with U1 and U2 RNP by binding an epitope common to the U1-specific A protein and the U2-specific B" protein (see also below). This serum contains also antibodies that bind to the U2-specific A' protein and which have been designated Ya antibodies (37). Whether these specificities are associated with particular rheumatic syndromes has to await further cross-sectional studies.

Antibodies against at least one of the low-molecular-weight snRNP proteins E, F and G have been described in few instances (33,38). In the collection of sera described in this report one was found which exhibits anti-F specificities (Table 1). In an independent study, it proved possible to demonstrate the presence of specific antibodies against the snRNP proteins E, F and G in five out of 32 sera chosen because of their high titre of anti-Sm antibodies (Fig. 4) (Reuter, Habets, van Venrooij and Lührmann, unpublished results).

The immune system of patients with connective tissue diseases appears therefore in principle to be able to produce antibodies against any of the major protein components of the U snRNPs.

## V. Immunological relationships among snRNP proteins

Inspection of the 34 sera whose snRNP-specific antibodies are given in Table 1 shows that the sera always react either with both B' and B proteins or with neither. The 12 sera that bind to the D protein react always with the B' and B protein. One is therefore led to ask whether antibodies against these three proteins frequently occur in association or whether there is immunological relation between the major Sm antigenic polypeptides and cross-reacting antibodies occur in the sera. This question is easily answered by a simple re-blotting procedure (39), once monospecific antibodies to one of the antigens have been produced. Antibodies from autoimmune sera are allowed to react with their antigenic proteins, which have been separated by SDS-polyacrylamide electrophoresis and transferred to nitrocellulose; the specific antibodies are then eluted from the protein bands. If, for example, the antibodies of the serum E.K., which react in an immunoblot with the B', B and D proteins, are eluted from the three bands and then allowed to react with the proteins of the snRNPs U1 to U6 on a blotting filter strip, each of the eluted antibodies binds to all three proteins B', B and D (Fig. 5A). Thus the Sm antibodies - at

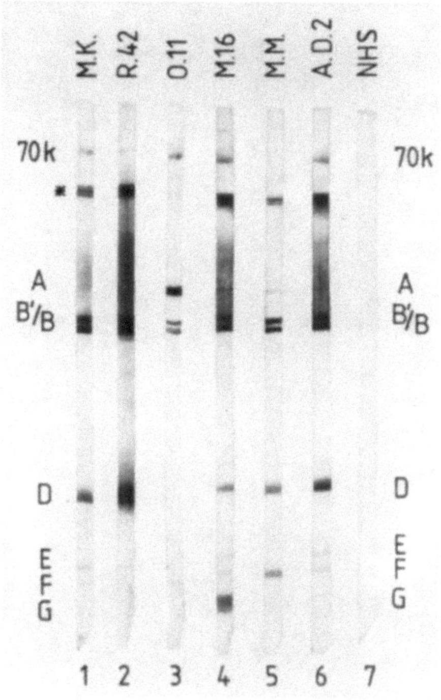


Fig. 4
Detection of antibodies against the snRNP proteins E, F and G in the sera of patients with rheumatic diseases by immunoblotting. Immunoblot was carried out as described in Fig. 3B. Staining of protein bands in the immunoblot indicated by an asterisk is due to the reaction of anti-B'/B antibodies with residual U2 RNA-protein complexes resistant to dissociation by SDS (see ref. 21 for details).

least those of this serum - cross-react with all three proteins, as was observed for various monoclonal antibodies (33,40,41). On the other hand, sera have also been described which contained autoantibodies against B' and B that did not show cross-reaction with the D protein and vice versa (31). Serum G.E. is an example which contains antibodies that only react with B' and B but not with D. Proteins B'/B and D thus also possess characteristic epitopes.

Interestingly, no sera have been described so far that react only with B' or B proteins. The reason for this was recently shown by comparative partial proteolysis of the two proteins:

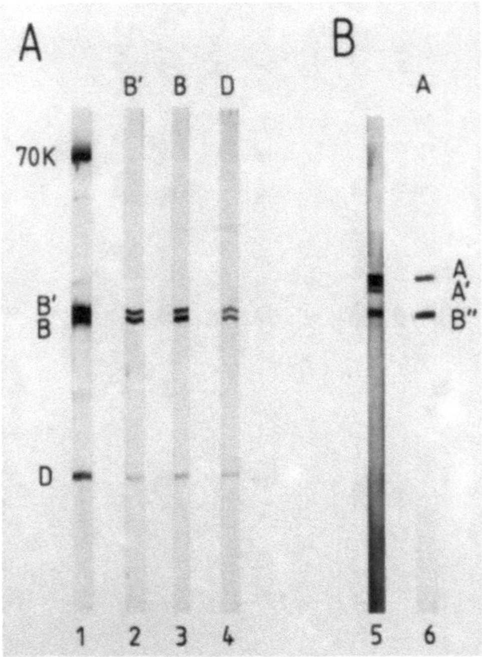


Fig. 5
Detection of immunological relationships among (A) proteins B', B and D as well as (B) proteins A and B" with affinity-purified human autoantibodies. Proteins from snRNPs U1-U6 were blotted and incubated either with the anti-Sm serum E.K. (lane 1) or with the anti-(U1/U2)RNP serum R.Sch. (lane 5). Antibodies were eluted from (A) the protein bands B', B and D, respectively or (B) the protein band A. These fractions were reused for immunoblotting on total snRNP proteins as indicated on the top of each lane.

they are highly homologous and produce identical peptide patterns with the exception of one distinct fragment each (42). They are thus so closely related immunologically that the autoantibodies do not distinguish the individual proteins.

Polypeptide D is clearly structurally distinct from the proteins B'/B, as is shown by the distinctive peptide patterns produced by partial protease digestion of proteins B' or B and D. Their common features may thus be restricted to the epitope(s) to which the cross-reacting patient sera and monoclonal antibodies bind.

By the re-blotting method it could also be shown that the anti-(U1/U2)RNP sera contain antibodies that cross-react with the U1-specific A and the U2-specific B" proteins (43). The re-blot in Fig. 5B shows the cross-reactivity of the two proteins with antibodies from serum R.Sch. that were eluted from the A band of the initial immunoblot obtained with this serum. That proteins A and B" share at least one epitope was recently formally proved by the isolation of cross-reactive monoclonal antibodies (40,44). Thus, protein A must bear at least two distinct antigenic epitopes reacting with autoantibodies, one that is recognized by anti-A specificities associated with anti-RNP sera and the other that is recognized by anti-(U1/U2)RNP antibodies.

## VI. Investigation of the anti-Sm/RNP autoimmune response by competition studies with monoclonal antibodies

The next step in the analysis of the autoantibody profile produced by patients with connective tissue diseases against snRNPs is the definition of the antigenic regions on individual snRNP proteins with which the autoantibodies react. This is made possible by a method which, through the measurement of the competition between monoclonal antibodies and autoantibodies for a single snRNP polypeptide, can differentiate between epitopes on the antigenic protein. Monoclonal antibodies directed towards snRNP polypeptides can be obtained from certain mouse strains, such as MRL/l, which are prone to develop SLE-like autoimmune diseases, or from patients suffering from SLE. Alternatively, such antibodies may be derived from genetically non-autoimmune mice immunized with isolated snRNPs or snRNP polypeptides. The latter procedure has the advantage that the role of the snRNPs as an immunogen in the anti-Sm/RNP autoimmune response can be studied by comparing the antibody patterns produced by the mice against the exogeneous snRNP antigens with the anti-snRNP antibody profile produced spontaneously by SLE patients.

Isolation and characterization of monoclonal antibodies

A series of mice, showing no pre-disposition for autoimmune disease (C57bl/6), was immunized with purified native U1 snRNPs from HeLa cells. In view of the high structural conservation of snRNP proteins during evolution, the high proportion of these mice (8 out of 10) that gave a positive immune response was at first surprising. When the immune response against individual snRNP polypeptides was studied by immunoblotting a remarkable similarity between the antibody spectrum of the immune sera of the mice and the antibody population expressed by patients with connective tissue diseases was observed (Fig. 6). The antibodies of the mice were directed against the same snRNP proteins as were the human autoantibodies and with roughly equal frequency, i.e. antibodies were found that were primarily directed against the polypeptides 70k, A, B', B, less against C or D. No antibodies against proteins E, F and G were detected in any of these eight mice (Fig. 6).

Hybridoma cell lines expressing antibodies specific for snRNP polypeptides were obtained by fusion of spleen cells of two of the immunized mice and the cell line X63Ag8.653. Activity for snRNPs was established initially by an enzyme-linked immunosorbent assay (ELISA) performed with anti-$m_3G$ affinity-purified snRNPs U1 to U6 (40). To verify reactivity with snRNPs immunoprecipitation analysis was performed using radiolabeled extracts from HeLa cells (Fig. 7A). Specificity of the monoclonal antibodies for individual snRNP polypeptides was analysed by immunoblotting using total proteins from purified snRNPs as antigenic material (Fig. 7B).

Five monoclonal (mAbs) were obtained, which can be assigned to four categories according to their properties in the above-mentioned assays.

i) Two mAbs, H386 and H399, specifically precipitate U1

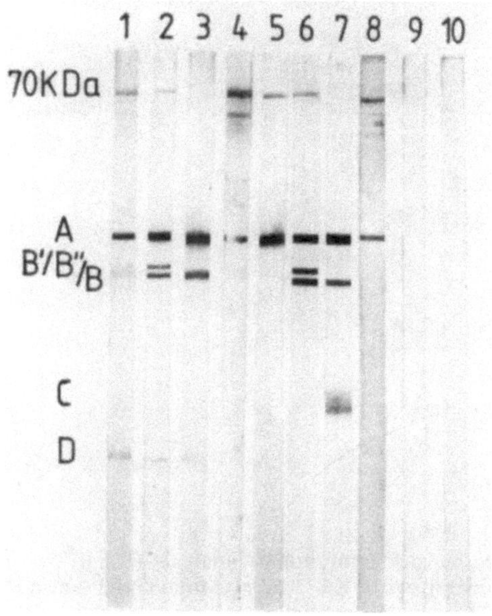


Fig. 6
Immunoblot analysis of the sera from C57bl/6 mice immunized with purified U1 snRNPs. Total snRNP proteins were used for the immunoblot to detect snRNP-specific antibodies in dilutions of sera from the mice M1-M9 (lanes 1-9) and one control mouse (lane 10) which was treated with complete Freund's adjuvant only.

snRNPs (Fig. 7A, lanes 4 and 5) and bind solely to the 70k protein of this particle (Fig. 7B, lanes 4 and 5).
ii) The mAb H304 reacts specifically with the A protein of U1 snRNP (Fig. 7B, lane 2).
iii) The mAb H126 cross-reacts with the three proteins B', B and D, showing that these have at least one common epitope (Fig. 7B, lane 1). Since these three proteins are common to all nucleoplasmic snRNPs, all these particles are precipitated by the monoclonal antibody (Fig. 7A, lane 1).
iv) One mAb, H356, selectively precipitates U1 and U2 snRNPs (Fig. 7A, lane 3) and reacts with the U1-specific A protein as well as with the U2-specific B" protein in the immunoblot (Fig. 7B, lane 3). In these properties mAb H356 resembles human anti-(U1/U2)RNP autoantibodies (see above).

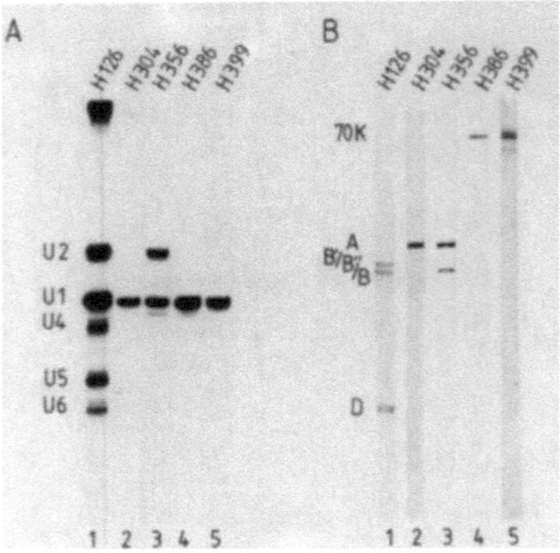


Fig. 7
Radioimmunoprecipitation (A) and immunoblot analysis (B) of the monoclonal antibodies derived from mice immunized with U1 RNPs. Monoclonal antibodies were isolated as described previously (40). Part A. HeLa nuclear extracts, labeled with $^{32}$P-phosphate, were used as antigen sources for radio-immunoprecipitation with the monoclonal antibodies H126 (lane 1), H304 (lane 2), H356 (lane 3), H386 (lane 4) and H399 (lane 5). The autoradiogram shows the positions of radioactively labeled RNA, after polyacryl-amide gel electrophoresis, contained in the immune precipitates. Part B. Immunoblot of the various monoclonal antibodies on proteins from anti-$m_3$G-purified snRNPs U1 to U6.

When the five monoclonal antibodies were further characterized by cytoimmunofluorescence (Fig. 8), they all showed the nuclear "speckled" staining pattern typical of snRNP-specific antibodies (1). They all react similarly with human cells (HeLa), mouse cells (3T3, not shown) and rat kangaroo cells (PtK2), which indicates a high degree of conservation during evolution of the structures to which they bind and against which the immunized mice have made antibodies (Fig. 8).

In contrast to the mAbs H126, H304 and H356, the mAbs H386 and H399 show fluorescence that is not confined to the nucleoplasm (Fig. 8). As well as a very weak cytoplasmic signal they give a signal from the nuclear membrane. Whether staining of the nuclear membrane is owing to cross-reactivity

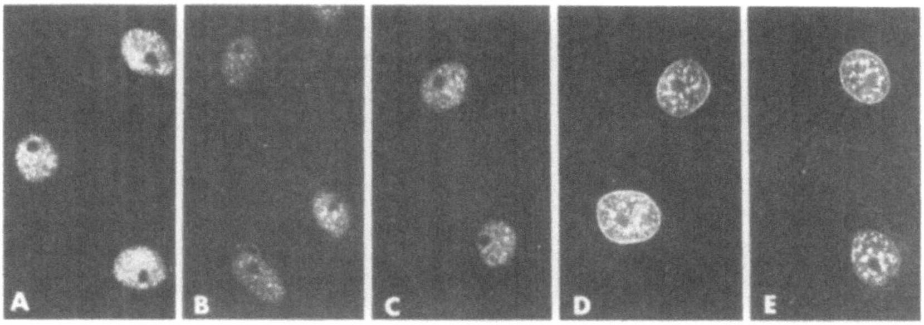


Fig. 8
Indirect immunofluorescence of monoclonal antibodies against snRNP proteins on PtK2 rat kangaroo cells. Immunofluorescence was performed essentially as described previously. (45). Panels A to E exhibit PtK2 cells reacted with monoclonal antibodies H126 (A), H304 (B), H356 (C), H386 (D) or H399 (E).

of the two mAbs with one of the nuclear lamins, has not yet been clarified. When mAbs H386 and H399 are blotted on total nuclear proteins from HeLa cells, they cross-react with two proteins of apparent molecular weights around 60k as well as with the 70k protein of the U1 snRNPs (40).

## Competition between monoclonal antibodies and patient autoantibodies for single snRNP polypeptides

The monoclonal antibodies described above made possible the definition of the regions on individual snRNP polypeptides with which the autoantibodies of the patients react, and the comparison of the human autoantibodies at the level of distinct antigenic determinants. For this purpose the mAbs H126, H304, H356 and H399 were labeled with $^{35}$S-methionine *in vivo* and isolated from the supernatant of the hybridoma cells. They were then used for a solid-phase antibody binding assay (SABA), in which various concentrations of a given human serum with anti-snRNP activity competed with the radioactively labeled mAb for binding to a purified single

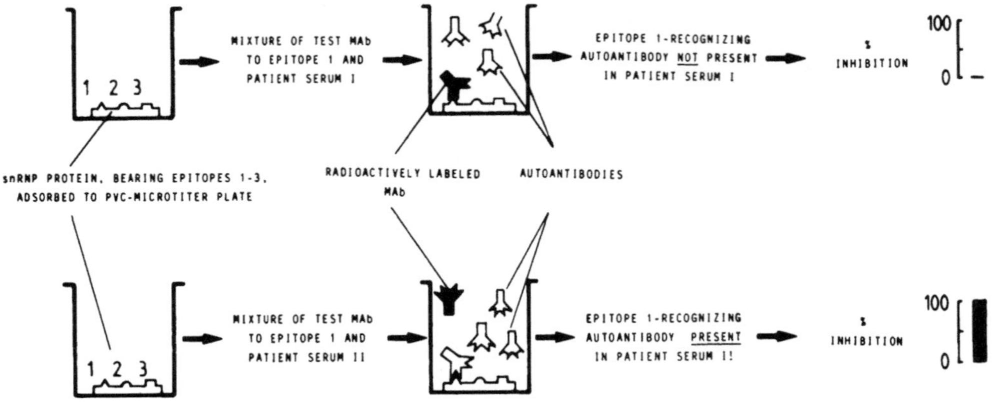


Fig. 9
Principle of the competition assay of monoclonal antibodies with human autoantisera for the binding to purified snRNP proteins.

snRNP polypeptide adsorbed to the wells of PVC microtiter plates. The principle of this SABA is shown schematically in Fig. 9. The various snRNP polypeptides were isolated by electroelution after separation of the total proteins of purified snRNPs U1 to U6 on SDS-polyacryamide gel electrophoresis (42). The anti-snRNP specificities of the human autoantisera were characterized by immunoblotting (Table 1).

The competition experiments are summarized in Table 2. The important result of these studies is the finding that the binding of each of the four monoclonal antibodies to their antigenic proteins is inhibited by at least one of the human sera that react with these proteins on the immunoblot. This means that the antibodies appearing in the mice in response to the exogenic U1 snRNPs recognize the same regions on the snRNP proteins as do the autoantibodies of the patients. Thus, for example, all sera containing autoantibodies against the A protein inhibit the binding of the mAb H304 to this protein. mAb H304 therefore defines a major antigenic determinant on the polypeptide A recognized by anti-RNP

Table 2: Competition of human autoantisera with monoclonal antibodies for binding to purified snRNP proteins.

| Patient sera | Anti-snRNP autoantibody specificities present in individual patient sera | | | | | Inhibition of binding of test mAb by patient sera | | | |
|---|---|---|---|---|---|---|---|---|---|
| | 70k | B/B' | D | A | B" | H399 | H126 | H304 | H356 |
| G.C.1 | + | + | + | - | - | + | + | - | - |
| B.A. | + | + | + | + | - | + | + | + | - |
| G.C.2 | + | + | - | + | - | + | - | + | - |
| L.E. | + | + | + | + | - | + | + | + | - |
| A.G. | + | + | + | + | - | - | + | + | - |
| L.L. | + | + | + | + | - | - | + | + | - |
| O.T. | + | + | + | + | - | - | + | + | - |
| R.M.2 | - | + | - | + | - | - | - | + | - |
| R.Sch. | - | - | - | + | + | - | - | + | + |
| G.P. | - | + | + | - | - | - | + | - | - |
| S.B. | + | - | - | + | - | + | - | + | - |

Electroeluted individual snRNP proteins were adsorbed to the wells of polyvinyl chloride microtiter plates. $^{35}S$-labeled monoclonal antibodies were tested for competition with autoantibodies in human autoantisera as follows: H399 vs. 70k, H126 vs. B', H304 and H356 vs. A protein. Competitive sera (+), noncompetitive (-). Autoantibodies specific for snRNPs were determined by immunoblot as described in Fig. 3.

associated anti-A autoantibodies. Similarly, autoantibodies against the regions on the B' and the 70k proteins, to which the mAbs H126 or H399 bind, appear to occur very frequently in the patient sera (Table 2). The binding of the mAb H356 to the A protein, however, is prevented only by anti-(U1/U2)RNP-antibody-containing serum (R.Sch.). As this serum also clearly competes with H304, it seems to contain both types of A-specific autoantibodies.

Autoantibody formation against several distinct epitopes does not seem to be restricted to the A protein. Thus, three out of nine sera that react with the 70k protein in immunoblotting, do not compete with the mAb H399 (A.G., L.L., O.T.). Clearly, they bind to at least one further epitope on this protein, which is not in the region that reacts with the two 70k-protein-specific monoclonal antibodies. Similarly, the sera G.C.2 and R.M.2 contain antibodies against the proteins B' and B that recognize a region other than that recognized by H126, since they fail to inhibit the binding of the latter to B' (Table 2). It is of interest that these two sera are the only ones in the whole selection that react in immunoblots with B' and B but not with D. This suggests that the patients can make antibodies against at least two epitopes on proteins B' and B: against one that is common to B', B and D and is also recognized by the mAb H126 and against another, that is found only on B' and B and that binds to the autoantibodies of the sera G.C.2 and R.M.2.

The competitive SABA used with isolated monoclonal antibodies, thus allows two conclusions: (i) the experimentally produced antibodies against U snRNPs recognize the same regions on the proteins as do the patients autoantibodies, and (ii) the autoimmune response is more complex than was assumed originally on the basis of immunoblotting results. The SABA further offers the possibility of defining immunodominant epitopes on individual snRNP polypeptides, against which patients primarily produce autoantibodies. It will be of great interest to find out how much the appearance of clinical symptoms of rheumatic diseases is connected with the appearance of autoantibodies against particular epitopes.

## VII. Distribution of epitopes on the snRNP proteins

The results of the analysis of the diversity of autoantibody populations in sera of patients with connective tissue diseases obtained by immunoblotting and competitive radioimmunoassays allow one to draw a provisional picture concerning the number and distribution of distinct antigenic determinants on the snRNPs (Fig. 10). At least 12 different epitopes on the snRNPs can so far be distinguished, two residing each on the proteins 70k, A, B'/B and D; the

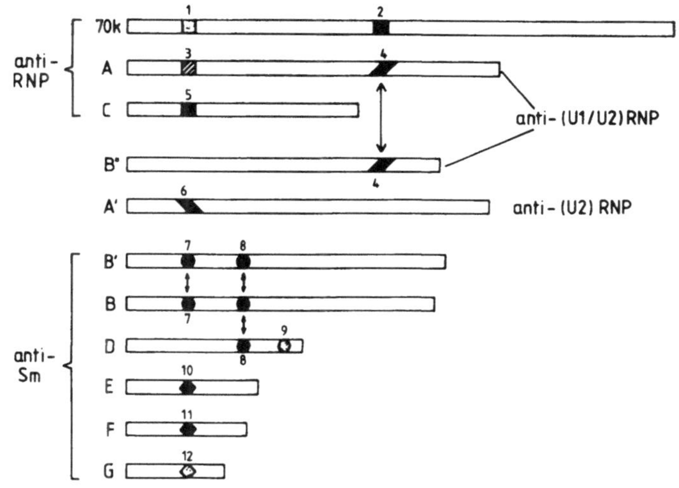


Fig. 10
Distribution of the antigenic determinants reacting with the autoantibodies on the U snRNPs. The numerals give the epitopes known to bind autoantibodies from SLE or MCTD patients. For further explanation see text.

polypeptides A', C, E, F and G contain at least one unique epitope each. Since some of the epitopes are shared by two or three proteins (Fig. 10) and some polypeptides such as A and D are probably present in the snRNP particles in at least two copies (46), the total number of independent antigenic regions on U1 snRNPs exceeds twenty. The complex protein structure of the U snRNPs is thus mirrored by an even more complex multiplicity of autoantibodies that bind to them.

## VIII. The nature of the immune response against U snRNPs in patients with connective tissue diseases

The observations made in the analysis of the anti-RNP and anti-Sm autoimmune response and of the experimental immune response of mice against U1 snRNPs suggest strongly that the anti-RNP/Sm autoantibody production in patients with connective tissue diseases arises from direct antigenic stimulation of their immune system by the endogeneous snRNPs. (i) The U snRNPs appear to be recognized as a physical entity by the immune system rather than the individual polypeptides, since antibodies against various snRNP proteins mostly occur together in individual patient sera. (ii) The immune response of the patients is subject to great individual variation, both qualitatively and quantitatively. (iii) The particles are immunogenic in the experiment with the mouse, and they result in the production of mouse antibodies against the same major RNP and Sm antigenic polypeptides against which the patients preferentially produce autoantibodies. (iv) Indeed, the immune system of the mouse reacts with the same regions of the snRNP proteins as does the immune system of the patients. (v) The introduction of snRNPs into the mouse induces autoreactive snRNP-specific antibodies. This means that in a normal organism B-cell clones are present that can react with the appropriate particles, and is no tolerance of snRNPs on the level of B cells.

At the same time, the findings speak against a possible spontaneous and unspecific endogeneous activation of the patients' B cell repertoire. In view of the great diversity of distinct antigenic determinants against which autoantibodies are produced it is not very likely either that these autoantibodies could have their origin via cross-reacting viral or bacterial neoantigens. Furthermore, the snRNP-specific antibodies derived from mice immunized with U1 snRNPs would recognize regions on the proteins of the particles other than those recognized by natural autoimmune sera, if the latter were raised against fortuitously crossreactive microbial antigens. This was in fact not observed.

While the snRNPs appear to play a central role as immunogen in the anti-Sm/RNP autoimmune response, the initial event that triggers off the autoantibody production is not yet clear, however. It is possible that an immune response can be provoked by a metabolic defect or a particular kind of lesion of the tissue, which would provide for the exposure of large quantities of U snRNPs to the immune system. For example, the UV irradiation of epidermal cells in vitro leads to the presentation of antigenic components of the U snRNPs on the cell membranes of living keratinocytes (47). Whether this is sufficient to initiate an immune response is not known. A major unanswered question is whether snRNP-specific B cells proliferate under T-cell control. In autoimmune MRL/Mp-+/+ mice, an animal model for SLE, Sm-specific T cells were detected that might have a helper function in the anti-Sm autoimmune response (48,49). In addition, the anti-Sm and anti-RNP autoimmune response of SLE patients is restricted to isotype IgG1 of the heavy chain, which suggests strongly a control by T cells of antibody production (50).

One might also speculate that a structural change in one component in the U snRNPs makes it appear to be foreign in the T-cell compartment, and that the T cells then induce antibody formation by B cells with specificity for the unchanged components of the U snRNPs, so that finally an immune response against the entire particle comes about. Such a change could be the expression of a variant structural gene for particular snRNP polypeptides towards which the immune system is not tolerant, or posttranslational modifications of snRNP proteins. With procedures at hand for the purification of snRNPs and the analysis of their proteins by two-dimensional polyacrylamide gel electrophoresis, these possibilities can now be tested.

Acknowledgements

We are indebted to Prof. Dr. H.H. Peter for his continual interest and support. We thank Drs. P. Woolley and B. Bennett for their input into this review and H. Markert for the preparation of the manuscript. Work from our laboratory was supported by grants from the Deutsche Forschungsgemeinschaft (to R.L.). R.R. was supported by a fellowship from the DECHEMA.

References

1. Tan, E.M. 1982. Adv. Immunol. 33, 167-240.

2. Tan., E.M., A.S. Cohen, J.F. Fries, A.T. Masi, D.J. McShane, N.F. Rothfield, J.G. Schaller, N. Talal, R.J. Winchester. 1982. Arthritis Rheum. 25, 1271-1277.

3. Sharp, G.C., W.S. Irvin, E.M. Tan, R.G. Gould, H.R. Holman. 1972. Am. J. Med. 52, 148-159.

4. Lerner, M.R., J.A. Steitz. 1979. Proc. Natl. Acad. Sci. USA 76, 5495-5499.

5. Budman, D.R., E.B. Merchant, A.D. Steinberg, B. Doft, M.E. Gershwin, E. Lizzio, J.P. Reeves. 1977. Arthritis Rheum. 20, 829-833.

6. Bresnihan, B., H.E. Jasin. 1977. J. Clin. Invest. 59, 106-116.

7. Steinberg, A.D., H.R. Smith, C.A. Laskin, B.J. Steinberg, B.S. Smolen. 1983. In: Advances in Systemic Lupus Erythematosus (J.P. Hayslett and J.A. Hardin, eds.). Grune & Straaton, New York, pp. 5-14.

8. Theofilopoulos, A.N., F.J. Dixon. 1985. Adv. Immunol. 37, 269-390.

9. Reddy, R., H. Busch. 1983. Progr. Nucleic Acids Res. Mol. Biol. 30, 127-162.

10. Steitz, J.A., S.L. Wolin, J. Rinke, I. Pettersson, S.M. Mount, E.A. Lerner, M. Hinterberger, E. Gottlieb. 1983. Cold Spring Harbor Symp. Quant. Biol. XLVII, 893-900.

11. Brunel, C., J. Sri-Widada, P. Jeanteur. 1985. Progr. Molec. Subcell. Biol. 9, 1-52.

12. Hashimoto, C., J.A. Steitz. 1984. Nucleic Acids Res. 12, 3283-3293.

13. Bringmann, P., B. Appel, J. Rinke, R. Reuter, H, Theissen, R. Lührmann. 1984. EMBO J. 3, 1357-1363.

14. Padgett, R.A., P.J. Grabowski, M.M. Konarska, S. Seiler, P.A. Sharp. 1986. Ann. Rev. Biochem. 55, 1119-1150.

15. Maniatis, T., R. Reed. 1987. Nature 325, 673-678.

16. Sharp, P.A. 1987. Science 235, 766-771.

17. Bringmann, P., J. Rinke, B. Appel, R. Reuter, R. Lührmann. 1983. EMBO J. 2, 1129-1135.

18. Lührmann, R., B. Appel, P. Bringmann, J. Rinke, R. Reuter, S. Rothe, R. Bald. 1982. Nucleic Acids Res. 10, 7103-7113.

19. Bringmann, P., R. Reuter, J. Rinke, B. Appel, R. Bald, R. Lührmann. 1983. J. Biol. Chem. 258, 2745-2747.

20. Hinterberger, M., I. Pettersson, J.A. Steitz. 1983. J. Biol. Chem. 258, 2604-2613.

21. Bringmann, P., R. Lührmann. 1986. EMBO J. 5, 3509-3516.

22. Bringmann, P., R. Lührmann. 1987. FEBS Letters 213, 309-315.

23. Lelay-Taha, M.-N., I. Reveillaud, J. Sri-Widada, C. Brunel, P. Jeanteur. 1986. J. Mol. Biol. 189, 519-532.

24. Tazi, J., C. Alibert, J. Temsami, I. Reveillaud, G. Cathala, C. Brunel, P. Jeanteur. 1986. Cell 47, 755-766.

25. Gerke, V., J.A. Steitz. 1986. Cell 47, 973-984.

26. Kessler, S.W. 1975. J. Immunol. 115, 1617-1624.

27. Towbin, H., T. Staehelin, J. Gordon. 1979. Proc. Natl. Acad. Sci. USA 76, 4350-4354.

28. Billings, P.B., R.W. Allen, F.C. Jensen, S.O. Hoch. 1982. J. Immunol. 128, 1176-1180.

29. Conner, G.E., D. Nelson, R. Wisniewolski, R.G. Lahita, G. Blobel, H.G. Kunkel. 1982. J. Exp. Med. 156, 1475-1485.

30. Matter, L., K. Schopfer, J.A. Wilhelm, T. Nyffenegger, R.F. Parisot, E.M. De Robertis. 1982. Arthritis Rheum. 25, 1278-1283.

31. Guldner, H.H., H.-J. Lakomek, F.A. Bautz. 1983. J. Imm. Meth. 64, 45-59.

32. Fisher, D.E., N.H. Reeves, G.E. Conner, G. Blobel, H.G. Kunkel. 1984. Proc. Natl. Acad. Sci. USA 81, 3185-3189.

33. Pettersson, I., M. Hinterberger, T. Mimori, E. Gottlieb, J.A. Steitz. 1984. J. Biol. Chem. 259, 5907-5914.

34. Habets, W.J., J.H.M. Berden, S.O. Hoch, W.J. van Venrooij. 1985. Eur. J. Immunol. 15, 992-997.

35. Pettersson, I., G. Wang, E.I. Smith, H. Wigzell, E. Hedfors, J. Horn, G.C. Sharp. 1986. Arthritis Rheum. 29, 986-996.

36. Mattioli, M., M. Reichlin. 1983. J. Immunol. 110, 1318-1324.

37. Mimori, T., M. Hinterberger, I. Pettersson, J.A. Steitz. 1984. J. Biol. Chem. 259, 560-565

38. Fisher, D.E., G.E. Conner, W.H. Reeves, G. Blobel, H.G. Kunkel. 1983. Proc. Natl. Acad. Sci. USA 80, 6356-6360.

39. Olmsted, J.B. 1981. J. Biol. Chem. 256, 11955-11957.

40. Reuter, R., R. Lührmann. 1986. Proc. Natl. Acad. Sci. USA 83, 8689-8693.

41. Williams, D.G., M.R. Stocks, P.R. Smith, R.N. Maini. 1986. Immunology 58, 495-500.

42. Reuter, R., S. Rothe, R. Lührmann: Nucl. Acids Res. (in press)

43. Habets, W., M. Hoet, P. Bringmann, R. Lührmann, W. van Venrooij. 1985. EMBO J. 4, 1545-1550.

44. Reuter, R., C.F. Lehner, E.A. Nigg, R. Lührmann. 1986. FEBS Letters 201, 25-29.

45. Reuter, R., B. Appel, J. Rinke, R. Lührmann. 1985. Exp. Cell Res. 159, 63-79.

46. Kinlaw, C.S., B.L. Robberson, S.M. Berget. 1983. J. Biol. Chem. 258, 7181-7189.

47. LeFeber, W.P., D.A. Norris, S.R. Ryan, J.C. Huff, L.A. Lee, M. Kubo, S.T. Boyce, B.L. Kotzin, W.L. Weston. 1984. J. Clin. Invest. 74, 1545-1551.

48. Cohen, P.L., R.A. Eisenberg. 1982. J. Immunol. 129, 2142-2145.

49. Bernard, N.F., R.A. Eisenberg, P.L. Cohen. 1985. J. Immunol. 134, 3812-3818.

50. Eisenberg, R.A., K. Dyer, S.Y. Craven, C.R. Fuller, W.J. Yount. 1985. J. Clin. Invest. 75, 1270-1277.

# NEW ULTRASENSITIVE DETECTION SYSTEMS fOR BIOLUMINESCENCE-ENHANCED ENZYME IMMUNOASSAYS.

R. E. Geiger, W. Miska

Abteilung für Klinische Chemie und Klinische Biochemie in der Chirurgischen Klinik Innenstadt der Universität München, Nußbaumstraße 20, D-8000 München 2, FRG.

## Introduction

Since the introduction of the radioimmunoassay by Berson and Yalow nearly three decades ago (1, 2) this technique has become a widely used tool for scientists in medical research and routine diagnosis for determining and quantifying antigens such as peptide and non-peptide hormones, drugs, enzymes, viruses, bacterial and tumor constituents, and other organic substances of biological and environmental interest (2 - 6). Radioimmunoassays are relatively simple in handling, and have a remarkably high specificity and sensitivity (4, 5). Disadavantages are the contact with radioactive material, the often short shelf-life of the tracers, and the unstability (radiolysis) of labelled antigens (7, 8).
For these reasons immunoassays with non-isotopic labels (e.g. fluorescence quenchers, chemiluminescence precursors) have been investigated and developed in the last years (9, 10). In particular, enzyme immunoassays (11-16) have proven to be suitable alternatives to radioimmunoassays. The various ways in which enzymes have been employed, either directly or indirectly as immunoassay labels, have been discussed in detail elsewhere (4,17).

More recently, highly sensitive bioluminescent and luminescent detection systems have became available for the determination of ATP and many different analysts (18-21). Bioluminescence is a natural phenomenon found in many lower forms of life (22).

Modern Methods in Protein Chemistry, Vol. 3

Naturally ocurring bioluminescent systems differ with regard to the structure and function of enzymes and cofactors as well as in the mechanism of the light emitting reactions (19-29).

In this communication we describe and demonstrate highly sensitive enzyme immunoassays in which bioluminescence is used as detection system (Fig. 1).

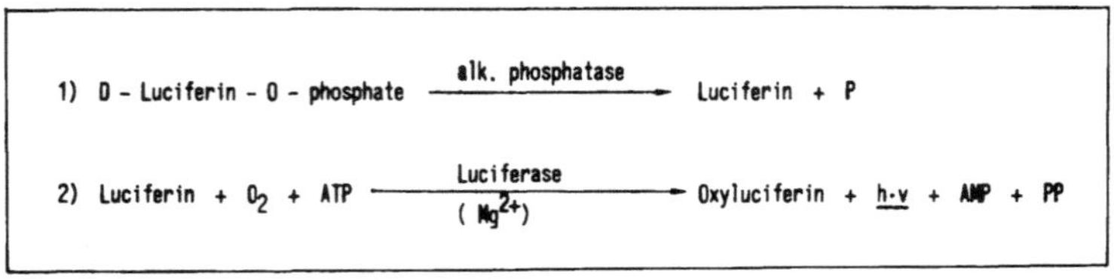


Fig.1 Scheme of bioluminescence reaction using firefly luciferase (Photinus pyralis).

The principle of these bioluminescence enhanced immunoassays is the following (Fig. 2): an antibody enzyme or antigen enzyme conjugate , respectively, which is bound in the enzyme immunoassay is enzymatically releasing D-luciferin from luciferin derivatives. Then D-luciferin is quantified in a luminometric detection system allowing a determination of antigen concentrations down to 5 pg/l.

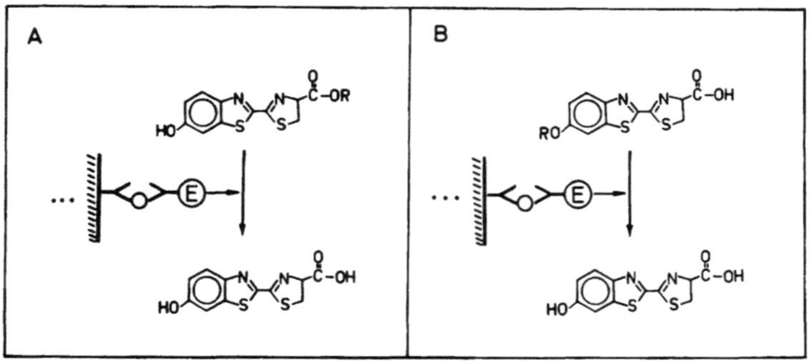


Fig.2 Scheme of the bioluminometric detection system for enzyme labelled antibodies. A, liberation of luciferin from luciferin derivatives substituted at the carboxyl group; B, release of luciferin from luciferin derivatives substituted at the hydroxyl group.

## Results

The structure of luciferin, a complex heterocyclic chemical compound was elucidated in 1963 by White et al. (30). Luciferin is very stabil at neutral pH solutions, whereas in acid solutions (pH below 2) a rapid degradation to D-cysteine and a benzthiazole derivative occurs. In alkaline solution luciferin is oxidized to dehydroluciferin which is a potent inhibitor of firefly luciferase (31). For these reasons synthesis strategies had to be developed which lead to the expected luciferin derivatives but did not destroy luciferin during synthesis. Luciferin derivatives (Fig. 3) were synthesized, purified by high performance liquid chromatography, characterized and proven for their use in enzyme immunoassays as described in ref. (32).

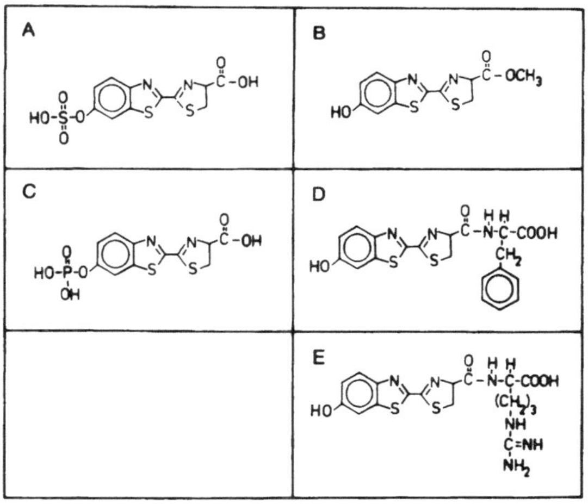


Fig.3 Synthetic D-luciferin derivaties:A, D-luciferin-O-sulfate; B, D-luciferin methyl ester; C, D-luciferin-O-phosphate; D, D-luciferyl-L-phenylalanine; E, D-luciferyl-L-N$^{\alpha}$-arginine.

Bradykinin, human urinary kallikrein (HUK) and immunoglobulin G were conjugated to carboxylic esterase, carboxypeptidases A and B, arylsulfatase and alkaline phosphatase according to established methods as described in ref. (33, 34). The enzymatic activities of the conjugates after synthesis were sufficiently high for the intended aim (36).

Using these enzyme conjugates, a number of different variants of immunoassays with different detection systems have been developed for the assay of bradykinin, human urinary kallikrein and human urinary kallikrein antibody-titer determinations as examples. The principles of the assays are shown schematically in Fig. 4.

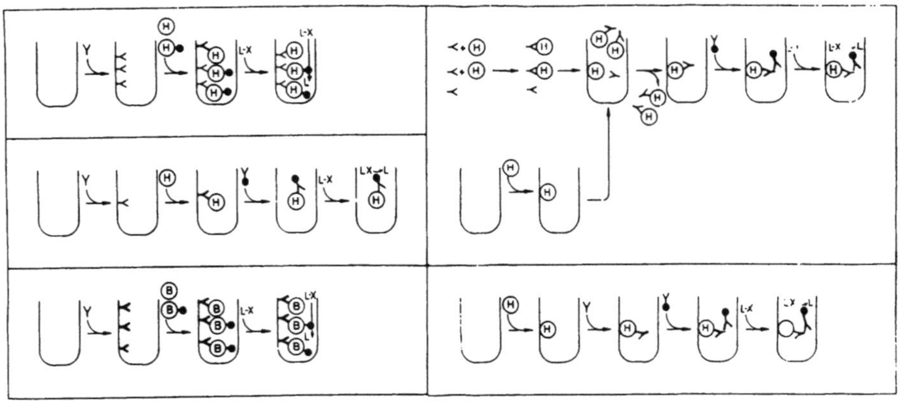

Fig.4 Scheme of the bioluminescence enzyme immunoassays.
A, competitive HUK assay, B, "indirect" HUK assay; C, HUK ELISA; D, competitive bradykinin assay; E, titer determination assay.
B, bradykinin; H, human urinary kallikrein; Y, immunoglobulin G; L-X, luciferin derivative; L, luciferin; Y-● , immunoglobulin G enzyme conjugate; B-● , bradykinin enzyme conjugate; H-● , HUK enzyme conjugate.

The bradykinin assay was performed as a competitive test (Fig. 4d), whereas for human urinary kallikrein a competitive (Fig. 4a), an immuno enzymometric assay (Fig. 4b), and a sandwich antigen assay (Fig. 4c) were established. Anti human urinary kallikrein titers were measured by sandwich antibody assay (Fig. 4e). The assays were performed as follows:

Buffer A: 15 mmol/l $Na_2CO_3$, 0.35 mol/l $NaHCO_3$, 0.2 g/l $NaN_3$, pH 9.6

Buffer B: 10 mmol/l $KH_2PO_4$, 15 mmol/l NaCl, 0.05 g/l, Tween 20 pH 7.4

Buffer C: 1.5 mmol/l $KH_2PO_4 x H_2O$, 0.14 mol/l NaCl, 0.5 g/l Tween 20, 2 mg/ml bovine serum albumin, pH 7.4

Competitive HUK enzyme immunoassay (Fig. 4a)

Microtiter plates were coated with rabbit anti-human urinary kallikrein IgG (10 ug/ml in buffer A; 0.2 ml) at 4°C over night. Thereafter the plates were washed (5 times) with buffer B. Than 0.15 ml of human urinary kallikrein standard or test samples and 0.05 ml of a human urinary kallikrein enzyme conjugate solution (see above) were added and the mixture was incubated for 24 h at 4°C. The plates were washed with buffer B and then 0.2 ml of the corresponding D-luciferin derivative solution was added and detected as described in the Table.

HUK Immuno-enzymometric assay (Fig. 4b)

Microtiter plates were coated with human urinary kallikrein (10 ug/ml buffer A; 0.2 ml per well) overnight at 4°C. Parallel but separately to coating 0.5 ml anti-human urinary kallikrein IgG solution (0.1 ng specific antibodies/ml buffer C) and 0.5 ml human urinary kallikrein solution (0.01-10 pg/ml buffer C) were incubated in capped polypropylene vials over night at 4°C. The microtiter plates were washed with buffer B (5 times). Thereafter 0.2 ml of each incubation mixture were transfered from the polypropylene vials to the corresponding microtiter plate wells and incubated at 37°C for 5 h. After washing of the plates (5 times), 0.2 ml of a goat anti-rabbit IgG enzyme conjugate solution in buffer C was added to each well and the plates were incubated again for 24 h at 4°C. The plates were washed with buffer B (5 times) and then the corresponding D-luciferin derivative solution was added (0.2 ml) to the wells. Bound IgG conjugates were detected as described in the Table.

HUK sandwich antigen assay (Fig. 4c)

Microtiter plates were coated with goat anti-human urinary kallikrein-IgG (10 ug/ml buffer A; 0.2 ml per well) at 4°C overnight. The plates were then washed (5 times) in buffer B. Human urinary kallikrein standard samples and test samples were diluted with buffer C. Of these samples 0.2 ml were added to the wells and the plates were incubated at 4°C for 24 h. After incubation, the plates were washed (5 times) with buffer B, 0.2 ml of rabbit anti-human urinary kallikrein-IgG enzyme conjugate solution was added to each well, and the plates were washed (5 times), 0.2 ml of the corresponding D-luciferin derivative substrate solution were added to the wells and the human urinary kallikrein bound IgG conjugates were detected as described in the Table.

Competitive bradykinin enzyme immunoassay (Fig. 4d)

Microtiter plates were coated overnight at 4°C with immunoselected anti-bradykinin IgG solution (10 ug/ml in buffer A; 0.2 ml per well). The plates were then washed with buffer B (5 times). Thereafter 0.15 ml bradykinin standard solution in buffer C and 0.05 ml bradykinin enzyme conjugate were added to the wells and incubated at 4°C for 24 h. Than the plates were washed with buffer B (5 times) and the corresponding D-luciferin derivative solution was added (0.2 ml) to the wells and the bound bradykinin conjugates were measured as described in the Table.

Enzyme immunoassay for titer determination (Fig. 4e)

Microtiter plates were coated with antigen (10 ug human urinary kallikrein/ml buffer A; 0.2 ml) at 4°C overnight. The plates were then washed with buffer B (5 times). Thereafter 0.2 ml of different antiserum dilutions (10 - 0.001 ul anti-human urinary kallikrein serum in 0.2 ml buffer C) were added and incubated for 3 h at 37°C. The plates were washed (5 times with buffer B), 0.2 ml of goat anti-rabbit IgG enzyme conjugate were added and incubated for 2h at 37°C. Then the plates were washed again with buffer B (5 times). The corresponding D-luciferin derivative solution was added (0.2 ml) to the wells and the bound conjugates were measured as described in the Table.

Assay systems have been optimized with respect to enzyme immunoassay conditions (binding of antigen/antibody to microtiter plates, antibody-antigen binding, etc.) as previously described (35), and to optimal detection system conditions (optimal concentration of substrates, temperatures, etc. (36)).

Table Scheme of the detection systems of the bioluminescence enhanced enzyme immunoassay.

| | Volume (ml) |
|---|---|
| Addition of D-luciferin derivative solution to the microtiter plate wells | 0.20 |
| | 60 min at 37°C* |
| Transfer to the luminometer | 0.10 |
| Additon of luminometric detection reagent (30 mmol/l HEPES, 6.6 mmol/l $MgCl_2$, 0.66 mmol/l EDTA, 0.1 mmol/l DTT, 5 mmol/l ATP, pH 7.75 containing 1 ug luciferase) | 0.40 |
| | light impulses were read for 10 s |

* in the AP-based detection system, the system has been incubated only for 30 min.

The luciferin derivative solutions used had the following concentrations: 2 umol/l D-luciferin methyl ester in 0.05 mol/l Tris/HCl, pH 7.5 (detection system I); 10 umol/l D-luciferyl-L-phenylalanine in 0.05 mol/l Tris/HCl, 3 g/l LiCl, pH 7 (detection system II); 10 umol/l D-luciferyl-L-$N^{\alpha}$-arginine in 0.05 Tris/HCl, 0.2 mol/l NaCl, pH 7.8 (detection system III); 10 umol/l D-luciferin-O-sulfate in 10 mmol/l sodium acetate, pH 5.0 (detection system IV) and 10 umol/l D-luciferin-O-phosphate in 10 mmol/l diethanolamine, 0.5 mmol/l $MgCl_2$, pH 9.8 (detection system V). Luciferin derivativescan be purchased by Novabiochem AG, CH-4448 Läufelfingen, Switzerland.

In Fig. 5 to Fig. 9 binding curves of different types of bioluminescence enhanced enzyme immunoassays for human urinary kallikrein, bradykinin and antibody titers (anti-human urinary kallikrein serum) are shown using different detection systems. The standard curves have been constructed by plotting the fraction of light impulses of the blank and the samples against the dose of antigens either in a logit-log mode (Fig. 6 and 9) or in a linear-log mode (Fig. 5, 7 and 8). Sensitivity and the slop of the curves are variing in dependency of the detection system and the type of enzyme immunoassay. The sensitivities of the bioluminescence enhanced enzyme immunoassays were defined according to the method described by Kaiser (37). The lowest concentration of human urinary kallikrein that produces a response greater than that caused in the absence of human urinary kallikrein was 1 fg per well in the immuno enzymometric assay (Fig. 8), this corresponds to 5 pg/l. For bradykinin the lowest detectable concentration was 1 fg per well corresponding to 5 pg/l (Fig. 5). Mass of antibodies which gave 50 % binding in the assay were 200 fg per well corresponding to 1 ng/l specific anti-human urinary kallikrein immunoglobulin G.

Fig.5 Dose response curves of bradykinin in the competitive bradykinin bioluminescence enzyme immunoassay.
▲, CE-based detection system I; O, CPA-based detection system II; ■, CPB-based detection system III; □, AS-based detection system IV; ●, AP-based detection system V.

Investigations performed concerning precision (intra and inter assay coefficient of variation), specificity and recovery of the assays lead to the same results as previously published for enzyme immunoassays with peroxidase as label (35, 38). The relative intraassay coefficient of variation (N between 15 and 25 for each assay; measured at different days) ranged from 2-6 % for all assays and detection systems. The intraassay coefficient was calculated for all standard curve concentrations (e.

g. Fig. 5; range $10^{-15}$ to $10^{-12}$ g/tube for ●). The interassay coefficient of variation of identical samples (N between 8 and 22) and detection systems containing all standard curve concentrations (e. g. Fig. 7; concentrations used for ●: $10^{-12}$, $10^{-11}$, $10^{-10}$, $10^{-9}$, $10^{-8}$ g/tube) was 4-12 %.

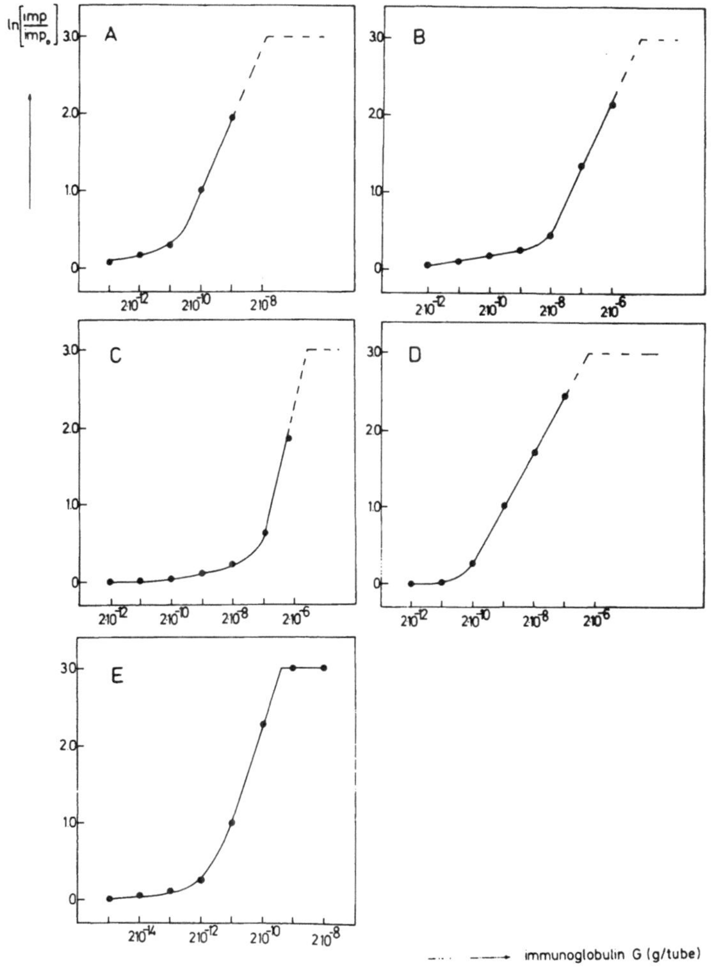


Fig.6 Binding curves of rabbit anti-HUK IgG in the HUK sandwich antibody assay.
A, CE-based detection system I; B, CPA-based detection system II; C, CPB-based detection system III; D, AS-based detection system IV; E, AP-based detection system V.

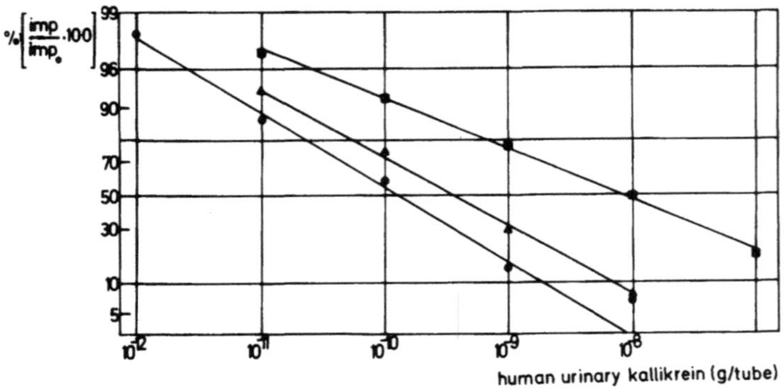


Fig.7 Binding curves of HUK in the competitive HUK assay. ■, CE-based detection system I; ▲, CPA-based detection system II; ●, CPB-based detection system III.

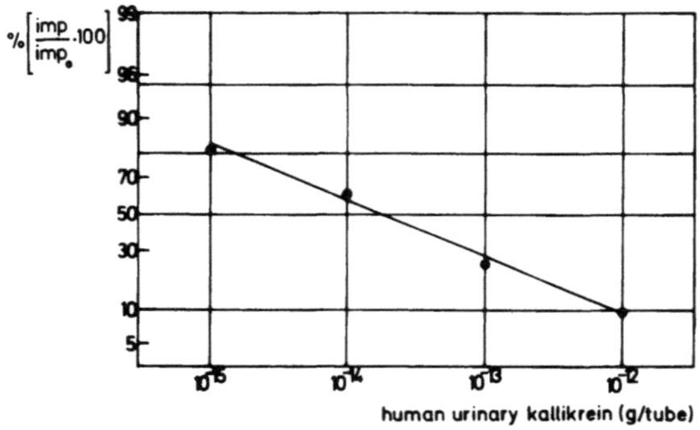


Fig.8 Binding curve of HUK in the immuno enzymometric HUK assay. ●, AP-based detection system V.

The sensitive bioluminescence enhanced enzyme immunoassay developed for peptides (bradykinin), proteins (human urinary kallikrein) and antibody titer determinations should be especially suitable for studies in which highly sensitive methods are needed (detection of monoclonal antibody synthesis in cell cultures, demonstration of in vitro translation products, etc.). By their use many problems in science and diagnosis could be solved more rapidly and at an earlier stage, as the quick determination of low concentrations of proteins or peptides produced by recombinant techniques in microorganisms or culture broth (29) or the determination of low antibody titers in the first stages of antibody formation, e.g.in diseases like AIDS. The high sensitivity of the bioluminescence-enhanced enzyme immunoassay (1 pg/l; in general radio and enzyme immunoassays reach a sensitivity of 10 to 100 ng/l; (39)) could be demonstrated by employing various bioluminescence detection systems for different types of enzyme immunoassay. Depending on the enzyme conjugates and substrates used, turnover rates of enzymes ranged from a $k_{cat}$ value of 11 l $mol^{-1} \cdot s^{-1}$ (carboxylic esterase (32)) up to 1010 $l \cdot mol^{-1} \cdot s^{-1}$ (alkaline phosphatase (32)). In Fig. 5 dose response curves obtained for detection system I to V in the competitive bradykinin bioluminescence enzyme immunoassay are shown. As can be seen, alkaline phosphatase conjugates gave a response curve with the lowest detection limit. Titer determinations (Fig. 6) and the different types of human urinary kallikrein immunoassays (Fig. 8, 9) showed similar good results. Alkaline phosphatase conjugates gave the best responses with the lowest limits of detection when compared with the other detection systems.

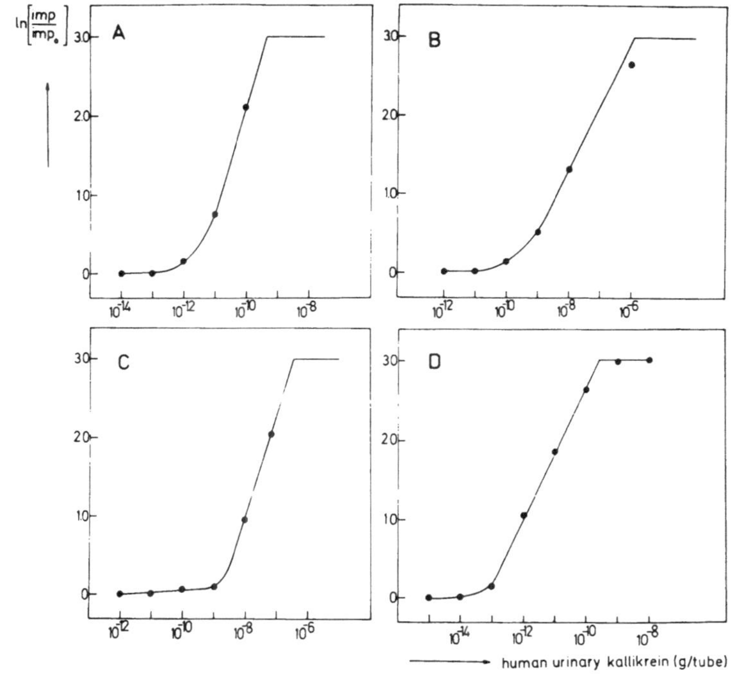


Fig.9 Binding curves of HUK in the HUK sandwich antigen assay. A, CE-based detection system I; B, CPA-based detection system II; C, CPB-based detection system III; D, AP-based detection system V.

The bioluminescence enhanced enzyme immunoassays described are useful for determination of antigens in a very low concentration range (down to $10^{-15}$ mol/l or 0.2 $10^{-19}$ mol/test). Especially conjugates with alkaline phosphatase may be used for work requiring highest sensitivity. The assay procedure is simple to perform, and the substrates are more than one year stable at -20°C in the dark. Substrate solutions can also be stored at -30°C for some weeks.

Acknowledgement

We wish to thank Prof. Dr. H. Fritz for his support during the work.

Abbreviations

CE, carboxylic esterase; CPA, carboxypeptidase A; CPB, carboxypeptidase B; AS, arylsulfatase; AP, alkaline phosphatase; HUK, human urinary kallikrein; HEPES, N-2-hydroxyethylpiperazine-N'-2-ethane sulfonic acid.

References

1. Berson, S. A., R. S. Yalow. 1957. J. Clin. Invest. 36, 873.

2. Yalow, R. S., S. A. Berson. 1959. Nature (London) 184, 1648.

3. Yalow, R. S. 1978. Science 200, 1236.

4. Vunakis, H. V., J. J. Langone. 1980. Meth. Enzymol. 70, Plenum Press, New York.

5. Langone, J. J., H. V. Vunakis. 1981. Meth. Enzymol. 74, Plenum Press, New York.

6. Schwalbe, M. 1984. Fresenius Z. Anal. Chem. 317, 596.

7. Mourier, G., L. Moroder, A. Previero. 1983. Z. Naturforsch. 39b, 101.

8. Yalow, R. S. 1982. Trends Anal. Chem. 1, 128.

9. Schall jr., R. F., H. J. Tenoso. 1981. Clin. Chem. 27, 1157.

10. Oellerich, M. 1980. J. Clin. Chem. Clin. Biochem. 18, 197.

11. Borrebaeck, C., B. Mattiasson. 1979. J. Solid-phase Biochem. 4, 57.

12. O'Sullivan, M. J., J. W. Bridges, V. Marks. 1979. Ann. Clin. Biochem. 16, 221.

13. Engvall, E., E. Ruoslahti. 1979. In: Immunoassays in the Clinical Laboratory (R. M. Nakamura, W. R. and E. S. Tucker, eds.). Alan R. Liss Inc., New York, p. 89.

14. Scharpe, S. L., W. M. Cooreman, W. J. Blomme, G. M. Laekeman. 1976. Clin. Chem. 22, 733.

15. Ngo, T. T., H. M. Lenhoff. 1981. Appl. Biochem. Biotechnol. 6, 53.

16. Marks, V. 1981. In: Therapeutic Drug Monitoring (A. Richens and V. Marks, eds.). Churchill Living stone, Edinburgh, London, Melbourne, New York, p. 155.

17. Blake, C., B. J. Gould. 1984. Analyst 109, 533.

18. Lundin, A., A. Rickardsson, A. Thore. 1976. Anal. Biochem. 75, 611.

19. Wood, W. G., H. Fricke, L. von Klitzing, C. J. Strasburger, P. C. Scriba. 1982. J. Clin. Chem. Clin. Biochem. 20, 825.

20. Fricke, H., C. J. Strasburger, W. G. Wood. 1982. J. Clin. Chem. Clin. Biochem. 20, 91.

21. Wood, W. G. 1984. J. Clin. Chem. Clin. Biochem. 22, 905.

22. Ward, W. W. 1985. In: Chemi- and Bioluminescence (J. G. Burr, ed.). Marcel Dekker, New York, p.321.

23. McCapra, F. M., K. D. Perring. 1985. In: Chemi- and Bioluminescence (J. G. Burr, ed.). Marcel Dekker, New York, p. 359.

24. McElroy, W. D., M. Deluca. 1985. In: Chemi- and Bioluminescence (J. G. Burr, ed.). Marcel Dekker, New York, p. 387.

25. Lee, J. 1985. In: Chemi- and Bioluminescence (J. G. Burr, ed.). Marcel Dekker, New York, p. 321.

26. Chappelle, E. W., G. V. Levin. 1968. Biochem. Med. 2, 41.

27. Thore, A. 1979. Science Tool 26, 30.

28. Slawinska, D., J. Slawinski. 1985. In: Chemi- and Bioluminescence (J. G. Burr, ed.). Marcel Dekker, New York, p. 533.

29. Kricka, L. J., P. E. Stanley, G. H. G. Thorpe, T. P. Whitehead. 1984. Analytical Applications of Bioluminescence and Chemiluminescence. Academic Press, New York.

30. White, E. H., F. McCarpa, G. F. Field. 1963. J. Amer. Chem.Soc. 85, 337.

31. Gish, D., F. H. Carpenter. 1963. J. Amer. Chem. Soc. 75, 5872.

32. Miska, W., R. Geiger. 1987. J. Clin. Chem. Clin. Biochem. 25,23.

33. Tae, H. 1983. Meth. Enzymol. 91, 580.

34. Carlsson, J., H. Drevin, R. Axen. 1978. Biochem. J. 173, 723.

35. Franke, M., S. Rohrschneider, R. Geiger. 1982. J. Clin. Chem. Clin. Biochem. 20, 621.

36. Geiger, R., W. Miska. 1987. J. Clin. Chem. Clin. Biochem. 25, 31.

37. Kaiser, H. 1965. Fresenius Z. Anal. Chem. 209, 1.

38. Geiger, R., W. Miska. 1986. Adv. Exp. Med. Biol., 198B, 531, 1986.

39. Yalow, R. S. 1982. Trend Anal. Chem. 1, 128.

# PRACTICAL CONSIDERATIONS IN BIOLUMINESCENCE-ENHANCED ENZYME IMMUNOASSAYS ILLUSTRATED BY A NEUTRAL PROTEINASE INHIBITOR OF PORCINE LEUKOCYTES

R. E. Geiger, G. Trefz

Abteilung für Klinische Chemie und Klinische Biochemie in der Chirurgischen Klinik Innenstadt der Universität München, Nußbaumstraße 20, D-8000 München 2, FRG.

## Introduction

Leukocytes have been shown to contain proteolytic as well as inhibitory activities (1). Proteolytic activity which is mainly found in granules and lysosomes of leukocytes is due to neutral serine proteinases such as elastase and cathepsin G (2, 3) and to thiolproteinases such as cathepsin B, H, and L (4, 5). In contrast, serine and thiolproteinase inhibitors are present in the cytosol of these cells (6, 7).
The physiological function of these inhibitors is not yet clearly unterstood. They may on the one side act as part of an intracellular defense system to regulate undesired proteolytic activities inside the cells. On the other side such inhibitors could assist the overall inhibitory potential of the organism e.g. in inflammatory processes (8), in which leukocytes release high amounts of proteinases by degranulation or during desintegration (9). The inhibitors might function by elimination of the proteases by complex formation thus preventing degradation of tissues and humoral factors as well as further activation of inflammatory cells.
In this communication we describe the evaluation of a very sensitive bioluminescence-enhanced enzyme immunoassay of porcine leukocyte neutral proteinase inhibitor (LNPI) which can be used for the determination of the inhibitor in porcine blood during inflammatory processes.

Modern Methods in Protein Chemistry, Vol. 3

Results

One of the initial steps in the development of a new bioluminescence-enhanced enzyme immunoassay is the selection of the type of assay that should be applied. Bioluminescence-enhanced enzyme immunoassays can be established for all types of enzyme immunoassays. In this illustration we decided to establish an immuno enzymometric assay of LNPI. The principle of the assay is shown in Fig. 1.

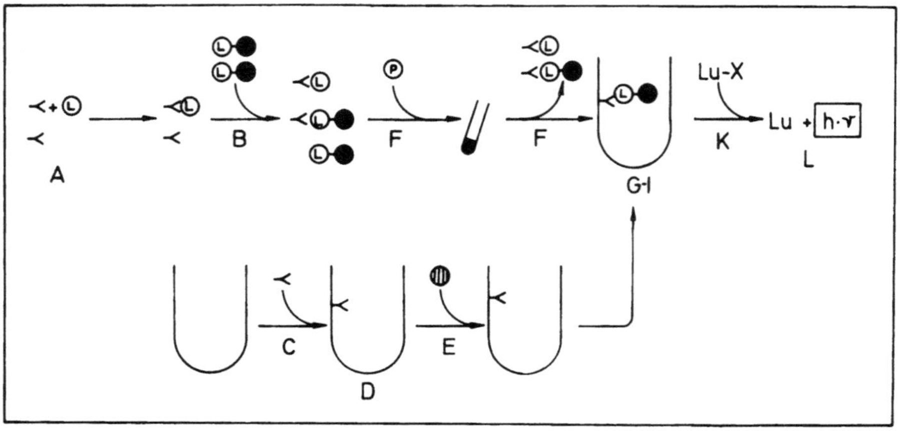


Fig.1 Scheme of the bioluminescence-enhanced enzyme immunoassay of LNPI.
A, preincubation of different amounts of LNPI with a constant amount of immunoselected antibodies (8 h at 4°C); B, addition of LNPI alkaline phosphatase conjugate and incubation over night; C, incubation of anti-LNPI IgG for 12 h at 4°C; D, binding of anti-LNPI IgG to surface of microtiter plates; E, additon of bovine serum albumin and incubation for 3 h at 4°C; F, addition of protein A Sepharose, incubation (2h at 4°C), centrifugation (10 min at 10,000 g) and transfer of supernatant of incubation mixture to microtitre plate wells coated with anti-LNPI IgG; G, incubation of supernatant for 3 h at 4°C; H, wash step; I, binding of LNPI alkaline phosphatase conjugate to surface fixed anti-LNPI IgG; K, addition of D-luciferin-O-phosphate and incubation for 30 min at 37°C; L, oxidation of luciferin by luciferase and measurement of light. Lu-X, D-luciferin-O-phosphate; Lu, D-luciferin; L, LNPI; L-●, LNPI alkaline phosphatase conjugate.

LNPI was purified from porcine PMN leukocytes by gel filtration on Sephadex G-75 superfine and ion exchange chromatography on Mono S and characterized as described in l.c. (6).

Antibodies against LNPI were raised in rabbits according to l.c. (10). The increase of antibody titers of LNPI was followed by an enzyme immunoassay procedure based on anti-rabbit IgG peroxidase conjugate as described by Geiger (11). At plateau levels of titers rabbits were exsanguinated.

IgG fraction of antisera were isolated according to Steinbuch and Audran (19).

To increase the sensitivity of the bioluminescence-enhanced enzyme immunoassay and to avoid unspecific binding LNPI-specific IgGs were isolated using LNPI-Sepharose (synthetized as described in l.c. (13)) according to (14). The LNPI-specific IgG solution had a protein concentration of 0.19 mg/ml.

LNPI alkaline phosphatase conjugate was prepared as described by Carlsson using SPDP$^R$ (15).

The next step to do in the initial phase of the development of a new bioluminescence-enhanced enzyme immunoassay was the precise determination of antibody titer of immunoselected IgG solution. Furthermore optimal assay concentrations of immunoselected immunoglobulin G and of the LNPI alkaline phosphatase conjugate had to be elucidated.

Titer determination of the immunoselected immunoglobulin G solution was performed by the Pansorbin$^R$ method as shown in Fig. 2. The test was performed as follows: 0.02 ml LNPI solution (0.02 mg/ml) and 0.01 ml of decreasing amounts of immunoselected anti-LNPI antibody solutions in buffer (0.1 mol/l Tris/HCl, 0.01 mol/l $CaCl_2$, pH 7.8) were incubated at 37°C for 1 h. Thereafter 0.05 ml Pansorbin$^R$ suspension (100 mg/ml) were added and incubated for 1 h at 37°C. The reaction mixture was centrifuged and chymotrypsin inhibition of the supernatant was tested as described in l.c. (12). The measured titer of the immunoglobulin solution was 0.048 mg/ml.

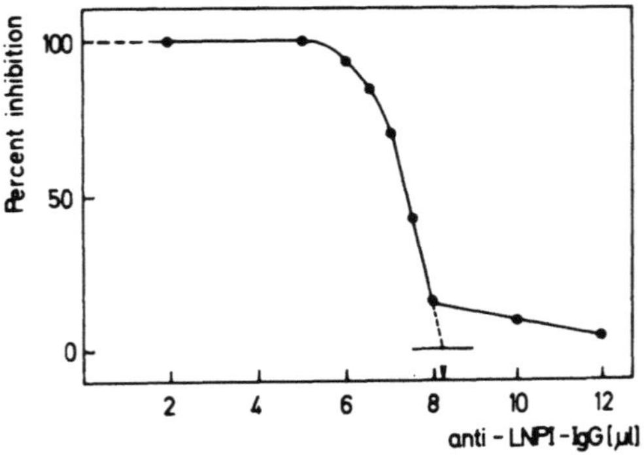


Fig.2 Titration of immunoselected anti-LNPI antibody solution using the Pansorbin$^R$ method; ..▼.., extrapolated equivalent point.

The optimal concentrations of immunoselected IgG and of LNPI alkaline phosphatase conjugate in the bioluminescence-enhanced immunoassay were elucidated as shown in Fig. 3. The tests were performed as follows:

Buffer A: 15 mmol/l $Na_2CO_3$, 0.35 mol/l $NaHCO_3$, 0.2 g/l $NaN_3$, pH 9.6

Buffer B: 100 mmol/l $K_2HPO_4$, 1.5 mol/l NaCl, 0.05 g/l, Tween 20, pH 7.4

Buffer C: 2.7 mol/l $KH_2PO_4xH_2O$, 0.14 mol/l NaCl, 0.5 g/l Tween 20, 10 mg/ml bovine serum albumin, pH 7.4

Buffer D: 2.7 mol/l $KH_2PO_4xH_2O$, 0.14 mol/l NaCl, 0.5 g/l Tween 20, pH 7.4

Microtiter plates were coated with anti-LNPI IgG fraction (2 ug IgG/ml buffer A; 0.2 ml) at 4°C overnight. The plates were then washed with buffer B (5 times). Thereafter 0.2 ml of 2 % (w/v) bovine serum albumin in buffer D were added and incubated for 2h at 4°C. Parallel to coating 0.05 ml of immunoselected antibody solutions (1, 5, 10, 20, 50 and 100 pg/ml) and 0.05 ml of decreasing LNPI alkaline phosphatase conjugate dilutions were incubated at 4°C over night. Then 0.1 ml buffer C and 0.1 ml Pansorbin$^R$ suspension (100 mg/ml) were added and incubated for 1 h at 37°C. After incubation the reaction mixture was centrifuged (5 min at 10,000 g), the supernatant was transfered to microtiter plate wells and incubated for 3 h at 37°C. Then 0.2 ml D-luciferin-O-phosphate solution (10 μmol/l D-luciferin-O-phosphate (Novabiochem AG, CH-4448 Läufelfingen, Switzerland) in 10 mmol/l diethanolamine, 0.5 mmol/l $MgCl_2$, pH 9.8) were added to each well and incubated for 30 min at 37°C. Thereafter, 0.1 ml of the incubation mixture and 0.4 ml of the test buffer (30 mmol/l HEPES, 6.6 mmol/l $MgCl_2$, 0.66 mmol/l EDTA, 0.1 mmol/l DTT, 5 mmol/l ATP, pH 7.75 containing 1 μg luciferase) were transfered to a luminometer (Biolumat, Fa. Berthold Typ 9500 T, Wildbad, FRG) and light impulses were monitored for 10s.

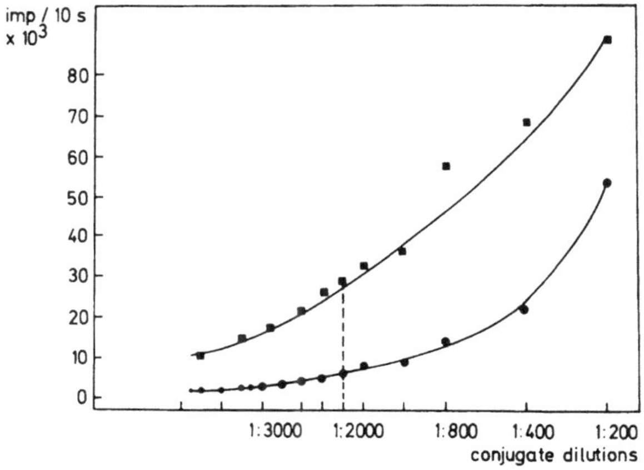


Fig.3 Binding of different amounts of LNPI alkaline phosphatase conjugate to surface-fixed anti-LNPI antibodies after preincubation with immunoselected anti-LNPI IgG. ■ , blank (buffer was used instead of immunoselected antibodies); ●, binding curve after preincubation of immunoselected antibodies (20 pg/ml) and decreasing amounts of LNPI alkaline phosphatase conjugate. Binding curves obtained after preincubation with 1, 5 and 10 pg/ml of immunoselected IgG and decreasing amounts of conjugate were between the curves (■/●); curves obtained for 50 and 100 pg/ml of immunoselected antibodies were identical to the upper curve (●).

As can be seen from data obtained in Fig. 3, a final dilution of LNPI alkaline phosphatase conjugate of 1:2000 and an immunoselected IgG solution of 20 pg/ml showed a sufficient response of about 30,000 light impulses for the blank (only conjugate, no antibodies preincubated) and about 4,000 light impulses for the preincubation mixture of conjugate and immunoselected IgG.

Using an immunoglobulin concentration of 20 pg/ml and a final LNPI alkaline phosphatase dilution of 1:2000 (enzymatic activity of the conjugate: 0.12 kU/l (20, 21)) a bioluminescence-enhanced enzyme immunoassay of LNPI was established. The assay system was optimized with respect to enzyme immunoassay conditions (incubation temperature and incubation time) as previously described (16). The bioluminescence-enhanced enzyme immunoassay was performed as follows: microtiter plates were coated with 0.2 ml IgG fraction (2 ug/ml buffer A) per well at 4°C over night. Thereafter 0.2 ml bovine serum albumin solution (20 mg/ml buffer A) were added to the microtitre plate wells and incubated at 4°C for 3 h to prevent unspecific binding.
Parallel to coating 0.05 ml of immunoselected LNPI antibody solution (20 pg/ml), 0.05 ml buffer C and 0.05 ml LNPI standard solutions and samples were incubated in siliconized polypropylene tubes at 4°C for 8 h. Then 0.05 ml LNPI alkaline phosphatase conjugate was added and incubatated over night at 4°C. After incubation 0.01 ml of a protein A-Sepharose suspension (50 mg/ml) were added and incubated for 2 h at 4°C. Then 0.06 ml buffer C was added to the tubes, centrifuged for 10 min at 10,000 g and 0.2 ml of supernatant was transfered to washed microtiter plates. The plates were incubated for 3 h at 37°C. After washing 0.2 ml of D-luciferin-O-phosphate solution (10 umol/l D-luciferin-O-phosphate in 10 mmol/l diethanolamine, 0.5 mmol/l $MgCl_2$, pH 9.8) was added and incubated for 30 min. Then 0.1 ml of the incubation mixture and 0.4 ml of the detection solution (30 mmol/l HEPES, 6.6 mmol/l $MgCl_2$, 0.66 mmol/l EDTA, 0.1 mmol/l DTT, 5 mmol/l ATP, pH 7.75 containing 1 ug luciferase) were mixed and light impulses were monitored for 10 s in a luminometer.

A dose response curve of LNPI in the LNPI bioluminescence-enhanced enzyme immunoassay is shown in Fig. 4.

Investigations performed concerning precision (intra and inter assay coefficient of variation), specificity and recovery of the assays lead to the same results as previously published for enzyme immunoassays with peroxidase as label (16, 17). The relative intraassay coefficient of variation (N = 10) ranged from 2-4 %. The intraassay coefficient was calculated for all standard curve concentrations (0.5 to 50 pg/ml). The interassay coefficient of variation of identical samples (N = 10) containing all standard curve concentrations was 4-10 %. The recovery of LNPI in the LNPI bioluminescence-enhanced enzyme immunoassy was about 90 to 95%.

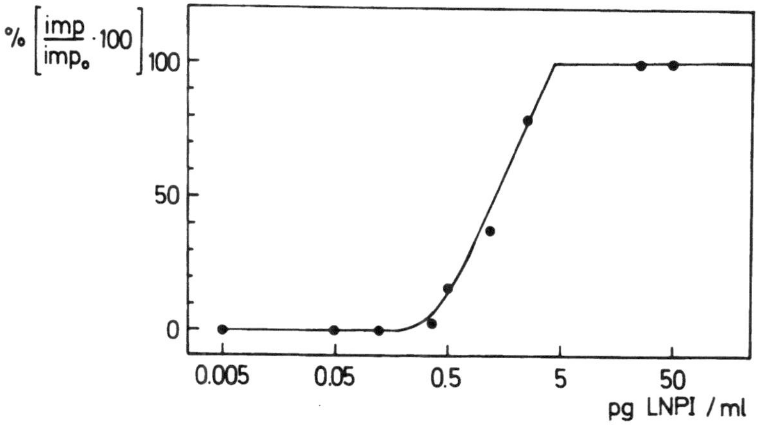


Fig.4 Binding curve of LNPI in the LNPI bioluminescence-enhanced enzyme immunoassay.

The sensitivities of the immunoassay was defined according to the method described by Kaiser (18). The lowest concentration of LNPI that produces a response greater than that caused in the absence of LNPI was 300 fg per well, this corresponds to 1.5 pg/ml. This is about 100fold more sensitive than measured with enzyme immunoassays using peroxidase as a label (6).

Acknowledgement

We wish to thank Prof. Dr. H. Fritz for his support during the work.

Abbreviations

AP, alkaline phosphatase; HEPES, N-2-hydroxyethylpiperazine-N'-2-ethane sulfonic acid; PMN, polymorphonuclear; LNPI, leukocyte neutral proteinase inhibitor; SPDP, 3-(2-pyridyldithio)propionic acid N-hydroxysuccinimide ester; IgG, immunoglobulin G.

References


1. Janoff, A., J. Blondin. 1971. Proc. Soc. Exp. Biol. Med. 136, 1050.

2. Fritz, H., H. Schiessler, R. Geiger, K. Ohlsson, K. Hochstraßer. 1978. Agents and Actions 8, 57.

3. Janoff, A., J. Blondin, R. A. Sandhaus, A. Mosser, C. Malemud. 1975. In: Proteases and Biological Control (E. Reich, D. B. Rifkin, E. Shaw, eds.), Cold Spring Harbour Laboratory, New York, p. 603.

4. Turk, V., Lj. Vitale. 1981. Proteinase and their Inhibitors, Mladinska Knjiga/Pergamon Press, Ljubljana/Oxford.

5. Katunuma, N., E. Kominami. 1983. Curr. Top. Cell. Reg. 22, 71.

6. Geiger, R., G. Trefz. 1987. Biol. Chem. Hoppe Seyler. in press.

7. Turk, V. 1986. Cysteine Proteinases and Their Inhibitors, Walter de Gruyter, Berlin.

8. Bieth, J. 1978. Front Matrix Biol. 6, 1.

9. Fritz, H., M. Jochum, K. K., Duswald, H. Ditmer, H. Kortmann, H. Neumann, H. Lang. 1984. In: Selected Topics in Clinical Enzymology, Vol. 2 (D. M. Goldberg, M. Werner, eds.), Walter de Gruyter Verlag, Berlin, p. 305.

10. Mann, K., R. Geiger, W. Göring, W. Lipp, E. Fink, B. Keipert, H. J. Karl. 1980. J. Clin. Chem. Clin. Biochem. 18, 395.

11. Geiger, R., L. Moroder, E. Wünsch. 1984. In: Peptides 1984 (U. Ragnarsson, ed.), Almquist and Wiksell International, Stockholm, p. 451.

12. Geiger, R., A. Junk, M. Jochum. 1985. J. Clin. Chem. Clin. Biochem. 23, 821.

13. Geiger, R., H. Fritz. 1981. Meth. Enzymol. 80, 466.

14. Geiger, R. 1886. In: Meth. Enzym. Anal. Vol. 11 (H. U. Bergmeyer, ed.) VCH Verlagsgesellschaft, Weinheim, p. 476

15. Carlsson, J., H. Drevin, R. Axen. 1978. Biochem. J. 173, 723.

16. Franke, M., S. Rohrschneider, R. Geiger. 1982. J. Clin. Chem. Clin. Biochem. 20, 621.

17. Geiger, R., W. Miska. 1986. Adv. Exp. Med. Biol., 198B, 531.

18. Kaiser, H. 1965. Fresenius Z. Anal. Chem. 209, 1.

19. Steinbuch, M., R. Audran. 1969. Arch. Biochem. Biophys. 134, 279.

20. Miska, W., R. Geiger. 1987. J. Clin. Chem. Clin. Biochem. 25,23.

21. Geiger, R., W. Miska. 1987. J. Clin. Chem. Clin. Biochem. 25, 31.

A NEW, VERY SENSITIVE, BIOLUMINESCENCE-ENHANCED DETECTON SYSTEM fOR PROTEIN BLOTTING (WESTERN BLOT)

R. E. Geiger, R. Hauber

Abteilung für Klinische Chemie und Klinische Biochemie in der Chirurgischen Klinik Innenstadt der Universität München, Nußbaumstraße 20, D-8000 München 2, FRG.

## Introduction

Protein blot analyses are now extensively used for research and for diagnosis of infectious diseases, e.g. hepatitis or AIDS (1-6). The first practical methods for transfer of proteins from gel electropherograms to cellulose or nitrocellulose were published in 1979 (7-9) and since that time there has been an explosive increase in applications of these techniques (for a review see 10). In the beginning, radioactive labels were used to increase the sensitivity of antigen detection (7, 8, 11, 12). During the last years non-radioactive labels such as horse radish peroxidase or alkaline phosphatase have been introduced (13-21), which allow a comparable sensitivity of detection. As little as 1 to 10 ng of protein can be detected under optimal conditons.

In this communication we describe the development of a bioluminescence-enhanced detection system with a much increased sensitivity of detection for protein blotting.

## Results

We report here an alternative, sensitive, bioluminescent-enhanced detection method of protein blotting. The principle of bioluminescent protein blotting is shown in Fig. 1.

Modern Methods in Protein Chemistry, Vol. 3

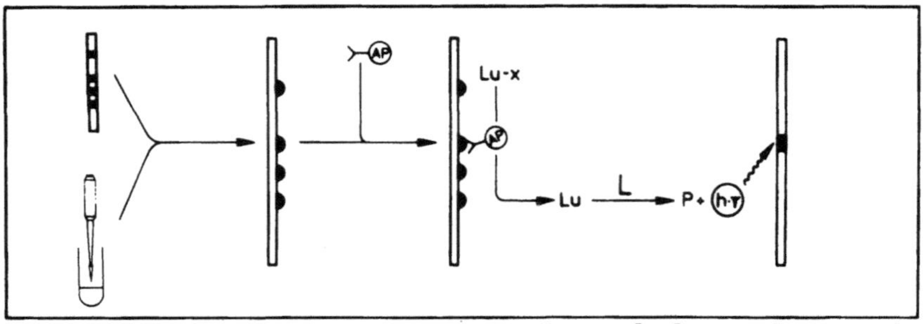


Fig.1 Scheme of the bioluminescent-enhanced detection system of protein blotting.

After application of proteins to a nitrocellulose filter, either by direct application or by electric transfer, an antibody alkaline phosphatase conjugate is added. Protein-bound alkaline phosphatase conjugate liberates luciferin from the substrate luciferin-O-phosphate, and luciferin is oxidized by luciferase (Photinus pyralis) under light emission. Emitted light exposes a sensitive photographic film. Thus specifically bound alkaline phosphatase conjugate can be visualized after development.

In the detection step, alkaline phosphatase and luciferase (Photinus pyralis) have to be active both in the same buffer solution. The pH optima of these enzymes, pH 7.8 for luciferase (Photinus pyralis) and pH 9.6 for alkaline phosphatase are different. Based on the results depicted in Fig. 2, a buffer of pH 8.0 was chosen, where alkaline phosphatase and luciferase are still sufficiently active.

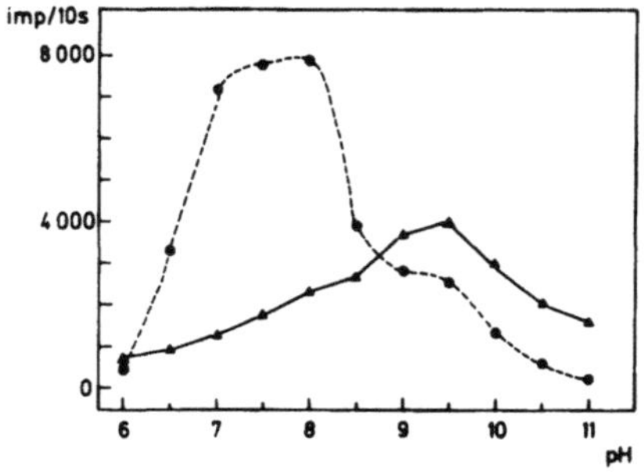


Fig.2 pH optima of calf intestine alkaline phosphatase ( ▲ ) and firefly luciferase (●, Photinus pyralis).
For details, see ref. 24 and 29.

Because of commercially non-available detection equipment, the detection has been performed using a self-constructed device as demonstrated in Fig. 3.

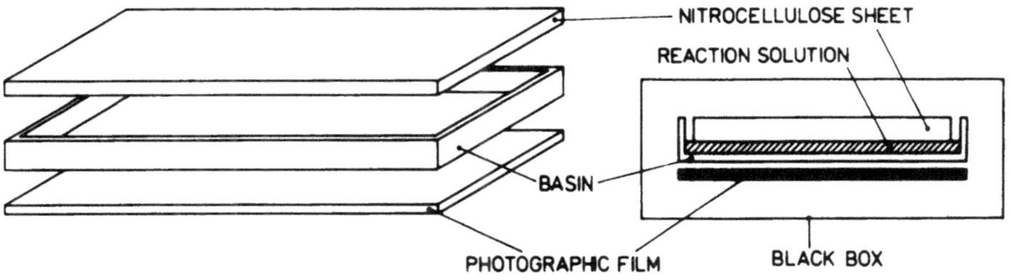


Fig.3 Scheme of the device used for the visualization of the bioluminescenence-enhanced detection system.

Fig. 4 shows an example of a photographic film, where even spots corresponding to 5 pg protein (rabbit immunoglobulin G) can be detected.

The detection system has been successfully used for the following antigens: rabbit immunoglobulin G, human urinary kallikrein and porcine leukocyte elastase. The detection system is at present not optimized with respect to sensitivity of photographic film and exposure time, whereas enzyme activities have been stabilized for two hours using the detection buffer described above. Limits of detection determined for the present method are between 5 to 5000 pg (corresponding to 30 x $10^{-18}$ mol of rabbit immunoglobulin G, to $10^{-16}$ mol of human urinary kallikrein (22) and to 7 x $10^{-16}$ mol of porcine leukocyte elastase (23)) at an exposure time of 2 hours. Taking into account that the measurement of light is performed under non-optimal conditions (only exposure of photographic film, no use of electronical photomultipliers or scanners) the detection limits can be lowered by powers of ten even at a shorter exposure time of the film. By the use of sensitive electronical detection systems, e.g. a photomultiplier or a luminometric scanner, protein samples could be measured by the same principle in the femtogram range ($< 10^{-18}$ mol) as demonstrated recently for bioluminescence-enhanced immunoassays (24, 25).

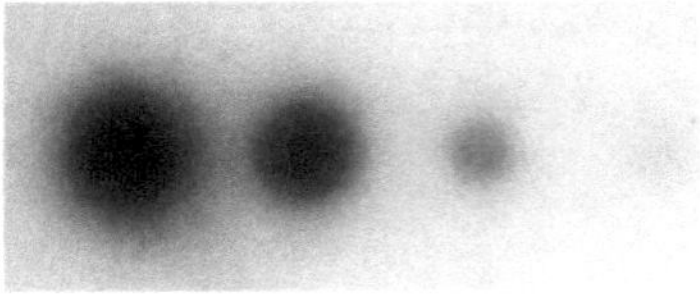

a b c d e

Fig.4 Photographic detection of a protein blot on nitrocellulose filter. a, 5 ng; b, 500 pg; c, 50 pg; d, 5 pg; e, control, no transfer of protein.
Protein (rabbit immunoglobulin G) was applied to buffer saturated nitrocellulose either by direct application or by electric transfer according to known methods (7, 26, 27). Nitrocellulose was then blocked with 5 % (w/v) bovine serum albumin in phosphate buffer (6.5 mmol/l $Na_2HPO_4$, 1.5 mmol/l/ $KH_2PO_4$, 137 mmol/l NaCl, 2.7 mmol/l KCl, pH 8.0) for 2 h at 25°C. Then nitrocellulose was incubated for 2 h at 25°C in phosphate buffer containing anti-rabbit IgG alkaline phosphatase conjugate (1:500 dilution of conjugate in phosphate buffer; Sigma Chemicals, St. Louis, U.S.A.). Thereafter the filter was washed 3 times with phosphate buffer without conjugate and transfered to a plastic basin containing about 6 ml of the detection solution (40 mmol/l HEPES buffer, 1.6 mmol/l diethanolamine, 6 mmol/l $MgCl_2$, 0.54 mmol/l EDTA, 3.4 mmol/l DTT, 2.6 mmol/l ATP, 2 mmol/l luciferin-O-phosphate, 0.15 mg luciferase, pH 8.0). The device was placed under light exclusion onto a photographic film (Kodak Tri X pan, 380 ASA), which was exposed for 2 h. After development and fixation of the film dark spots could be detected. Luciferin-O-phosphate was synthetized as described in ref. (24) or can be purchased by Novabiochem AG, CH-4448 Läufelfingen, Switzerland.

Very sensitive detection systems are of special interest and need today. By their use many problems in science and diagnosis could be solved more rapidly and at an earlier stage, as the quick determination of low concentrations of proteins or peptides produced by recombinant techniques in microorganisms or culture broth (28) or the determination of low antibody titers in the first stages of antibody formation, e.g.in diseases like AIDS.

## Acknowledgement

We wish to thank Prof. Dr. H. Fritz for his support during this work.

## Abbreviations

>-AP, antibody alkaline phosphatase conjugate; Lu-x, luciferin-O-phosphate; Lu, luciferin; P, Oxyluciferin.

## References

1. Gershoni, J. M. 1985. TIBS 10, 103.

2. Gupta, R. C. 1982. Clin. Exp. Immunol. 49, 543.

3. Erlich, H. A., G. Rodgers, P. Vaillancourt, F. G. Araujo, J. S. Remington. 1983. Infect. Immun. 41, 683.

4. Hall, W. W., P. W. Choppin. 1981. N. Engl. J. Med. 304, 1152.

5. Gordon, J., H. Towbin, M. Rosenthal. 1982. J. Rheumatol. 9, 247.

6. Kühnl, P., S. Seidl, G. Holzberger. 1986. Vox Sang. 51, 15.

7. Towbin, H., T. Staehelin, J. Gordon. 1979. Proc. Natl. Acad. Sci. USA 76, 4350.

8. Renart, J., J. Reiser, G. R. Stark. 1979. Proc. Natl. Acad. Sci. USA 76, 3116.

9. Bowen, B., J. Steinberg, U. K. Laemmli, H. Weintraub, H. 1980. Nucleic Acids Res. 8, 1.

10. Towbin, H., J. Gordon. 1984. J. Immunol. Meth. 72, 313.

11. McMichael, J. C., L. M. Greisinger, I. Millman. 1981. J. Immunol. Meth. 45, 79.

12. Billings, P. B., S. O. Hoch, P. J. White, D. A. Carson, J. H. Vaugham. 1983. Proc. Natl. Acad. Sci. USA 80, 7104.

13. Hawkes, R. 1982. Anal. Biochem. 123, 143.

14. Kyhse-Andersen, J. 1984. J. Biochem. Biophys. Meth. 10, 203.

15. Dao, M. D. 1985. J. Immunol. Meth. 82, 225.

16. O'Connor, C. G., L. K. Ashman. 1982. J. Immunol. Meth. 54, 267.

17. Jalkanen, M., S. Jalkanen. 1983. J. Clin. Lab. Immunol. 10, 225.

18. Brada, D., J. Roth. 1984. Anal. Biochem. 142, 79.

19. Hsu, Y. H. 1984. Anal. Biochem. 142, 221.

20. Turner, B. M. 1983. J. Immunol. Meth. 63, 1.

21. Moeremans, M., G. Daneels, J. de Mey. 1985. Anal. Biochem. 145, 315.

22. Geiger, R., U. Stuckstedte, H. Fritz. 1980. Hoppe Seyler's Z. Physiol. Chem. 361, 1003.

23. Geiger, R., A. Junk, J. Jochum. 1985. J. Clin. Chem. Clin. Biochem. 23, 821.

24. Miska, W., R. Geiger. 1987. J. Clin. Chem. Clin. Biochem. 25, 23.

25. Geiger, R., W. Miska. 1987. J. Clin. Chem. Clin. Biochem. 25, 31.

26. Lin. W., H. Kasamatsu. 1983. Anal. Biochem. 128, 302.

27. Bjerrum, O. J., Schafer-Nielsen. 1986. In: Electrophoresis 1986 (M. J. Dunn, ed.). VCH Weinheim, Bergstraße.

28. R. Geiger, R. Hauber, unpublished results.

29. R. Geiger, W. Miska. 1987. Anal. Biochem., in press.

# THE USE OF COLLOIDAL METAL PARTICLES IN PROTEIN BLOTTING

Marc Moeremans, Guy Daneels, Marc De Raeymaeker,
Bart De Wever and Jan De Mey
Laboratory of Biochemical Cytology
Division of Cellular Biology and Chemotherapy
Department of Life Sciences
Janssen Research Foundation
B-2340 Beerse, Belgium

## I. OVERALL STAINING OF PROTEIN BLOTS

### Introduction

Blotting has proven to be a valuable tool to study the binding activities of various kinds of ligands (e.g. lectins, hormones, enzymes, antibodies...). For reviews see reference 1-3.

The transfer of proteins from an electrophoresis matrix to a membrane is a critical and variable step. This is the main reason to visualize the overall protein pattern on an immobilizing matrix. The migration itself and, in case of migration, the subsequent adsorption onto the immobilizing matrix are not always evident. Therefore staining is valuable for judging transfer efficiency. In addition it facilitates the correlation of immunodetected bands with the entire protein pattern. This implicates that overall staining is at least as sensitive as the used immunodetection technique.

Modern Methods in Protein Chemistry, Vol. 3

Impressed by the optical and physicochemical characteristics of colloidal gold we, and independently ohters as well, started investigating the possibility of using these metal particles for overall protein staining of nitrocellulose blots.[4-5] This AuroDye staining, of which the sensitivity is comparable with silver staining of polyacrylamide gels, matches the sensitivity of the different immunostainings.

Although widely used, nitrocellulose membranes are sometimes replaced by nylon based membranes.[6] The physico-chemical characteristics of the latter prevent the use of anionic dyes like e.g. Coomassie Brilliant Blue, Amido Black and also AuroDye. Encouraged by the results obtained with negatively charged gold particles we developed a method based on the use of positively charged iron (hydrous) oxide sols (FerriDye).[7] The accumulation of these particles results in a weak yellowish-brown staining of the protein bands or spots. However, the amplification of this signal with potassium hexacyanoferrate (Perls' reaction) results in deeply-blue stained proteins. FerriDye can be used on all types of membranes (NC: nitrocellulose, PVDF: polyvinylidene difluoride, nylon and positively charged nylon based membranes. AuroDye is limited to NC and PVDF.

## A. AuroDye

### Materials

AuroDye forte (colloidal gold stain) and Tween 20 were obtained from Janssen Life Sciences Products. All other chemicals were of analytical reagent grade and double distilled water was used throughout. Nitrocellulose membranes were obtained from BioRad and polyvinilydene difluoride (PVDF) membranes from Millipore, filter paper (No. 604A-24343) from Schleicher and Schüll or ,(No. 3030917) from Whatman.

Methods

AuroDye forte is a highly sensitive, direct, protein stain. In order to reduce background staining, which forms on proteinaceous substances, several precautions have to be taken before performing AuroDye.

Protein impurities which may already originate during sample preparation and electrophoresis (e.g. dirty glass plates, dirty electrophoresis tank) will result in a disturbing, irreversible background staining. Therefore it is essential to work under very clean conditions not only during blotting and subsequent staining but also during all the foregoing steps.

In view of its high sensitivity it is recommended when, performing electrotransfer in a buffer tank, to put an additional membrane at the cathodic side of the gel. This is very helpful in reducing background staining which originates from the buffer tank and sandwich set up. The additional membrane probably adsorbs impurities out of the system that would otherwise adsorb onto the protein blot. (See fig 1.)

A common source of protein contaminants are the Scotch-Brite pads. It is advisable to replace them frequently and to rinse them thoroughly with distilled water before and after each transfer. The filter paper used in the sandwich must be of high quality too (see materials).

Another important point which has to be considered is the protein load. Most people are used to very insensitive staining methods like e.g. Amido black, as a consequence they use rather heavy protein loads. This is absolutely not necessary for AuroDye staining because excessive protein loads will prevent the detection of individual bands. Amounts, that give satisfactory results when silver staining polyacrylamide gels, perform at least equally well in AuroDye staining.

Background staining may also be caused by incorrect handling of the transfer membrane. Manipulations of the membrane must be kept to a minimum and use clean tweezers to handle the blots by their edges. Even contact with gloves must be avoided. Incubations should be preferentially done in clean plastic disposable polycarbonate trays on a rotary shaker.

Stock solutions

- Phosphate buffered saline (PBS)

| | | |
|---|---|---|
| -1.47 mM potassium dihydrogen phosphate | | 200 mg |
| -8.1 mM disodium hydrogen phosphate | | 1150 mg |
| -136.9 mM sodium chloride | | 8.00 g |
| -2.68 mM potassium chloride | | 200 mg |
| -bidistilled water | ad | 1000 ml |

- PBS-0.3 % Tween 20

| | | |
|---|---|---|
| -0.3 % (v/v) Tween 20 | | 3.0 ml |
| -PBS | ad | 1000 ml |

Procedure

Proteins can be blotted in conventional tanks using low ionic strength buffers supplemented or not with SDS. Besides several other advantages the recently introduced semi-dry blotting apparatus obviates the need of using an additional blotting membrane.[8]

After transfer, the membrane to be stained with AuroDye, is extensively washed with PBS-0.3 % Tween 20 for 30 min at 37 °C. Subsequently the blot is washed three times 5 minutes in the same buffer, at room temperature on a rocking table. Volumes are not that important but at least 0.4 ml per $cm^2$ blotting membrane must be used.

The blot is rinsed briefly in excess bidistilled water, drained off and incubated in AuroDye forte. For optimal results, incubation times are in the order of 1 to 4 hours. Prolonged staining does not cause increased background staining, nor overstaining. The blot is washed with excess bidistilled water and air-dried.

AuroDye staining is extremely sensitive (subnanogram amounts of protein/$mm^2$). However there may be some special applications were a higher sensitivity is required.

Pretreatment of the membrane with 1 % KOH for 5 min at room temperature increases signal intensity. After washing with PBS, AuroDye staining can be started as explained above.

The sensitivity of the AuroDye signal can also be substantially increased by silver enhancement. After incubation in AuroDye forte the membrane is washed 3 x 5 min in excess bidistilled water. Subsequently the membrane is incubated in IntenSE II as explained in the immunogold/silver staining section.

Remarks

In order to obtain an optimal visualization of specific antigen-antibody interactions, the amount of protein that will be loaded on the gel must be carefully chosen. Depending on the antigen concentration in the sample, sometimes rather high protein loads have to be used. However they may result in protein leakage of the membrane. This will cause agglutination of the gold particles which is demonstrated by a colour shift from red to purple-blue. In this case the blot is immediately rinsed with excess bidistilled water and subsequently incubated in a new solution of AuroDye. The use of membranes with a stronger binding capacity e.g. PVDF-membranes has proven to be very useful in these cases.

In addition these high protein loads may consume a large number of gold particles. Therefore, in cases where AuroDye forte becomes colourless, it has to be replaced by a new solution.

B. FerriDye

Materials

The FerriDye kit, including potassium ferrocyanide, was obtained from Janssen Life Sciences Products. ZètaProbe membranes were obtained from BioRad, polyvinylidene difluoride (PVDF) based membranes from Millipore. All other chemicals used were of analytical reagent grade and double distilled water was used throughout.

Methods

FerriDye is a direct protein stain. Therefore it is also in this case necessary to work under clean conditions. They are not as critical as the ones described for AuroDye because the method is about eight times less sensitive. However it is also here advantageous to include an additional membrane at the cathodic side in the sandwich set-up.

FerriDye contains cacodylate and the amplification involves the use of potassium ferrocyanide in acidic medium. Therefore in order to minimize possible health risks it is advisable to wear disposable gloves and to work in a fume hood when handling solutions.

Methods

Stock solutions

- FerriDye

| | | |
|---|---|---|
| FerriDye A | | 1 ml |
| FerriDye B | | 1 ml |
| bidistilled water | ad | 10 ml |
| prepare just before use | | |

- Amplifier solution

| | |
|---|---|
| Potassium Ferrocyanide | 210 mg |
| bidistilled water | 30 ml |
| HCl 1 N | 20 ml |
| prepare just before use | |

Procedure

After transfer, the membrane is extensively washed three times 10 minutes with excess bidistilled water.

Subsequently, the blot is drained off and incubated in actual FerriDye for 1 hour at room temperature. The protein bands will stain yellowish-brown at this stage.

Excess FerriDye is removed by washing three times 2 minutes in excess bidistilled water. The colloidal iron signal is enhanced by incubation in amplifier solution for about 1 minute (in a fume hood). The yellowish-brown bands will turn blue.

The membrane is washed 5 minutes in excess bidistilled water and air-dried.

## II. IMMUNOGOLD/SILVER STAINING (IGSS) OF PROTEIN BLOTS

### Introduction

Visualization of specific interactions on the membrane can be done in a number if ways.[1-3] Although possible, in most cases the primary antibody itself is not labelled. The common practice is to use a labelled secondary antibody or protein A. However instead of a two step method sometimes a three step method is used e.g. in the case were a biotinylated secondary antibody is visualized with the aid of labelled streptavidin.

Probably due to the success of related techniques e.g. immunoprecipitation where radiolabelled antigens are visualized with autoradiography, iodinated probes (antibodies or pA) are still widely used. Besides the possible health risks, long exposure times are needed to visualize bands.

Since the introduction of this immuno-replica technique one has tried to adapt immunocytochemical and/or Elisa procedures to detect specific antigen-antibody interactions. In this view enzymatic techniques like e.g. indirect peroxidase, PAP, ABC and alkaline phosphatases are well known. These methods use a chromogene as substrate and some are said to be as sensitive as those where radiolabelled antibodies are used.[3]

It is obvious that when immunogold reagents were introduced as coloured marker in light microscopy that their applicability was also studied in immunoblotting. Its intrinsic colour characteristics results in a pinkish red signal and provided that a rather high concentration of antigen is present.[9-12]

However, the capacity of gold particles to catalyze the reduction of silver ions can intensify an immunogold signal in such a way that even an initially invisible band will show up.[9]

In this chapter the reader will be provided with detailed technical information how to use this label in order to obtain optimal results.

## Materials

Nitrocellulose and ZètaProbe membranes were obtained from Biorad, polyvinylidene difluoride (PVDF) membranes from Millipore. Bovine serum albumin, gelatin (IGSS qualified), biotinylated secondary antibody and normal goat serum were obtained from Janssen Life Sciences Products. AuroProbe BL Plus (secondary antibody, protein A and streptavidin) reagents and Initiator and Enhancer solutions (silver enhancement products, Intense II) were also obtained from Janssen Life Sciences Products. All other chemicals were of analytical reagent grade and double distilled water was used throughout.

## Methods

Samples can take many different forms (e.g. total cell extracts, membrane fractions, etc.). Depending on the antigen studied its concentration within a sample may vary considerably. It is practically impossible to give an optimal dilution for a particular antigen. It is, in most cases, necessary to perform a few pilot experiments in order to optimize the final dilution of both the sample and the primary antibody. For the latter, a concentration of 1 to 2 µg/ml for purified antibodies gives in most cases satisfactory results.

However it has been described that incubation for 1 hour with a $10^{-9}$ M (= 0.15 μg/ml for IgGs) is usually sufficient.[2] In case of antisera a dilution of 1/100 or 1/500 is advisable. Culture supernatants should be diluted 1/5 to 1/10, ascites about 100 times. These figures are only indicative and for optimal results several dilutions have to be tested.

Although not as critical as for overall protein staining it is necessary to work under clean conditions. Metal impurities will influence silver enhancement and can cause background staining. Therefore blots have to be manipulated with clean, preferentially plastic, tweezers and the use of disposable polycarbonate trays is advisable.

## Methods

### Stock Solutions

- Tris buffered saline (TBS, pH 8.2)

| | | |
|---|---|---|
| 20 mM Tris(hydroxymethyl)-aminomethane | | 2.42 g |
| 150 mM sodium chloride | | 8.77 g |
| distilled water | ad | 900 ml |
| adjust the pH with 1 N HCl | | |
| 20 mM sodium azide | | 1.3 g |
| bidistilled water | ad | 1000 ml |

- Blocking solution

| | | |
|---|---|---|
| 5 % BSA | | 5 g |
| TBS, pH 8.2 | ad | 100 ml |

- Wash buffer

| | | |
|---|---|---|
| BSA | | 1 g |
| TBS, pH 8.2 | ad | 1000 ml |

- TBS-0.1 % BSA supplemented with gelatin

| | | |
|---|---|---|
| gelatin (IGSS qualified) | | 1 ml |
| TBS-0.1 % BSA | ad | 20 ml |

- Silver enhancement mixture

| | |
|---|---|
| Enhancer | 10 ml |
| Initiator | 10 ml |

prepare just before use

Methods

After electrophoresis (e.g. 1D or 2D) and subsequent transfer to the membrane, excessive binding sites are saturated by incubation in blocking solution. For nitrocellulose and PVDF membranes incubation in 5 % BSA-Tris at 37 °C for 30 min is sufficient. However, in the case of nylon based membranes, stronger conditions are required i.e. overnight incubation in 10 % BSA-Tris at 37 °C. Except for this blocking step all other steps are carried out at room temperature and are the same for any kind of membrane be it nitrocellulose, PVDF or nylon based.

After blocking the membrane is washed three times for 5 min in 0.1 % BSA-Tris and subsequently incubated in primary antibody solution. This primary antibody solution (in 0.1 % BSA-Tris) is supplemented with 1 % normal serum (of the same species as the secondary antibody) in case of detection with secondary antibody gold probes.

Antibodies, especially affinity purified antibodies, are precious products. Therefore the incubation volume must be kept to a minimum. This can be done by using heat sealed tailored plastic bags and by limiting the volume of antibody used to about 1 ml per 10 $cm^2$ blotting membrane. Avoid the entrapment of air-bubbles as much as possible and incubate on a tilting apparatus (e.g. Labquake, Labindustries, Berkeley, Calif., Cat. No. T400-220). Membrane strips can be easily incubated in stoppered plastic tubes.

After incubation in the primary antibody, the blot is washed three times with excess 0.1 % BSA-Tris for 10 min. This is best done in a large Petri-dish on a rocking platform. Subsequently the membranes or strips are incubated in the AuroProbe BL Plus reagent for 2 hours.

Depending on the probe used, proceed as described in A or B. The volumes used can also here be substantially reduced by using tailored plastic bags. In this case about 1 ml for every 10 $cm^2$ will be sufficient.

A. AuroProbe BL Plus secondary antibody or pA

After washing excess AB, the membrane is incubated in the gold reagent (either secondary AB or pA) for 2 hours. The gold reagent is diluted 100 times with 0.1 % BSA buffer supplemented with 1:20 v/v of IGSS qualified gelatin.

B. Auroprobe BL plus streptavidin

After washing excess AB, the membrane is incubated with a biotinylated secondary AB (2 µg/ml in 0.1 % BSA-Tris) for 1 hour. Subsequently, after rinsing 3 times 5 min in 0.1 % BSA-Tris, the membrane is incubated in the 100 x diluted (in 0.1 % BSA-Tris supplemented with gelatin) gold reagent for 2 hours.

After immunogold detection the membrane is washed twice with 0.1 % BSA-buffer for 5 min after which it can be air-dried or processed for silver intensification as described below.

Silver enhancement amplifies the signal significantly and although the gold signal itself can be quite satisfactory we routinely perform intensification as described below.

After washing the membrane in 0.1 % BSA-Tris it is washed twice in excess bidistilled water for 1 min. Equal volumes of Intensifier and Enhancer are poured into a clean Petri-dish and the washed membrane is immediately incubated in this solution for about 15 to 20 min. The reaction mixture is not very susceptible to light. There is a save margin between a typical enhancement time and the start of self nucleation of the reagent. However, both, the reaction velocity and stability of the enhancer are influenced by temperature variations. This is shown in table 1. In view of these figures one has to adapt silver enhancement according to laboratory temperature conditions.

After enhancement the membrane is washed with excess bidistilled water and air-dried. If necessary the enhancement can be prolonged (to 25 min at 22 °C) or repeated in order to improve the sensitivity. In this case the membrane is subjected to silver intensification for 20 min (22 °C), rinsed in excess distilled water and immediately incubated in fresh silver enhancement mixture. After enhancement, the blot is washed with excess bidistilled water and dried between filter paper.

table 1

| Temperature °C | Enhancement time min | Stability min |
|---|---|---|
| 16 | 15-25 | >50 |
| 18 | 12-21 | >45 |
| 20 | 11-19 | >35 |
| 22 | 10-18 | >30 |
| 24 | 9-15 | >25 |

General Discussion

The results show that colloidal metal particles prove to be a valuable tool for both non-specific (overall protein staining) and specific staining (immunodetection).

Nonspecific staining of protein blots on nitrocellulose or PVDF membranes is based on the capacity of gold particles to adsorb to all kind of macromolecules. Part of this bond seems to be of hydrophobic nature and is relatively pH independent. However binding is significantly enhanced at low pH's were the negatively charged gold particles are attracted in a strong, irreversible way by the cationized proteins. The accumulation of a large number of gold particles results in a highly sensitive red to dark red staining of the protein bands or spots.

The most essential feature of this staining method is its matching sensitivity with the different immuno-overlay assays. In addition the technique is simple, straightforward, doesn't require multiple manipulations or special equipment and AuroDye forte is commercially available.

The anionic character of AuroDye forte limits its use to nitrocellulose- or PVDF-membranes. Although not so widely used, nylon based membranes have several reported advantages.[6] For overall protein staining on these membranes FerriDye has demonstrated to be a valuable alternative for AuroDye, in spite of its lower sensitivity (about 8 times). It is a two step method but it is simple and fast. In addition it performs equally well on nitrocellulose-, PVDF-, and nylon based membranes. To our knowledge it is the only direct protein stain on nylon- or positively charged nylon based membranes.

The reason to use highly sensitive techniques for immunodetection are twofold. Not only have minor components to be visualized in complete cell extracts but often the availability of such samples is limited.

Immunogold staining itself results in a weak pinkish signal (sometimes practically invisible) and in most cases it doesn't meet the required sensitivity mentioned above. However the possibility to amplify this signal by silver enhancement makes of this marker a competitive immunovisualization technique. It sounds perhaps paradoxal that the silver amplification is no longer light sensitive. But, like mentioned in the text there is a clear gap between a typical enhancement time and stability time. Of course this makes the IGSS method much more practical to use. In addition the newest generation of silver enhancement reagents have several other interesting features. The reaction occurs at neutral pH (= 7.3) and the reagents are not very sensitive to most silver precipitating ions. The possibility to use readily available solutions is yet another adavantage of IntenSE II.

Where AuroDye is limited to nitrocellulose or PVDF-membranes there is no limitation whatsoever for immunogold/silver staining. However, besides the differences in blocking conditions, we must once again draw your attention on the buffer system used. We tried several buffer systems and the only one that is fully compatible with IGSS is the one described in the text.

A last general consideration which has to be made is: How to present unequivocal blotting data? Each item described below plays an essential role and presenting immunoblotting results in this way should be encouraged.

A critical and sometimes variable step in an immunoblotting experiment is the transfer. In order to be able to judge the transfer efficiency results have to include a stained gel (e.g. silver) and an overall stained membrane replica (e.g. AuroDye). The latter facilitates also the correlation of immunodetected bands with the total protein pattern. To complete the data a negative control (where the primary antibody is omitted or where a non related primary antibody is used) has to be shown. In order to be clear, these figures, a silver stained gel, the overall stained membrane replica, the immunoblot and the negative control have to be shown very well aligned to each other. Data presented in this way will be very clear to the reader who can now easy interprete the results.

References


1. Gershoni, J.M. and Palade, G.E., Anal. Biochem., 1983, 131, 1-15.
2. Towbin, H. and Gordon, J., J. Immunol. Methods, 1984, 72, 313-340.
3. Beisiegel, U., Electrophoresis, 1986, 7, 1-18.
4. Moeremans, M., Daneels, G. and De Mey, J., Anal. Biochem., 1985, 145, 315-321 .
5. Rohringer, R. and Holden, D.W., Anal. Biochem., 1985, 144, 118-127.
6. Gershoni, J.M. and Palade, G.E., Anal. Biochem., 1982, 124, 396-405.
7. Moeremans, M., De Raeymaeker, R., Daneels, G. and De Mey, J., Anal. Biochem., 1986, 153, 18-22.
8. Kyshe-Anderson, J., J. of Biochem. and Biophys. Methods, 1984, 10, 203-209.
9. Moeremans, M., Daneels, G., Van Dijck, A., Langanger, G. and De Mey, J., J. Immunol. Methods, 1984, 74, 353-360.
10. Brada, D. and Roth, J., Anal. Biochem., 1984 ,142, 79-83.
11. Hsu, Y.H., Anal. Biochem., 1984, 142, 221-225.
12. Surek, B., and Latzko, E., Biochem. Biophys. Res. Commun., 1984, 121, 284-289.

# AMINO-ACID SEQUENCE ANALYSIS OF PROTEINS SEPARATED BY ONE- AND TWO-DIMENSIONAL GEL ELECTROPHORESIS AND ELECTROBLOTTED ON POLYBASE-COATED GLASS-FIBER SHEETS

G. Bauw, M. Puype, M. Van Montagu, J. Vandekerckhove

Laboratory of Genetics, State University Gent, Ledeganckstraat 35, B-9000 Gent, Belgium

J. Van Damme

Plant Genetic Systems N.V., Plateaustraat 22,B-9000 Gent, Belgium

## Summary

Glass microfiber sheets to which quaternized ammonium polybases are adsorbed, were used as immobilizing membranes for protein electroblotting from sodium dodecylsulfate polyacrylamide gels. We have studied blotting efficiencies for various proteins, using different types of polybases and transfer buffers. We found that coating the glass-fiber sheets with poly (4-vinyl-N-methylpyridinium iodide) and carrying out the electrotransfer in Tris (hydroxymethyl) aminomethane-boric acid buffer (50 mM each) provides the best conditions found so far. Until now, the $NH_2$-terminal sequence of about one hundred different glass-fiber immobilized proteins, polypeptides and glycoproteins, with molecular weight ranging from 10,000 to 150,000, have been successfully determined using classical gas-phase sequence analysis. The technique was found to be especially suitable for partial sequence analysis of proteins which were first enriched before final purification on gel.

Moreover, we demonstrate that it is possible to determine the $NH_2$-terminal amino acid sequence of the major proteins present in a total cellular

Modern Methods in Protein Chemistry, Vol. 3

extract , after separating them by 2D-PAGE followed by electroblotting on coated glass-fiber. This allows one to analyse phenotypic alterations in cells or organisms at the level of the primary structure of the proteins. Based on the protein-sequence, a DNA-probe can be synthesized, which will allow the identification and isolation of the corresponding gene. This procedure was used to analyse the major proteins present in a total cellular extract of Nicotiana plumbaginifolia. Of the 10 proteins selected for analyses, eight yielded a sequence, ranging from 10 to 38 residues. Three of them are very homologous with other eukaryotic proteins: triosephosphate isomerase, Mn-superoxide dismutase and 1,3-beta-glucanase.

## Introduction

Proteins can be efficiently electroblotted from sodium dodecylsulfate-polyacrylamide gels (SDS-PAGE) onto nitrocellulose or positively charged nylon membranes (for review see 1). These supports, are not directly suitable as protein immobilizers in the gas-phase sequencing technology (2), since they are not chemically resistant against the chemicals and solvents used in the Edman chemistry. Glass microfilter (GF) paper provides a suitable alternative, provided it can be coated with positively charged groups. Such a coating can be obtained in several ways. First, by chemical derivatization of the silanol groups on the glass-fiber using an appropriate aminosilane derivative (e.g. aminopropyltrimethoxysilane) followed by quaternization (3). Second, by adsorption of positively charged polymers (4). The first approach was followed by Aebersold et al. to develop activated GF-paper sheets (5). Using the second method, we have allowed quaternized polybases to adsorb onto GF-paper in order to form a tightly bound positively charged layer. In previous papers (6-8), we have used Polybrene (1,5 dimethyl-1,5 diazaundecamethylene polymethobromide) coated GF-sheets as immobilizing supports for the electroblotting of proteins from SDS-gels and it is shown that immobilized proteins can be sequenced using a gas-phase sequenator and subjected to acid hydrolysis yielding their composition.

Here, we report on our further experience with this technique after successful sequence analysis of more than one hundred electroblotted proteins. We demonstrate that higher blotting efficiencies can be obtained by changing the transfer buffer and the type of polybases. The improved procedure is suitable as a routine method for $NH_2$-terminal sequence analysis not only for proteins separated by one-dimensional, but also by two-dimensional gel electrophoresis. The latter technique is particularly interesting since it now allows phenotypic analysis of cells at the primary structure level of their constituent proteins.

## Experimental Procedures

### Glass-fiber modification

Polybrene was obtained from Janssen Chimica (Belgium). Poly(4-vinyl N-methyl pyridinium iodide) (P4VMP) was synthesized by radical polymerization of the monomer 4-vinylpyridine, followed by methylation. Details on the synthesis and chemical characterization of the polybase will be reported elsewhere (Bauw et al. in preparation). GF-sheets (Whatman GF/C glass microfiber filters), previously cut at the size of the gel , were coated with the polybases by soaking then for 15 minutes in a 3 mg/ml solution of the polybase in deionized water. At least 5 sheets (14 x 15 cm) can be coated with 100 ml of the polybase solution. The sheets were air dried in a dark place and stored until use. Before being mounted in the blotting-sandwich, they were washed twice for 5 min, in 100 ml water to remove excess of non-adsorbed polybase.

### Plant material and protein extraction

Cell suspensions were initiated from homozygous Nicotiana plumbaginifolia plants using the procedure of Flick et al. (9). Cultures were grown for two weeks in the dark in Murashige and Skoog (MS) medium (10), supplemented with naphtalene acetic acid (0.1 mg/ml). Proteins were extracted from the cells by the procedure of Hurkman et al. (11). One gram of cell pellet was frozen in liquid nitrogen and ground to a fine powder in a dry-ice precooled mortar. The material was suspended in 5 ml extraction buffer (0.7 M sucrose, 0.5 M Tris-base, 30 mM HCl, 50 mM EDTA, 0.1 M KCL, 2 % 2-mercaptoethanol, 2 mM phenylmethylsulfonylfluoride, and 0,2 % polyvinylpolypyrrolidone). The solution was shaken for 10 minutes at 4°C, and the proteins were extracted with 5 ml of water-saturated phenol. Proteins were recovered from the phenol layer by overnight precipitation at -20°C with 5 volumes of 0.1 M ammonium acetate in methanol. The resulting protein pellet was washed once with 5 ml precipitation buffer, once with 5 ml of 80% acetone in water, and finally redissolved in buffer containing 9.8 M urea, 2 % Nonidet-P40, 2 % ampholines (pH 3.5-10), and 0.1 M dithio threitol. Final protein concentrations were 1 mg/ml as measured by the protein-dye binding assay (12).

Polyacrylamide gel analysis

Proteins were separated on 1.5 mm thick one-dimensional gels as described by Laemmli (13). Two-dimensional polyacrylamide gel separations were carried out essentially as described by Bravo (14). The non-equilibrium pH gel electrophoresis (NEPHGE) was performed in tubes with 1.5 mm inner diameter. One hundred microliters of protein solution were loaded for the first dimension. SDS-containing gels, serving as the separation medium for the second dimension, were casted in 2-mm-thick slab gels. The composition of the separation gel was 15 % acrylamide/0.08 % bisacrylamide. The gel buffer was 0.375 M Tris-HCl (pH 8.7), and 0.1 % SDS. The stacking gel was composed of 5 % acrylamide, 0.10 % bisacrylamide, 0.125 M Tris-HCl (pH 6.8) and 0.1 % SDS. The electrode buffer contained 0.05 M Tris, 0.19 M glycine, and 0.1 % SDS (pH 8.3).

Protein blotting on polybase-coated glass-fiber

Protein electroblotting was carried out as described previously (6), using various transfer buffers and polybase coatings (see Results and Discussion). In order to avoid excessive swelling of high percentage gels (more than 12.5 % acrylamide) during the blotting procedure with transfer buffers

depleted of methanol, the gels were pre-equilibrated. This step was carried out by gently shaking the gel for 2 hours at room temperature in NaOH-neutralized 50 mM boric acid (pH 8) containing 0.1 % SDS. The electro-transfer was carried out using a Bio-Rad blotting apparatus, set at a constant voltage of 35 V. After blotting, the GF sheets are washed at least three times with 100 ml of 20 mM NaCl, 10 mM boric acid, pH 8, in order to eliminate ionically bound glycinate.
Proteins can be visualized by dipping the dried glass-fiber sheets for 2 or 3 seconds in a dilute fluorescamine solution (1 mg fluorescamine per 750-1000 ml HPLC grade acetone). The immobilized proteins are then visible as fluorescent spots upon illumination with a U.V. lamp (mineral light lamp, UVL56, Ultraviolet Products Inc., San Gabriel USA). Use of higher fluorescamine concentrations or longer incubations in the solution should be avoided, since it gives rise to a higher degree of $NH_2$-terminal blocking.

Gas-phase sequencing of immobilized proteins

Protein bands of 1D-gels were cut out from the blots,if longer, shortened to 12 mm long pieces and mounted in the reaction chamber of the 470A gas-phase sequenator (Applied Biosystems Inc., USA). In order to have sufficient material, the same protein spot was collected from three identical 2D-blots. These glass-fiber pieces were then evenly placed in the reaction chamber of the gas-phase sequenator. The sequenator was run without modifications of the normal program. The stepwise liberated PTH-amino acids were identified with an on-line 120A PTH-amino acid analyser (Applied Biosystems Inc.)

## Results and Discussion

In previous communications, we have shown that proteins could be electro-eluted in NaOH-neutralized boric acid buffer (pH 8.0) and adsorbed onto GF/C glass-fiber sheets coated with Polybrene (6-8). During further application of this technique, we realized that these conditions were not always optimal for certain proteins. Since the technique would gain in value if the blotting could be efficiently applied to a large variety of proteins, we started a more systematic study of both the elution properties of proteins and the binding capacities of the GF support (G. Bauw et al., in preparation). These studies demonstrate that conditions for optimal elution were generally conditions of low binding capacity and vice versa. Two extremes could be seen : (i) the addition of 0.05 % SDS to the transfer buffer always leads to complete elution of all proteins, but fails to give efficient binding of the proteins on the polycationic glass-fibers. (ii), when methanol (20%) was added to the transfer buffer, most proteins remained in the gel, but those which could be eluted were immobilized in very high yields.

These counteracting effects could be explained as follows. Since blotting takes place in no SDS-containing transfer buffer, SDS is only present in the 1.5 mm thick gel. Thus, during their electromigration, proteins are surrounded by a continuously decreasing concentration of free SDS molecules resulting in a "steady" release of bound SDS and a continuous formation of new complexes with less bound SDS (15). This effect may be influenced (i) by the type and concentration of the ionic buffer. (ii) by SDS-impurities, notably C16 alkylsulfates which bind more tightly, and (iii) by the presence of organic solvents (such as methanol). As a result of this release of SDS, protein-SDS complexes will migrate more slowly and some of the proteins may even become insoluble and precipitate in the gel.

Protein binding on cationic GF is probably due to a coulombic attraction, initially between the sulfate head of SDS and the positively charged quaternary ammonium groups. These ionic attractions are then reinforced by hydrophobic interactions between the side chain of the detergent, the polybase, and the proteins. With such a situation, one may expect that SDS-poor protein complexes will neutralize fewer positive charges than complexes which contain higher amounts of bound SDS. Consequently, proteins with diminished SDS-coats will be bound more efficiently by the positively charged support than proteins which still contain high amounts of detergent. In the extreme case, SDS containing transfer buffers will saturate the polybase with monomeric SDS before the protein is able to reach the sheet.

Our efforts were therefore directed toward combining electrotransfer buffer systems exhibiting less "SDS-stripping" activity with polybases of higher charge density. Polybrene, the polybase which was used originally, contains quaternary ammonium ions alternating with hydrophobic hexamethylene and trimethylene bridges. Other potentially useful polybases should contain a higher charge density balanced by groups with enough hydrophobic character. Of the various polybases studied, poly (4-vinyl-N-methylpyridinium iodide) (P4VMP) exhibited the highest binding capacities. The difference in protein-binding capacity with polybrene-coatings was especially relevant when Tris-borate (a better protein eluting buffer) was used. Here, polybrene showed a drastic decrease of its capacity, while P4VMP maintained its original binding capacity.

Table 1.

Protein-Blotting and Binding Efficiencies

| Transfer buffer | elution efficiency | elution time | Protein binding Polybrene | P4VMP |
|---|---|---|---|---|
| 50 mM boric acid pH 8 | 0 | 20 h | < 50 % | 65 % |
| 50 mM boric acid 20% MeOH, pH 8 | - | 24 h | 70 % | 75 % |
| Tris-boric acid 50 mM each | + | 7 h | <50 % | >80 % |
| Tris-boric acid 50 mM each 20 % MeOH | - | 10 h | 60 % | 70 % |

Overall blotting efficiencies are given in a semiquantitative way and were measured on GF/C glass-fiber coated with either polybrene or poly(4-vinyl-N-methylpyridine) (P4VMP). Elution efficiencies given in a comparative way; - :bad elution, some proteins remain in the gel; 0 :all proteins elute, but not always complete; + : all proteins elute for more than 90 % out of gel.

Table 1 shows an evaluation of the different blotting systems assayed with the following proteins : actin, bovine serum albumin, carbonic anhydrase, chymotrypsinogen, human hepatitis B antigen, human gamma-interferon, myoglobin, ovalbumin, phosphorylase b, human tissue plasminogen activator and tubulin.

The best blotting conditions derived from Table 1 are those where glass-fiber sheets are coated with P4VMP and used in combination with : 50 mM Tris - 50 mM boric acid, as transfer buffer.

Extensive swelling of high percentage polyacrylamide gels, which results in deformed bands on the blots, can occur when methanol is not present in the transfer buffer. To avoid this, the gels can be pre-equilibrated. We found that equilibration of the gels with 50 mM boric acid/NaOH, pH 8 during 2 hours gave no detectable diffusion of the protein bands even when 0.1 % SDS is present.

Sequencing efficiencies of immobilized proteins

Initial coupling yields were determined for immobilized human serum albumin (from 2D-gels), chymotrypsinogen, human gamma-interferon, subunits S2 and S3 of Bordetella pertussis toxin (bound on polybrene coats) and for myoglobin bound on P4VMP-coated and polybrene-coated supports. Values of 50 %, 25 %, 40 %, 8 %, 20 %, 65 % and 50 % were found respectively. The yields were exceptionally low for the polypeptides starting with an $NH_2$-terminal serine ($\pm$ 10 %). This is best illustrated by a sequence analysis of the five dissimilar subunits of the Bordetella pertussis toxin (16) (Fig. 1A). Here, subunits S2 and S3, present in the toxin oligomer in equimolar amounts, are equally efficiently immobilized and show a very similar $NH_2$-terminal sequence : S-T-P-G-I-V-I-P for S2 and V-A-P-G-I-V-I-P for S3. In two independent sequence runs of both subunits, we consistently found lower yields for S2 than for the very similar subunit S3. Other polypeptides with $NH_2$-terminal serine (for example, some proteolytic fragments of vimentin) have consistently failed to give high initial yields (results not shown).

Unexpectedly, we also noticed a considerable difference when the same protein was electroblotted in different conditions. We have found that the initial yields of myoglobin are consistently (2 independent experiments) higher when the protein is electroblotted on P4VMP-sheets in Tris-borate (see Table 1) rather than on polybrene-coated supports in Na-borate (see Table 1). So far, it is not clear whether these differences are due (i) to the use in the polybrene blot of 20 % methanol which might contain contaminants which react with the $NH_2$-group, (ii) to the difference in electroblotting time, with the shorter one being more favorable, or (iii) to a small difference in the nature of binding of the protein due to differences in the transfer buffer or polybase which may produce different degrees of washing out in the first Edman cycle.

Repetitive cycle yields were determined for myoglobin, human serum albumin, and Bordetella pertussis toxin subunits S2 and S3. They all varied between 90 and 93 %, independent of the initial yield and of the polybase used.

The most important point is to know how much sequence information may be expected when a certain amount of a given protein is applied to the gel slot. The result will depend on a number of factors which can be summarized in the following points : 1, the "history" of the protein-purification which may result in partial $NH_2$-terminal blocking before the final gel separation; 2, the nature of the $NH_2$-terminal residue ($NH_2$-terminal serine gives low initial coupling yields); 3, the elution and binding properties of the protein; 4, the blotting conditions (see above). As a rule, we obtained better results when proteins were loaded on the gel early in the purification (by limiting the number of pre-purification steps) and by avoiding long boiling times in sample buffer. Using determined blotting conditions one can roughly calculate final sequence yields in the following way. One polybrene GF/C coated sheet adsorbs approximately 60 % when the protein in the gel is present in sub-saturating concentrations (below 20 micrograms/$cm^2$). One GF/C sheet coated with P4VMP will adsorb approximately 90 % of a protein concentration below 25 micrograms/$cm^2$. If we add to this recovery initial sequencing coupling yields varying between 10 % and 80 %, then the percentage of sequencable protein may vary between 6 % and 72 % of protein originally present in the gel. These results may be improved by using a second adsorbing layer. Note that when a minimal amount of 1 picomol of the phenylthiohydantoin-residue (PTH) can be measured, a sequence of 37 residues can be obtained starting from 50 pmol sequencable protein and with a repetitive yield of 90 %.

Applications of the protein-blotting technique

The technique of glass-fiber blotting is especially convenient when only partially purified protein samples are available. They can be further purified by one-step SDS-PAGE (a technique with high resolution and general applicability). Typically, one can use fractionated ammonium sulfate precipitation, an immuno-affinity or -precipitation step, or even total lysates of E. coli in which recombinant proteins are expressed at high levels (see for instance the separation of the E. coli lysate proteins expressing high levels of human gamma-interferon; a sample kindly provided by Dr. E. Remaut, Gent) (17) (Fig. 1B).

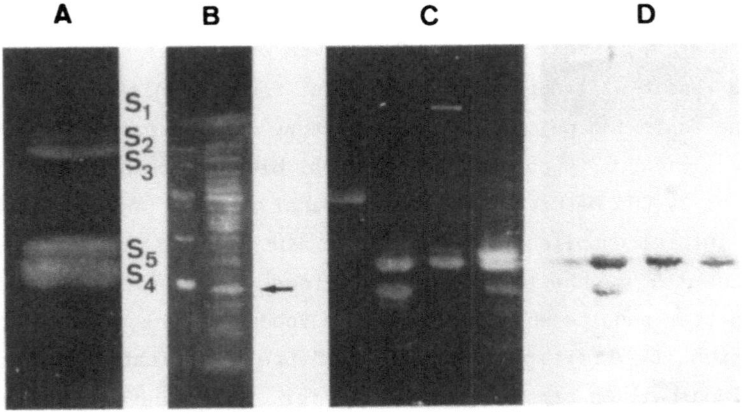


Fig. 1. Applications of the protein blotting technique. A, separation and electroblot of the five dissimilar subunits of the Bordetella pertussis toxin. B, separation of a total lysate of E. coli expressing recombinant human gamma interferon (indicated by arrow). C, chymotryptic resistent core fragments of various Bacillus thuringiensis strains, compared with reference proteins (left lane). D, the corresponding immunoblot on nitrocellulose using monoclonal antibodies raised against the toxic fragment.

The technique is also conveniently used for the analysis of individual monomers separated on PAGE from purified multisubunit protein complexes (an example is shown for the separation of the five dissimilar subunits of Bordetella pertussis toxin (16) (Fig. 1A).

Another application involves the protein-chemical identification of proteolytic fragments, which have been identified by an overlay-blotting technique to display specific binding activities. A specific case is the combination of the glass blot with the immuno-blot. In practice this is easily done by mounting a sheet of nitrocellulose or derivatized nylon on the back-side of the GF-sheet. The GF-sheet is used for total protein detection, while the organic sheet is used for classical immuno-detection. An example of this combination is shown in an experiment involving the endogenous crystal proteins formed upon sporulation of the gram-positive bacterium *Bacillus thuringiensis* (B.t.) strain Berliner consisting of three major proteins (MW : around 130,000). One of these proteins is toxic for the larvae of lepidopterean and dipterean species (18). Mouse monoclonal antibodies were raised against the cloned toxic protein and were shown to inhibit the toxic effect of the protein but also of its toxic 60,000 Mr chymotryptic fragment. Figs. 1C and 1D show the GF-blot of the mixture of chymotryptic fragments of the different crystal proteins and identifies one of them as the antigen. The $NH_2$-terminal sequence allowed mapping of the toxic 60,000 MW fragment within the native protein sequence (19).

Blotting and sequencing from 2D-gels

Combination of high capacity coated glass-fiber with transfer buffers allowing total elution of most proteins, provides the possibility to sequence proteins separated by classical, high-resolution 2D-PAGE. Figure 2 shows the fluorescamine stained glass-fiber blot of a 2D-gel separation (NEPHGE and SDS-PAGE) of a total protein extract of *Nicotiana plumbaginifolia* cell suspensions.

Proteins of plants can raise special problems since they are generally extracted from cells in the presence of high amounts of proteases. Phenol extraction followed by protein precipitation with methanol-$NH_4Ac$ results in clean preparations with no "trimming" at the $NH_2$-terminus, or proteolytic cleavage. We here demonstrate the possibility of routinely obtaining useful $NH_2$-terminal sequence information from proteins separated on 2D-gel. We analyzed 10 proteins from *Nicotiana plumbaginifolia*, which originated of the electroblotted 2D-gel of a total cellular extract.

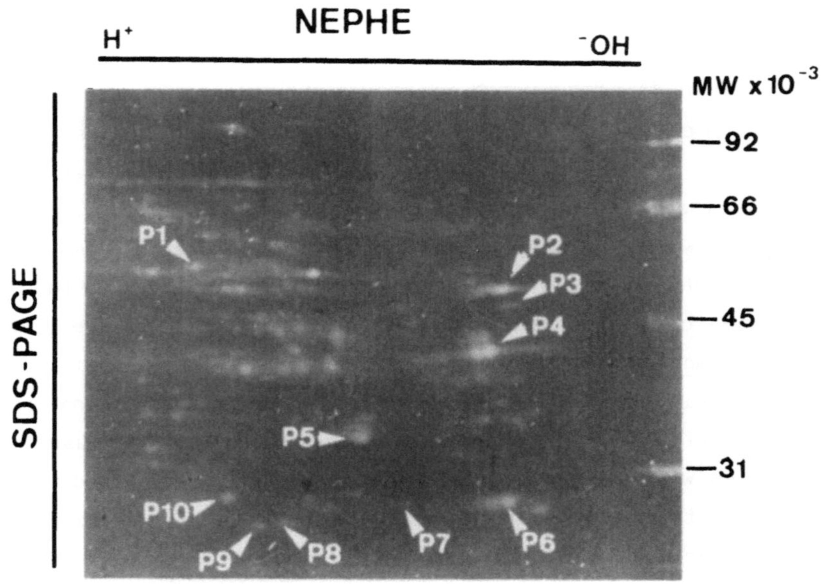


Fig. 2: Glass-fiber blots of Nicotiana plumbaginifolia proteins separated by 2D-gel electrophoresis. Total protein extracts are separated by nonequilibrium pH gel electrophoresis (horizontal direction) and SDS-PAGE (vertical direction). The separated proteins were electroblotted onto glass-fiber sheets coated with poly(4-vinyl-N-methylpyridinium iodide) and detected by fluorescamine. The molecular weight (MW) markers used in the second dimension are indicated at the right side of the figure. Proteins selected for sequence analysis are numbered in reference to the protein nomenclature of Table 2.

In order to have sufficient material of some less abundant but well separated proteins (eq. P7-P10), we collected the protein spots of three identical 2D-electroblots. The results of the sequence-analyses are summerized in Table 2. Only 2 of the 10 proteins showed a blocked $NH_2$-terminus (proteins P1 and P2), a number unexpectedly low. The remaining 8 spots could be sequenced for 10-38 cycles, yielding a total of 148 amino acids (an average of 18,5 amino acids per protein).

In order to illustrate the sequencing efficiencies obtained by the improved electroblotting procedure, we show some traces of the phenylthiohydantoin (PTH) amino acid analysis used to deduce the sequence of protein P7 (Figure 3). Repetitive sequencing yields, calculated from the recoveries of PTH-leucine at positions 5, 9, and 30, raise to 90 % for residues 5-30, and to 91 % between residues 9-30. The calculated repetitive yields for

PTH-valine (positions 10-29) are 93 %. These values are obtained with 19 picomoles of sequencable protein, corresponding with initial yields of 50 %. Similar repetitive yields were calculated from the sequence degradation of protein P9 (the Nicotiana plumbaginifolia Mn-superoxide dismutase) 98 % between residues 1-16 and 92 % between residues 9-16; results not shown). In the sequence of proteins P6 and P9, we noticed the existence of a modified amino acid residue, eluting with retention time between that of PTH-Asn and PTH-Ser (designated X in Table 2).

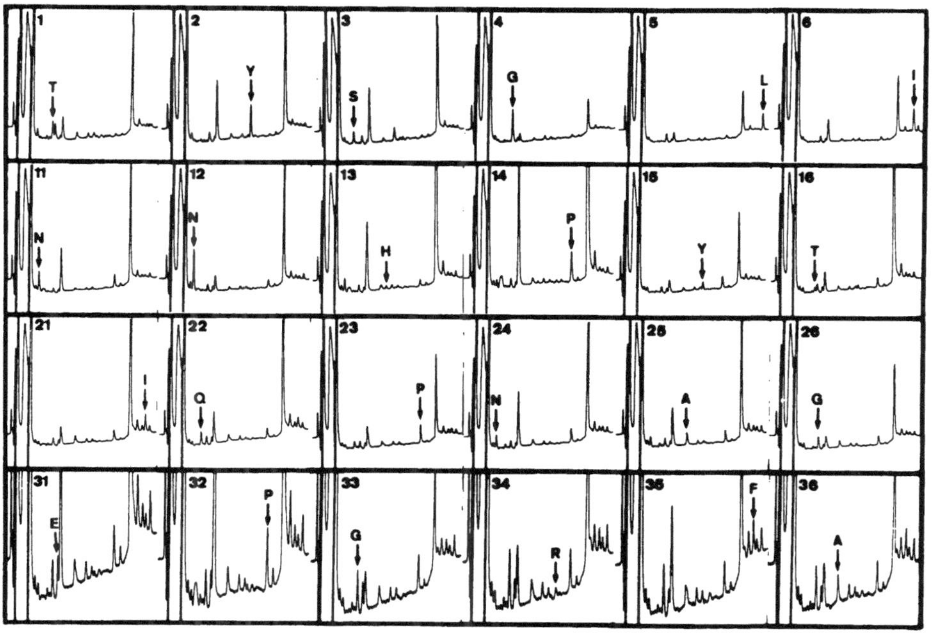


Fig. 3: Sequencing analysis of Nicotiana plumbaginifolia protein P7. HPLC traces of cycli 1-6, 11-16, 21-26 and 31-36 from an $NH_2$-terminal amino-acid sequence analysis are shown. Phenylthiohydantoin-amino acids were analyzed "on line" by chromatographing 40 % of the compound from each cycle. The positions of the phenylthiohydantoin-amino acids assigned in the traces are indicated by the one-letter notation. Traces for residues 1-6, 11-16 and 21-26 are recorded with attenuation 3, wile those for residues 31-36 are recorded with attenuation 1.

Recent DNA-sequence analysis of the cDNA clone encoding protein P9, revealed that this derivative is a post-translationally modified tyrosine residue (Aliotte T., De Loose M., Inze D., personal communication). In two other positions of the 8 sequenced proteins, we failed to identify the PTH residue, although the neighbouring residues could be characterised unambiguously. These cases are very likely to be modified amino acid residues of which plant proteins are known to be especially rich. Another complication in the sequence analysis is the occasional occurrence of micro-heterogeneity, indicating that some of the proteins are expressed in at least two different isoforms. We observed double amino acid residues on some positions of proteins P4 and P8.

All obtained protein sequences were compared with the information stored in the National Biomedical Research Foundation (NBRF) protein sequence data bank (release 9). Our search revealed significant homology between the $NH_2$-terminal sequences of three Nicotiana proteins and the $NH_2$-terminal region of previously identified proteins (Table 3).

(1) The $NH_2$-terminal sequence of protein P5, covering 21 amino acid residues, is completely identical in the region 10-21 with the amino acid sequence derived from the cDNA sequence of tobacco (1 - 3)-$\beta$-glucanase (Shinshi, cited in reference 20). The N. plumbaginifolia sequence is 66.7 % homologous to that of barley (1 - 3, 1 - 4)-$\beta$-glucanase isoenzyme II (20).

(2) The sequence containing residues 5-20 of protein P9, reveals 87.5 % homology with residues 3-18 of human Mn-superoxide dismutase (21) and 68.7 % homology with the corresponding enzyme of Saccharomyces cerevisiae (21).

(3) The 10 $NH_2$-terminal residues identified for protein P10 were sufficient to align this protein for 90 % with the corresponding sequence of Zea mays triosephosphate isomerase (22). Extensive homology was also found with the isomerase of rabbit (23) and S. cerevisiae (24).

Table 2. $NH_2$-Terminal Sequence Analysis of Ten Nicotiana plumbaginifolia Proteins Separated on Two-Dimensional Gels and Electroblotted onto Polybase-Coated Glass-Fiber Sheets

| | M.W. | Sequencable amount (pmol) | Sequence |
|---|---|---|---|
| P1 | (60) | | blocked |
| P2 | (52) | | blocked |
| P3 | (50) | 14 | V A G R S F V P I A |
| P4 | (45) | 50 | A I T A N V I V A $^{N}_{K}$ D G S K T Y |
| P5 | (34) | 30.8 | E S I G V C Y G M L G N N L P N H W E V I |
| | | | (β-1,3-glucanase) |
| P6 | (27) | 50 | A D P S P S P V L D T N G D K I Y A X ? N |
| P7 | (26) | 19 | T Y S G L I L T L V N N H P Y T I W P A I Q P |
| | | | N A G H P V L E P G R F A L |
| P8 | (24) | 14 | A I K $^{F}_{V}$ E G P P ? P |
| P9 | (24) | 12 | L Q T F S L P D L P Y D X G A L E P A I ? G D |
| | | | (Superoxide dismutase) |
| P10 | (27) | 10 | G R T F F V G G N W |
| | | | (Triosephosphate isomerase) |

Proteins are denoted (P1-P10) as shown in Fig. 2. Arabic numbers given in the second column indicate the protein molecular weights ($x\ 10^{-3}$). The sequencable amount indicates the amount amino acid liberated in the first cycle of the Edman degradation. The amino acid sequence is given in the one-letter notation. Positions where no PTH derivatives were identified are indicated by question marks. X denotes a modified residue with typical retention time in the PTH amino acid analysis system, which was observed in two proteins (P9 and P12). Two amino acids on the same position indicate microheterogeneity. Blocked proteins did not yield an amino acid sequence.

Table 3. Comparison of Three N. plumbaginifolia Protein Sequences (P5, P9, and P10) with Homologous Enzymes Identified in other Eukaryotes

**β(1 → 3)glucanase (P5)**

| | |
|---|---|
| N. plumbaginifolia | E S I G V C Y G M L G N N L P N H W E V I |
| N. tabacum | L G N N L P N H W E V I |
| Barley | (E S) I G V C Y G M S A N N L P A A S T V W |

**Mn-Superoxide dismutase (P9)**

| | |
|---|---|
| N. plumbaginifolia | L Q T F S L P D L P Y D X G A L E P A I ? G D |
| Human | K H S L P D L P Y D Y G A L E P H I N A Q |
| S. cerevisiae | K V T L P D L K W D F G A L E P Y I S G Q |

**Triosephosphate isomerase (P10)**

| | |
|---|---|
| N. plumbaginifolia | G R T F F V G G N W |
| Zea mays | M G R K F F V G G N W |
| S. cerevisiae | A R T F F V G G N F |
| Human | A P S R K F F V G G N W |

Amino acid residues are given by the one-letter notation. X denotes a post-translationally modified residue, also recognized in other plant proteins. The regions showing complete homology are shown in boxes.

## Conclusion

Electroblotting of proteins on a poly(4-vinyl-N-methylpyridinium iodide)-coated glass-fiber supports opens new perspectives for primary structure analysis of gel purified proteins. The major advantage of this technique resides in the possibility to combine PAGE, a general applicable separation method for proteins with high resolution, and the very sensitive gas-phase sequencing technology. So far, glass-fiber sheets which are either covalently (5) or non-covalently (6) modified, have been used for this purpose. Other protein binding supports, resistant to the chemicals and solvents used in the techniques, may become available in the near future. Protein blotting on glass-fibers coated with a polycation can be used for: i) analysis and confirmation of $NH_2$-terminal sequences of recombinant proteins, ii) $NH_2$-terminal mapping of epitopes and functional domains, iii) $NH_2$-terminal sequence analysis of gel-separated subunits of purified multi-subunit enzyme complex, iv) to provide amino acid sequence information, which can be used to synthesize DNA probes to isolate and identify the corresponding gene. The application of glass-fiber blotting for amino acid sequence analysis of two-dimensionally separated proteins demonstrated in this communication makes it possible to carry out phenotypic analysis of total cell lysates at the level of partial primary structures. The obtained information can then be used to isolate the corresponding gene for further characterisation of the protein.

## Acknowledgements

This research was supported by the Belgian National Fund for Scientific Research (N.F.W.O.). G.B is a fellow of the Belgian IWONL. J.V Research Associate of the N.F.W.O.

References

1. Biesiegel U. 1986. Electrophoresis 7, 1.

2. Hewick, R.M., M.W. Hunkapiller, L.E. Hood, W.J. Dreyer. 1981. J. Biol. Chem. 256, 7990.

3. Eltekov, Y.A., A.V. Kiselev, T.D. Khokhlova, T.S. Nikitin. 1973. Chromatographica 6, 187.

4. Pefferkorn, E., Q. Tran, R. Varoqui. 1981. J. Polym. Sci. 19, 27.

5. Aebersold, R.H., D.B. Teplow, L.E. Hood, S.B.H. Kent. 1986. J. Biol. Chem. 261, 4229.

6. Vandekerckhove, J., G. Bauw, M. Puype, J. Van Damme, M. Van Montagu. 1985. Eur. J. Biochem. 152, 9.

7. Vandekerckhove, J., G. Bauw, M. Puype, J. Van Damme, M. Van Montagu. 1986. In: New and Improved Methods in Protein Microsequence Analysis (B. Wittman-Liebold, J. Salnikov and V. Erdmann, eds.). Springer Press New York-Heidelberg, p. 179.

8. Vandekerckhove, J. 1986. In: Proteins : structure and function. Selected papers from the first Symposium of American Protein Chemists (L'Italien, J. ed.) Plenum Press, New York (in press)

9. Flick, C.E., D.A. Evans, W.R. Sharp. 1983. In: Handbook of Plant Cell Culture (D.A. Evans, W.R., Sharp, P.V. Ammirato and Y. Yamada, eds.). Macmillan Publishing Co., New York Vol 1, p. 13.

10. Murashige, T., F. Skoog. 1962. Physiol. Plantarum 15, 473.

11. Hurkman, W.J., C.K. Tanaka. 1986. Plant Physiol. 81, 802.

12. Bradford, M.M. 1976 Anal. Biochem. 72, 248.

13. Laemmli, U.K. 1970. Nature 277, 680.

14. Bravo, R. 1984. In: Two-dimensional Gel Electrophoresis of Proteins (J.E. Celis and R. Bravo, eds.). Academic Press, Orlando, p. 3.

15. Putnam, F.W., H. Neurath. 1945. J. Biol. Chem. 159, 195.

16. Capiau, C., J. Pétre,J. Van Damme, M. Puype, J. Vandekerckhove J. 1986. FEBS Lett. (in press)

17. Simons, G., E. Remaut, B. Allet, R. Devos, W. Fiers 1984. Gene 28, 55.

18. Dulmage, H.T. 1980. In: Microbial Control of Pests and Plant Diseases 1970-1980 (H.D. Burges, ed.). Academic Press, London, p. 193.

19. Höfte, H., H. de Greve, J. Seurinck, S. Jansens, J. Mahillon, C. Ampe, J. Vandekerckhove, H. Vanderbruggen, M. Van Montagu, M. Zabeau, M. Vaeck. 1986. Eur. J. Biochem. 161, 273.

20. Fincher, G.B., P.A. Lock, M.M. Morgan, K. Lingelbach, R.E.H. Wettenhall, J.F. Mercer, A. Brandt, K.K. Thomson. 1986. Proc. Natl. Acad. Sci. USA 83, 2081.

21. Harris, J.I., H.M. Steinman. 1977. In: Superoxide and Superoxide Dismutase (A.M. Michelson, J.M. McCord and Fridovich, eds.). Academic Press, London, p. 225.

22. Marchionni, M., W. Gilbert. 1986. Cell 46, 133.

23. Corran, P.H., S.G. Waley. 1975. Biochem J. 145, 335.

24. Alber, T., G. Kowasaki. 1982. J. Mol. Appl. Genet. 1, 419.

# PHOTOAFFINITY LABELING AND LOCALIZATION BY MICROSEQUENCING OF AN ION CHANNEL PROTEIN

F. Hucho, W. Oberthür[§]

Institut für Biochemie, Freie Universität Berlin,
Thielallee 63, 1000 Berlin 33.

F. Lottspeich

Max-Planck-Institut für Biochemie, Martinsried.

(§ - current address: Max-Planck-Institut für Biochemie, Martinsried.)

## Summary

Photoaffinity labeling of active proteins poses several problems. Various methods are reviewed, aimed at different scopes: identification of funtional sites, characterization of transient conformational states, localization of effector domains in the quaternary and primary structures. The nicotinic acetylcholine receptor is described here as a model system which has been investigated for many years using various photoaffinity labeling techniques. In connection with microsequencing methods detailed information concerning structure and function of this membrane protein has been obtained.

Modern Methods in Protein Chemistry, Vol. 3

## Introduction

Photoaffinity labeling of proteins is applicable in many cases where the specificity of affinity labeling with other reagents is not sufficient. Its advantage is based on the fact that the time point of the reaction can be determined by the experimentalist: in the dark the reagent is incubated with the protein and reversibly bound preferentially to the site having maximum affinity. After excess reagent has been removed (especially from low affinity binding sites), the light is turned on and the photoreaction is started.
Several factors determine the usefulness of a photolabel: i) The affinity, i.e., the specificity of reversible binding in absence of light. ii) The yield of photolysis, i.e., the concentration of the reactive intermediate created by irradiation of the reagent. iii) The lifetime of the reactive intermediate. A long lifetime affects specificity negatively because the reactive intermediate may dissociate from its primary binding site to react with distant sites of the protein. The latter problem can be diminished by adding 'scavengers', molecules which react with the photoactivated intermediates diffusing about, thereby preventing them from reacting with nonspecific sites on the protein. Primary amines are powerful scavengers of free radicals arising from photolysis of light sensitive compounds.
Disadvantages of photoaffinity labeling are: i) The usually low yiels, ii) The lack of group specificity with respect to the amino acids forming the ligand binding site, iii) The long irradiation time required for efficient labeling and the long duration of the photoaffinity labeling experiment (see below).
Azido or diazo derivatives of a ligand are most commonly applied as photoaffinity labels for a specific protein site, but preparing the derivatives is time consuming and because of the bulky photosensitive group the ligand may lose

affinity and specificity. Recently, therefore, in many cases the unmodified ligand has been used as a photolabel. A prerequisite for this is, of course, an electron system in the ligand which absorbs light and is shifted by the absorbed energy to an excited state reactive enough to cause covalent reaction with the protein.

*Application in receptor research* A successful application of this approach was the identification of benzodiazepine binding proteins in mammalian brain membranes (1). Benzodiazepines are tranquillizers which interact with GABA-receptors at inhibitory synapses in the brain. They act by enhancing the affinity of GABA-receptors for the inhibitory transmitter. Like the barbiturates, they function as allosteric effectors, not,as initially assumed, by binding to a receptor (a 'benzodiazepine receptor') of their own.

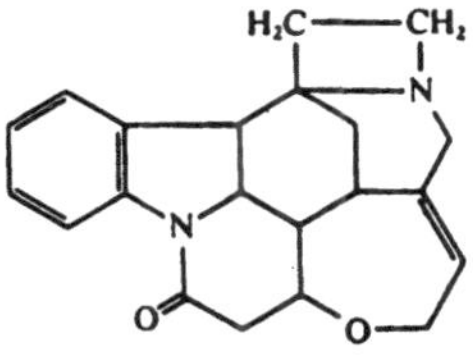


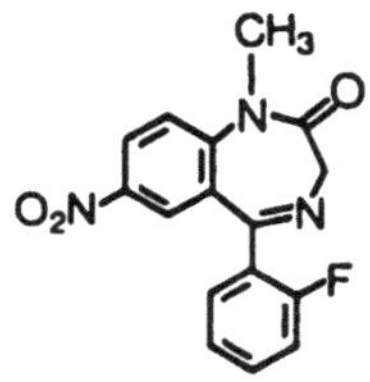


*Strychnine (left) and Flunitrazepam (right), two drugs applied as photoaffinity labels for identifying glycine- and GABA/benzodiazepine-receptors, respectively.*

The binding protein was subjected to photoaffinity labeling with the benzodiazepine $^{3}$H-flunitrazepam. Flunitrazepam, though not light-sensitive per se, becomes highly photoreactive upon binding to its binding protein in synaptic membranes. Among the many proteins present only one, a peptide of $M_r$ 50,000, was photolabeled in membranes from rat cerebral cortex. A similar approach led to the identification of another inhibitory neurotransmitter receptor of the vertebrate central nervous system (2). Strychnine is a potent inhibitor of the spinal cord glycine receptor. It is light sensitive and was used as a photoaffinity label for identifying the glycine receptor. A $M_r$ 48,000 peptide was identified which is part of the $M_r$ 250,000 glycine receptor.

Rapid photoaffinity labeling

As mentioned above, an important disadvantage of photoaffinity labeling is its slowness. The protein-photolabel complex is usually irradiated for minutes to ensure sufficient incorporation of the label. Therefore only stabile protein states can be investigated by this method; transient, short-lived states are not accessible. This is especially distressing where a key function of a protein rests just on such a transient state. Before we discuss an example of this in more detail, we would like to describe a method for photoaffinity labeling with millisecond time resolution (3). We combined a stopped-flow apparatus with a high-energy pulse laser (fig. 1c). The laser is triggered by a switch which is closed by the movement of the pistons of the syringes containing the protein and ligand, respectively. The trigger can be delayed electronically to start the photoreaction at variable time intervals after mixing protein and ligand in the mixing chamber of the stopped-flow apparatus. The delay time can be varied from 0 ms to 500 s. A detailed description of the experimental setup has been published elsewhere (3); first results from photoaffinity labeling experiments have proven its usefulness (4-6).

A different device for photolabeling after rapid mixing was developed by Karlin and his coworkers (7,8). In this case (fig. 1a) the main aim was to accomplish labeling of large amounts of protein by means of a continuous-flow, rapid-mixing technique. This resembled the experimental setup (fig. 1b) designed by Changeux and his coworkers (9) for photolabeling the nicotinic acetylcholine receptor with tritiated chlorpromazine and other drugs with similar properties. The main

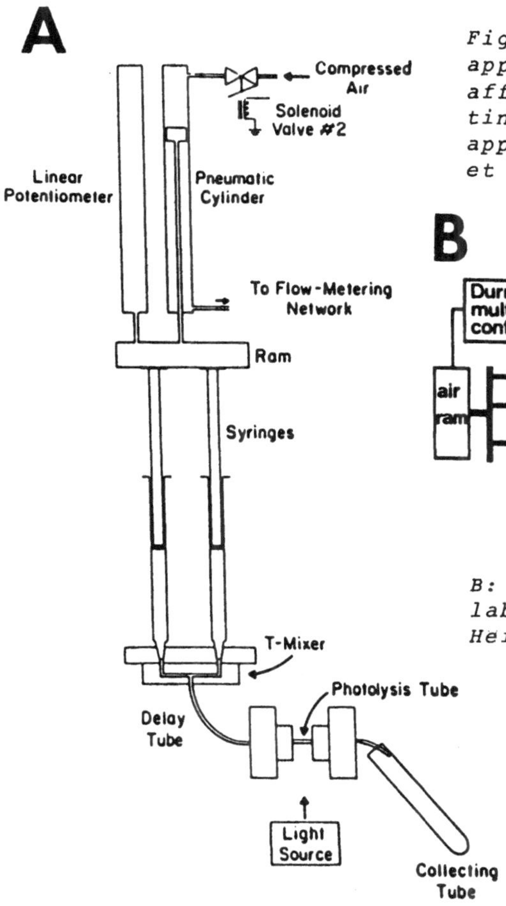


*Fig. 1. Three alternative apparatus for rapid photo-affinity labeling. A: continuous-flow rapid mixing apparatus according to Cox et al. (7).*

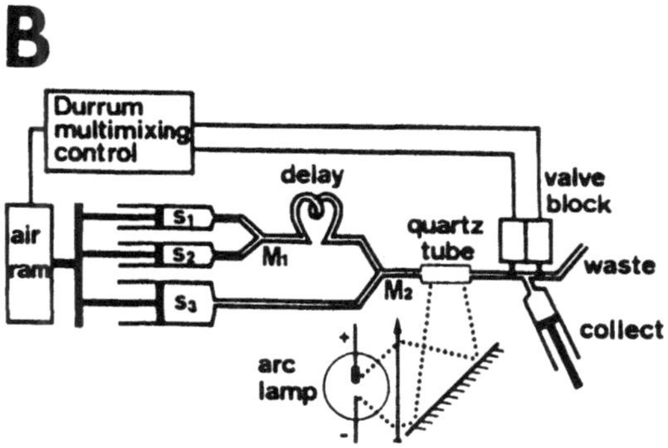


*B: Rapid-mixing photo-labeling apparatus used by Heidmann and Changeux (9).*

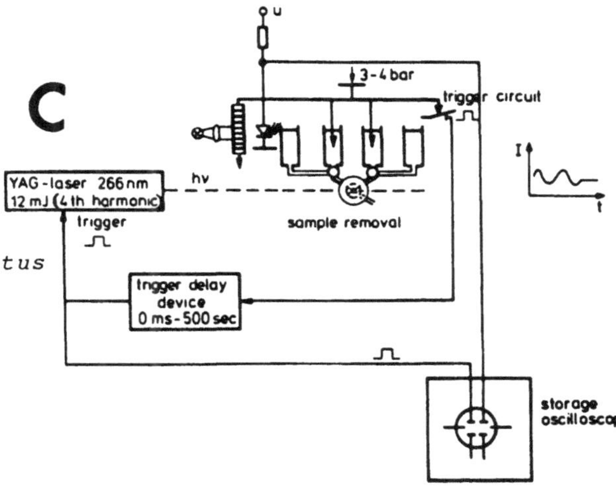


*C: Laser-flash apparatus for photoaffinity labeling with high time resolution, designed by Fahr & Hucho.*

difference between the three methods (3, 7, 9) is the time resolution. Photolabeling with normal UV light sources requires longer irradiation and the design of the mixing chamber and tubing system determines the delay time between mixing and photolysis. In the method by Cox et al, (7) 20 ms is the resolution obtainable. Heidman and Changeux (9) apply their apparatus in the 100 milliseconds to seconds range. The dead time of our apparatus (3) is 2.4 ms. Of course another important parameter is the yield, which depends largely on the reactants and which has not been compared for the three methods under similar conditions.

| | Cox et al. | Heidmann & Changeux | Fahr & Hucho |
|---|---|---|---|
| delay time (ms) | 20 | 50 | 2.4 |
| photolysis | 1 ms | < 20 ms | 5 ns |
| lower limit of time resolution (ms) | 20 | ~100 | ~ 2.4 |

Table 1. Comparison of three rapid photolabeling methods

It so happened that the three rapid photolabeling methods described above have been developed for the investigation of the same protein, the nicotinic acetylcholine receptor from *Torpedo* electric tissue. Even the same problem was tackled, the location of the binding site for small molecules blocking non-competitively the ion flux through the channel regulated by acetylcholine. The only, albeit important, difference was the chemical nature of the small molecule applied. In each case the localisation was attempted on the level of the quaternary and primary structures.

## Photoaffinity labeling of the nicotinic acetylcholine receptor (AChR)

*Classical methods* Photoaffinity labeling was successfully applied in various investigations of the AChR. These investigations may serve as model experiments for similar membrane proteins. In 1970 Kiefer et al. (10) introduced 4-azido-2-nitrobenzyltrimethylammoniumfluoroborate.

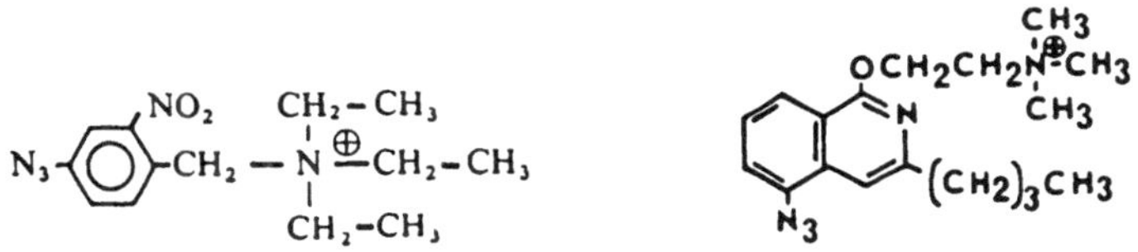

*Photoaffinity labels used in investigations of the nicotinic acetylcholine receptor: 4-azido-2-nitrobenzyltrimethylammonium (left) and 5-azido trimethisoquin (right).*

With its quaternary ammonium group it posesses some structural analogy to the cholinium groups of the cholinergic agonist acetylcholine and other agonists and antagonists. Accordingly, it was shown to inactivate (upon irradiation) AChR in frog satorius muscle. With the radioactive arylazide (11) we obtained evidence that the smallest (α-) polypeptide chains of AChR contained the binding sites for cholinergic antagonists and that the three large chains (β, γ, δ) contained binding sites for quaternary ammonium compounds, too. Later this was confirmed with arylazido derivatives of α-neurotoxins from snake venoms (12 - 14). Photoaffinity labeling in this case was found mainly in the α-chains, but some cross-linking of toxin to the δ-chain could also be obtained. This was interpreted as indicative of a quaternary structure placing the δ-polypeptide chain next to the α-subunit. Besides experiments of this kind, which were aimed at elucidating the function of the different polypeptide chains of the AChR and at the quaternary structure of the complex, several attempts have been made at gaining information concerning allosteric and other functional sites by photoaffinity labeling. In reviewing these I would like to first introduce the protein AChR and the problems. Then I will present our own experiments and the localisation of a channel blocking photolabel in the primary structure of the receptor protein. Finally, I shall draw conclusions by proposing a structural model of the channel of the acetylcholine receptor and by comparing our results with the results obtained by other laboratories.

*The biochemistry of the AChR* The nicotinic acetylcholine receptor (AChR) is at present the best characterized neurotransmitter receptor (15). It is a heteropentameric protein of $M_r$ 290,000, having the quaternary structure $\alpha_2\beta\gamma\delta$. All five polypeptides are glycosylated and span the membrane. $\beta$, $\gamma$, and $\delta$ are phosphorylated. The primary structures are known from the cDNA sequences of the subunit precursors. The receptor contains large amounts of $Ca^{2+}$, some of which is very tightly bound and is still found in the respective bands on the SDS polyacrylamide gel. The two $\alpha$ chains contain the binding sites for agonist and competitive antagonists; the functions of the other chains are unknown. The $\delta$-chain appears to be especially important for the conductivity of the ion channel which is regulated by the AChR.

The structure and functional mechanisms of the receptor's ion channel are unknown. Reconstitution experiments with the purified AChR have proven that both the ligand binding sites and the channel are integral parts of the $M_r$ 290,000 protein complex. In our view the pharmacology of the AChR provides the most direct access to elucidate on the molecular level the architecture of both functional components, the signal recognition (ligand binding) and the effector (ion translocation) site within the protein. There are three classes of pharmacologically distinguishable effectors: i) agonists, which bind and trigger opening of the channel, ii) competitive antagonists, which bind to the same sites as the agonists but have no effect other than preventing the agonist from binding, and iii) noncompetitive antagonists, which also block the response of the AChR to agonists, not by competing, but by binding to another binding site. This binding site for noncompetitive antagonists is thought to be directly related to the ion channel, blocking it either sterically or allosterically.

Noncompetitive antagonists are a very heterogeneous group of compounds comprising natural toxins like histrionicotoxin, local anesthetics like lidocain, detergents, neuroleptics like chlorpromazine and several more. The fluorescent noncompetitive antagonists have been used extensively for monitoring conformational changes in the AChR caused by agonist binding. Experiments of this kind helped to define the receptor as an allosteric protein (16). Attempts were made to localize binding sites for noncompetitive antagonists by means of covalently reacting derivatives. Again photoaffinity labeling was especially successful in this respect.

5-azido-[$^{3}$H]trimethisoquin was shown to selectively label the δ-chain. Since photolabeling was prevented by well-characterized noncompetitive antagonists, it was concluded that a site involved in the regulation of ion translocation had reacted with the label. Agonists affected the yield of the photoaffinity labeling, indicating that the reaction site senses structural changes related to AChR activity. Photolabeling with several drugs having no azido groups gave similar results (fig. 2). $^{3}$H-trimethisoquin, $^{3}$H-phencyclidine, $^{3}$H-perhydrohistrionicotoxin, $^{3}$H-chlorpromazine, all labeled upon irradiation with UV light sites within the AChR responding to various degrees to agonists (17). After the primary structures of the four receptor polypeptide chains were published it became an obvious goal to localize the label in the amino acid sequence.

We used the lipophilic cation $^{3}$H-triphenylmethylphosphonium ($^{3}$H-TPMP$^{+}$) (fig. 2) as the photoaffinity label (18). We showed it previously to be a blocker of the AChR ion channel without competing with the agonist acetylcholine for its binding sites on the α-polypeptide chains. This molecule has been shown to react to UV irradiation covalently with the AChR, in the absence of other cholinergic effectors, mainly with its α-polypeptide chain (19). The photochemistry of this reaction is completely unknown. The action spectrum and a characteristic species difference (the agonist-stimulated labeling of the δ-chain has been observed only with AChR from *Torpedo marmorata*, but not with *Torpedo californica*) hints at the possibility that it is actually an amino acid side chain and not the photolabel itself that is excited by UV light.

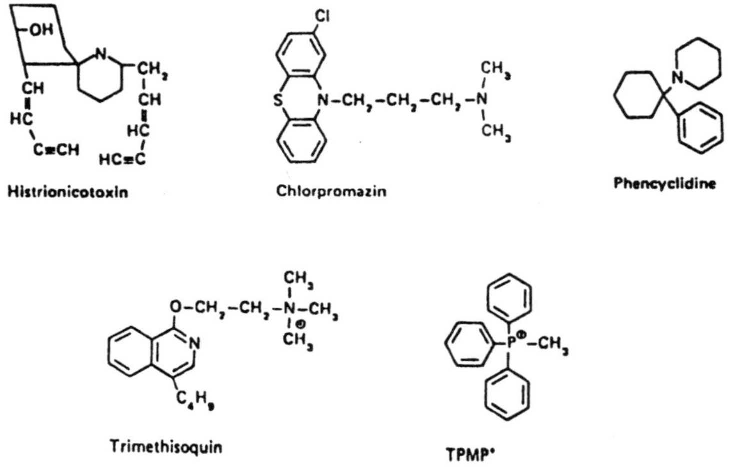


*Fig. 2: Noncompetitive antagonists used as photoaffinity labels of the nicotinic acetylcholine receptor.*

From our analysis of the kinetics of labeling (5) we distinguished three separate receptor states: i) resting receptor (in absence of any cholinergic effector except $^3H$-$TPMP^+$), ii) desensitized receptor (in presence of 0,1 mM carbamoylcholine, an agonist), and iii) AChR in an antagonist state (a conformation we observed in presence of 0,1 mM of the competitive antagonist flaxedil). Accordingly, we set out to localize the photolabel $^3H$-TPMP$^+$ in the primary structure and we performed three sets of experiments:

1) Localization of $^3H$-TPMP$^+$ in the primary structure of the δ-chain.
2) Comparison of the site of photolabeling in the α-, β-, and δ-polypeptide chains, (for unknown reasons no labeling of the γ-chain occurred under any circumstances).
3) Comparison of the site of photolabeling in the various states described above (i - iii).

After having worked out the conditions of the photoaffinity labeling (19), it was necessary to develop a method for separating quantitatively the labeled polypeptide chains. Labeling was exclusively performed with AChR-rich membranes isolated from *Torpedo marmorata* electric tissue (20). After labeling the membranes were dissolved in Triton X-100. The best method for obtaining the pure labeled polypeptide chains turned out to be preparative SDS-polyacrylamide gel electrophoresis (fig. 3). Especially the α- and δ-chains were easily purified with reasonable yield. Other methods (gel chromatography on TSK-columns, HPLC) turned out to be ineffective. Next, it was necessary to obtain smaller peptides from the labeled receptor subunits. AChR is fairly resistant against proteolysis, possibly because of the hydrophobic character of its membrane-spanning domains. Limited proteolysis with acids proved to be one method of breaking down receptor subunits: separation of the cleavage products on a TSK-3000 column (gel filtration) showed that after 1 h hydrolysis most of the δ-chain was partially hydrolysed (fig. 4b: as compared to a, the left peak, corresponding to the intact chain, disappeared). 2 h hydrolysis (fig. 4d) did not improve the result to a large extent, but an interesting effect was the shift of the $^3H$-TPMP$^+$-carrying peak to a larger apparent molecular size. This effect may be related to the hydrophobicity of the peptides formed. The result of acid proteolysis was similar to

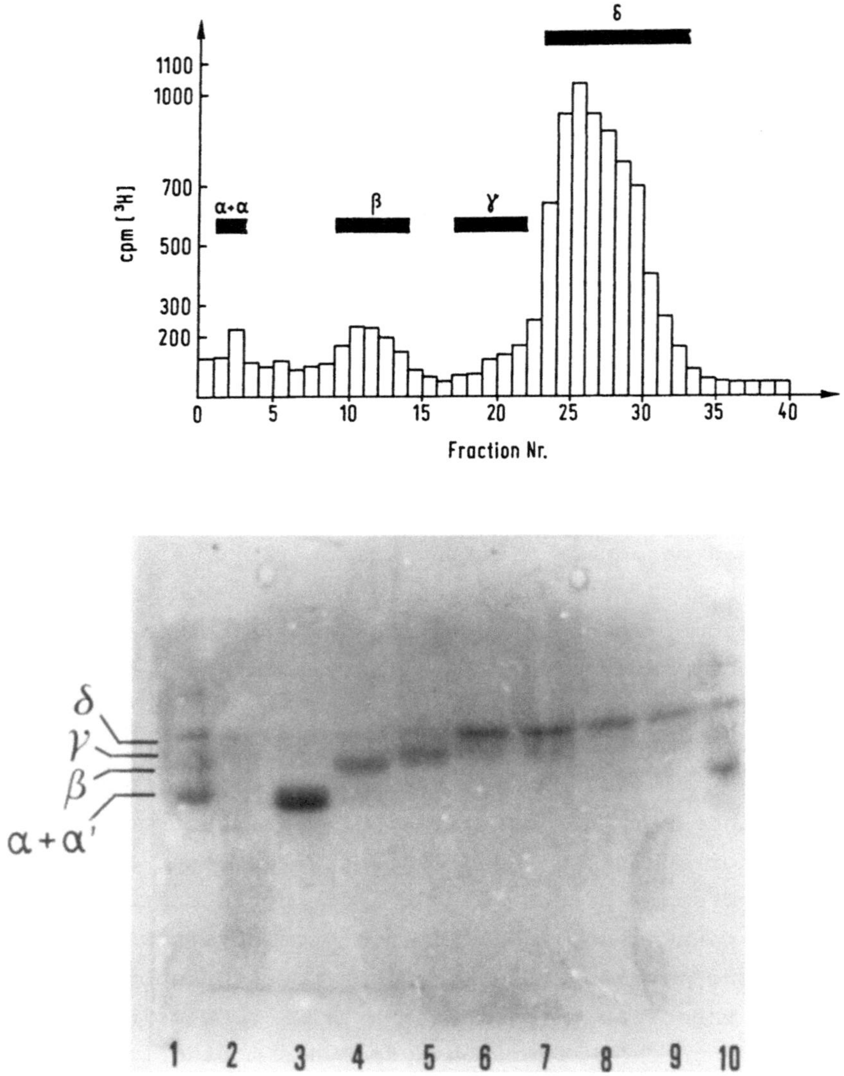


*Fig. 3:* *Preparative SDS-Polyacrylamide gel electrophoresis for separation of $^{3}H$-$TPMP^{+}$-labeled receptor subunits.*

*Separation was performed with an apparatus from BRL (Bethesda). Upper gel 3%, lower gel 7.5%, 6 mA/150 V. Elution buffer: 0.1% SDS in 0.576 M glycine, 0.075 M Tris/HCl, pH 8.3. Elution rate 15 ml/h, fractions were 3 ml.*

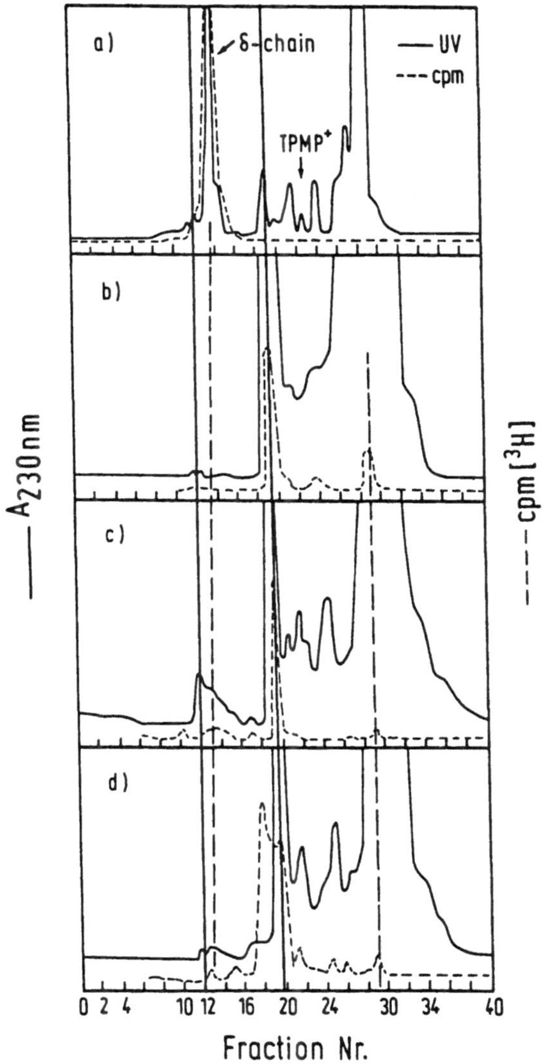


*fig. 4: Cleavage of AChR δ-chain; comparison of methods. a) Reference, chromatography of $^{3}H$-$TPMP^{+}$-labeled intact δ-chain on a TSK 3000 column, 600 x 4 mm, from LKB. The peaks to the right of the δ-chain peak stem from impurities caused by the preparative SDS electrophoresis. b) δ-chain after 24 h CNBr cleavage. c) δ-chain after 1 h cleavage with 90% formic acid at 108° C. d) as c, but after 2 h cleavage. Elution buffer: 0,25% TFA adjusted to pH 6.3 with $NH_4OH$, 0.1% SDS, 0.1% NaCl, 2.5% 2-propanole. Flow rate 0.2 ml/min.*

CNBr-cleavage (fig. 4b) which was used in the further experiments.

*Localization of $^3$H-TPMP$^+$ in a CNBr-peptide from δ-subunit (labeled in the resting state of AChR):* An important result of the cleavage experiments (fig. 4) was the proof that the label is truly stable. To localize it in the primary structure of the δ-chain we then separated the CNBr-peptides by HPLC (21, Organogen HP-Gel, RP-7 column). The fraction containing maximum radioactivity was subject to liquid- and gas-phase sequencing. Automated Edman degradation allowed us to clearly identify peptide CB5, amino acids 258 - 293 of the δ-chain (fig. 5). The maximum of radioactivity was released in the fifth Edman step, representing amino acid 262, a serin residue in the known sequence of the δ-chain from *Torpedo californica*.

*Localization of $^3$H-TPMP$^+$ in a trytpic peptide*: The peptide which we sequenced was not quite pure. CB1 and CB2 (corresponding to amino acids 1 - 58 and 59 - 160) could be identified as major impurities and very low amounts of even further peptides were visible. Since the yield of the photolabeling was only on the order of 1% (19) the radioactivity released in the fifth Edman cycle could well stem from one of these impurities. We therefore repeated the whole procedure using trypsin as the cleaving agent: we cleaved $^3$H-TPMP$^+$-labeled δ-chain with trypsin (10% w/w, 15 h digestion), separated the tryptic peptides by HPLC and sequenced the main radioactive fraction (fig. 6). Again this fraction contained more than one peptide, among which T21 (amino acid 257-277) dominated. The maximum of radioactivity was released in the sixth Edman step. Now we could localize the site of the photolabeling unambiguously: five positions toward the aminoterminus there must be a methionine (the site of CNBr cleavage), six steps away must be a lysine or arginine (the site of tryptic cleavage). Lysine (or arginine) and methionine must be located next to each other, being five, respectively six steps

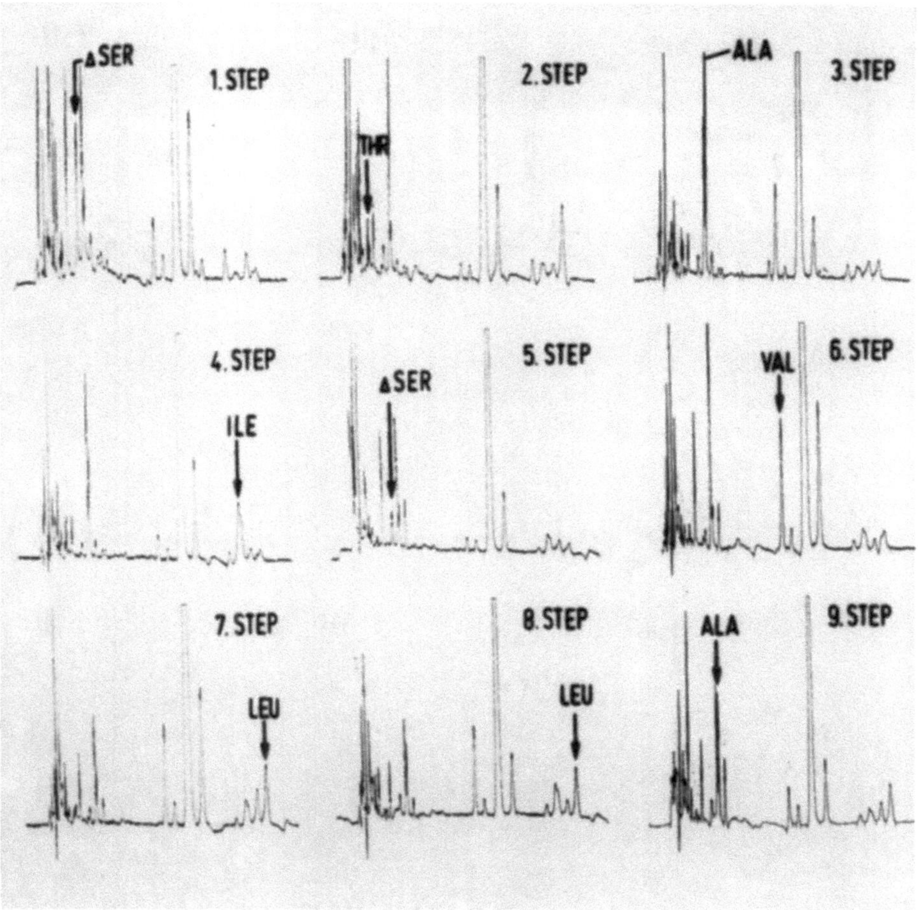


*Fig. 5: Sequencing of $^{3}H$-$TPMP^{+}$-labeled CNBr-peptide with a gas phase sequenator. Identification of PTH-amino acids by HPLC.*

| Step | Peptide CB5 |
|---|---|
| 1 | SER |
| 2 | THR |
| 3 | ALA |
| 4 | ILE |
| 5 | SER ← Maximum of Activity [$^{3}H$] |
| 6 | VAL |
| 7 | LEU |
| 8 | LEU |
| 9 | ALA |

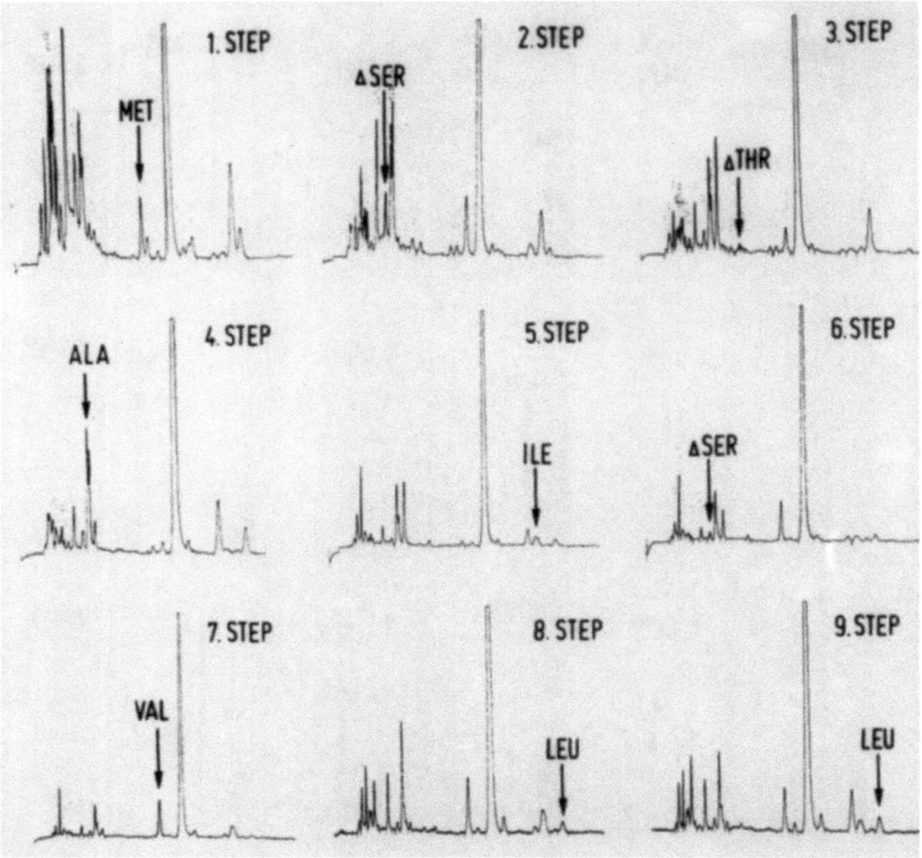


*Fig. 6: Sequencing of tryptic peptides: Identification of PTH-amino acids by HPLC. Three peptides could be identified (see text).*

| Peptide / Step | T43 | T19/20 | T21 | |
|---|---|---|---|---|
| 1 | ASN | LYS | MET | |
| 2 | ALA | PRO | SER | |
| 3 | TYR | LEU | THR | |
| 4 | ASP | PHE | ALA | |
| 5 | GLU | TYR | ILE | |
| 6 | GLU | VAL | SER (circled) | ← Maximum of activity [$^3H$] |
| 7 | VAL | ILE | VAL | |
| 8 | GLY | ASN | LEU | |
| 9 | ASN | PHE | LEU | |

'upstream' of the labeling site. Fortunately, the sequence -lys-met- occurs only once in the δ-chain, allowing the definitive conclusion that position 262 (probably a serine) carried the label.

*Localization of $^3H$-TPMP$^+$ in other chains of the AChR* This procedure was applied to localizing the site of photolabeling in the α- and β-chains of AChR as well. Tryptic cleavage and HPLC resulted in an interesting phenomenon: although the peptide patterns obtained were very different with the different polypeptide chains (as one would expect) the radioactivity eluted from the RP column at identical positions (fig. 7). Localization by microsequencing (22) showed that α-, β-, and δ were labeled in homologous positions. (α 248, β 254, δ 262).

*Model of the AChR-ion channel* Covalent labeling of homologous positions in the primary structures of several receptor subunits can only mean that the binding site for the channel-blocking label is formed by these subunits together. The most popular channel model (23) proposes that the AChR ion channel is made up of five homologous amino acid sequences believed to form amphipathic helices ('helix-A model') spanning the membrane contributed by the five receptor subunits. Amino acid δ 262 and the respective labeling sites in the other subunits are not located in this amphipathic helix A, but rather in another membrane spanning the helix, called helix M2 (according to the chain-folding prediction by Finer-Moore and Stroud, 23). We therefore propose a model (15, 22, 24, 25) in which the channel is formed by the five helices M2 of the respective receptor subunits (fig. 8).

*Labeling of AChR in different functional states* The AChR is an allosteric protein existing in various conformational and functional states. Three states can be easily defined (15): the R-state (resting state, in absence of any cholinergic effector molecule, channel closed), the A-state (active state, in presence of agonists, channel transiently open), and the

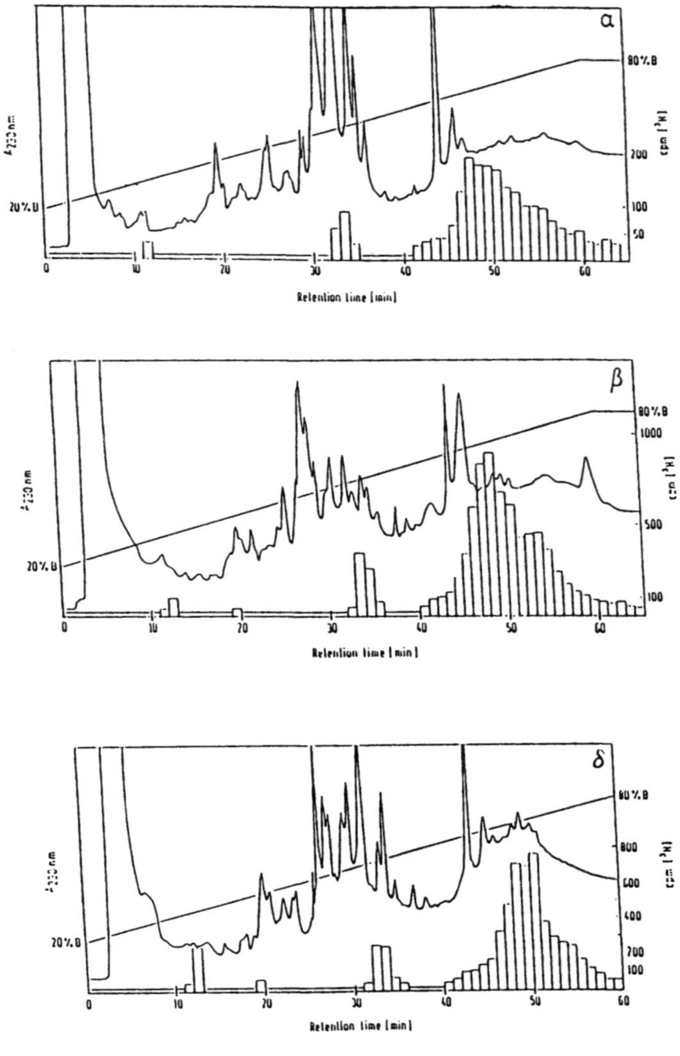


*Fig. 7: Homologous labeling of α-, β-, and δ-chains. HPLC separation of tryptic peptides. Note the different peptide composition and the similar distribution of radioactivity, indicating that homologous sequences in the different receptor subunits were labeled.*

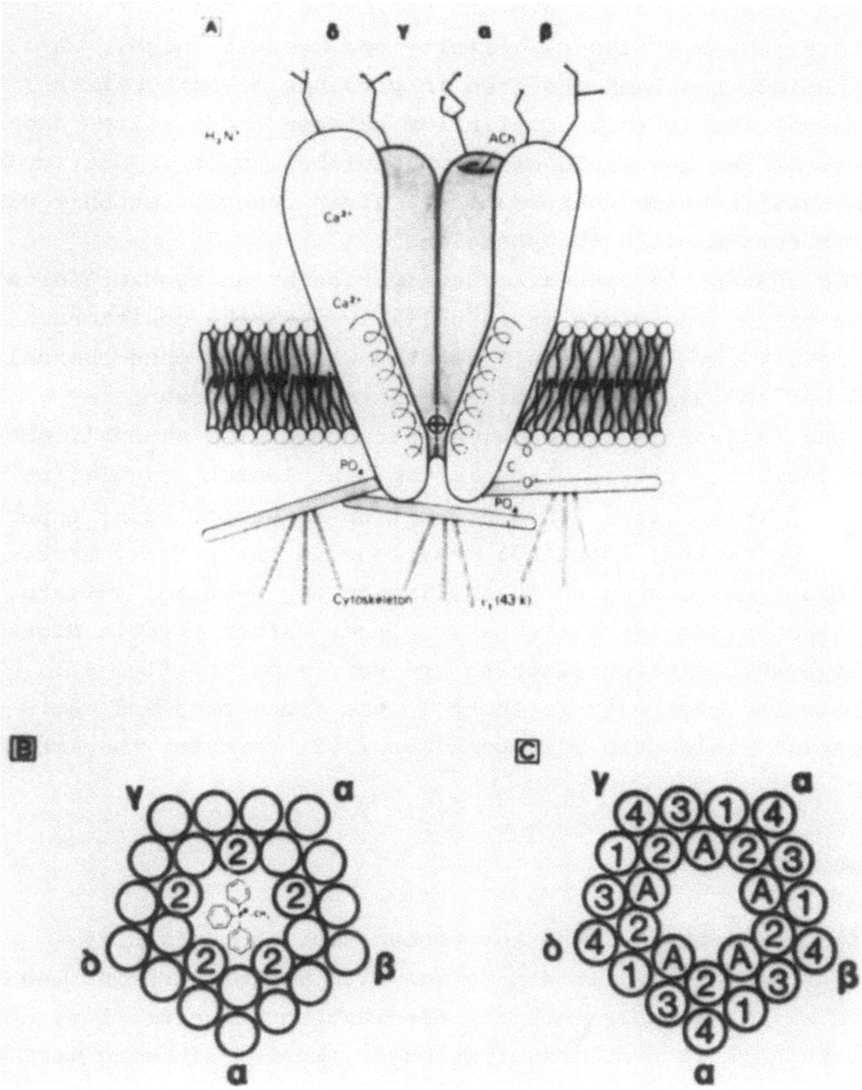


*Fig. 8: The helix-2 model of the AChR ion channel (15, 25).*

*a) longitudinal section*
*b) and c) cross sections.*

D-state (desensitized state, in prolonged presence of agonists, channel closed, affinity for agonists high). A fourth state has been observed in presence of antagonists (5). Because of the dynamic equilibrium between these states and because of the low yield of the photolabeling it is difficult to be sure in which conformation a given receptor subunit was when it reacted with the photolabel.
The ion channel is open after activation by an agonist for a few ms only. Therefore it is unlikely that the positions photolabeled by $^{3}$H-TPMP$^{+}$ are located within the open channel. We rather assume that the binding site for the channel-blocking reagent is in the wide entrance of the channel, above the selectivity filter which is the "bottleneck" regulating the ion permeability. This assumption is supported by experiments showing that identical positions in the primary structure are labeled when photolabeling in the A-state, D-state, or in the antagonist state takes place. After tryptic digestion the HPLC-elution patterns are very similar (fig. 9). The main radioactivity is found in the fractions, and microsequencing yields the same position (262) carrying the label.

## Conclusion

Experiments with the AChR have shown that photoaffinity labeling, expecially in connection with microsequencing techniques, is a valuable tool for elucidating structure-function relationships in proteins. The major problem limiting its success is the low yield of photoaffinity labeling, which hampers conclusions concerning the specificity of the photo reaction. In the case of the AChR our conclusions are supported by Giraudat et al. (26), who localized the reaction site of the noncompetitive antagonist chlorpromazine in position 262 of the δ-chain as well. Time-resolved photoaffinity labeling with the noncompetitive inhibitor quinacrine azide (7,8) gave very different results. This label reacted

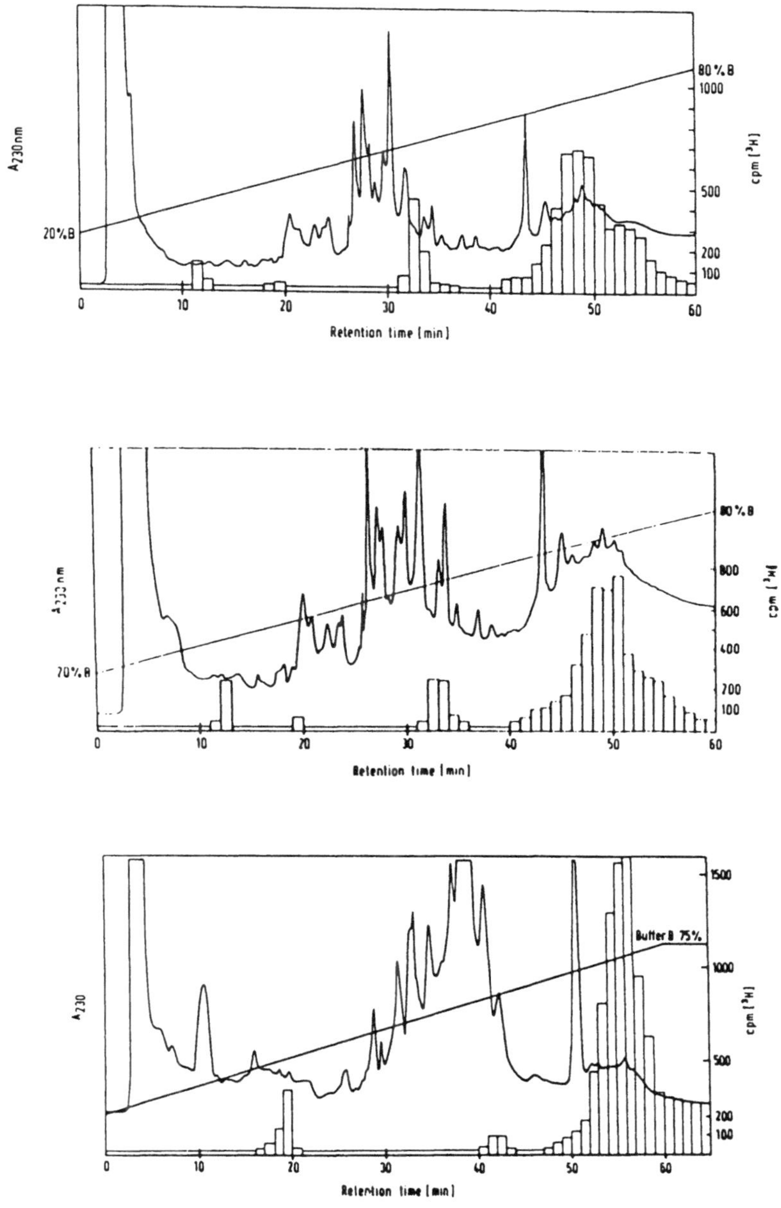


*Fig. 9: Comparison of labeling in different functional states. HPLC separation of tryptic peptides from δ-chains labeled with the receptor in the resting (top), desensitized (middle), and antagonist (bottom) state (24).*

predominantly with the α- and β-chains (27). Here the target was helix M 1 of the folding predicted in (23).
In any case, results from photoaffinity labeling should be supported by evidence obtained with different methods. A crucial role for helix M 2 was confirmed by site-directed mutagenesis experiments in combination with patch clamp single channel measurements (28).

Acknowledgements

The sequencing work described in this article would have been impossible without the support and encouragement of Dr. B. Wittmann-Liebold. Financial aid by the Deutsche Forschungsgemeinschaft (SFB 312) and the Fonds der Chemischen Industrie is gratefully acknowledged.

1. Moehler, H., Bathersby, M.K., Richards, J.G. 1980. Proc. Natl. Acad. Sci. USA 77, 1666-1670.

2. Graham, D., Pfeiffer, F., Betz, H. 1983. Eur. J. Biochem. 131, 519-525.

3. Fahr, A., Hucho, F. 1986. J. Neurosci. Meth. 16, 29-38.

4. Muhn, P., Fahr, A., Hucho, F. 1984. FEBS Lett. 166, 146-150.

5. Fahr, A., Lauffer, L., Schmidt, D., Heyn, M., Hucho, F. 1985. Eur. J. Biochem. 147, 483-487.

6. Fahr, A., Hellmann, S., Lauffer, L., Muhn, P., Hucho, F. 1985. 36. Mosbach Colloquium, 103-112.

7. Cox, R.N., Kaldany, R.-R., Brandt, P.W., Ferren, B., Hudson, R.A., Karlin, A. 1984. Anal. Biochem. 136, 476-486.

8. Cox, R.N., Kaldany, R.-R., Di Paola, M., Karlin, A. 1985. J. Biol. Chem. 260, 7186-7193.

9. Heidmann, T., Changeux, J.-P. 1984. Proc. Natl. Acad. Sci. USA 81, 1897-1901.

10. Kiefer, H., Lindstrom, J., Lennox, E.S., Singer, S.J. 1970. Proc. Natl. Acad. Sci. USA 67, 1688-1694.

11. Hucho, F., Layer, P., Kiefer, H., Bandini, G. 1976. Proc. Natl. Acad. Sci. USA 73, 2624-2628.

12. Witzemann, V., Muchmore, D., Raftery, M.A. 1979. Biochemistry 24, 5511-5518.

13. Hucho, F. 1979. FEBS Lett. 103, 27-32.

14. Tsetlin, V., Pluzhnikov, K., Karelin, A., Ivanov, V. 1983. In: Toxins as Tools in Neurochemistry (F. Hucho and Yu. A. Ovchinnikov, eds.). de Gruyter, Berlin, New York, p. 159-169.

15. Hucho, F. 1986. Eur. J. Biochem. 158, 211-226.

16. Changeux, J.-P. 1981. Harvey Lect. 75, 85-254.

17. Heidmann, T., Oswald, R.E., Changeux, J.-P. 1983. Biochemistry 22, 3112-3127.

18. Lauffer, L., Hucho, F. 1982. Proc. Natl. Acad. Sci. USA 79, 2406-2409.

19.Muhn, P., Hucho, F. 1983. Biochemistry 22, 421-425.

20.Schiebler, W., Hucho, F. 1978. Eur. J. Biochem. 85, 55-63.

21.Oberthür, W., Muhn, P., Baumann, H., Lottspeich, F., Wittmann-Liebold, B., & Hucho, F. 1986. EMBO J. 5, 1815-1819.

22.Hucho, F., Oberthür, W., & Lottspeich, F. 1986. FEBS Lett. 205, 137-142.

23.Finer-Moore, J., & Stroud, R. 1984. Proc. Natl. Acad. Sci. USA 81, 155-159.

24.Oberthür, W., Lottspeich, F., & Hucho, F. 1987. In: "Receptors and Ion Channels", (Yu. A. Ovchinnikov and F. Hucho, eds.), Walter de Gruyter, Berlin, New York.

25.Guy, R. & Hucho, F. 1987. TINS, in press.

26.Giraudat, J., Dennis, M., Heidmann, Th., Chang, J.-Y., & Changeux, J.-P. 1986. Proc. Natl. Acad. Sci. USA 83, 2719-2723.

27.Karlin, A., Kao, P., & Di Paola, M. 1986. TINS 7, 304-308.

28.Imoto, K., Methfessel, C., Sakmann, B., Mishina, M., Mori, Y., Komo, T., Fukuda, K., Kurasaki, M., Bujo, H., Fujita, Y., & Numa, S. 1986. Nature 324, 670-674.

# MICROCHARACTERIZATION OF PHOSPHOSERINE CONTAINING PROTEINS. LOCALIZATION OF THE AUTOPHOSPHORYLATION SITES OF SKELETAL MUSCLE PHOSPORYLASE KINASE

HELMUT E. MEYER, EDELTRAUT HOFFMANN-POSORSKE, CARL C. KUHN & LUDWIG M.G. HEILMEYER, JR.

Institut für Physiologische Chemie, Abteilung Biochemie Supramolekularer Systeme, Ruhr-Universität Bochum, D-4630 Bochum, FRG

## I. Introduction

Until recently, the localization of phosphorylation-sites inside proteins or even smaller peptides was a difficult task. Subfragmentation and purification of smaller peptides were necessary if an isolated phosphopeptide contained more than one serine residue and this was an even more tedious task if two or more serines were in series (1-5).
Suggestions to overcome these difficulties were made by several investigators (6-11). All of them used the ß-elimination of the phosphate group as the first step, as originally introduced by Downs and Pigman (12) for the ß-elimination of a carbohydrate-group at O-glycosylated serines or threonines. The dehydroalanine resulting from the phosphoserine is then derivatized using different addition reagents like methyl-

Modern Methods in Protein Chemistry, Vol. 3

amine, sulfite, methylmercaptane or activated hydrogen.

To solve the problems arising during microsequence amino acid analysis of phosphoserine containing proteins, we developed improved methods for amino acid analysis (13) and amino acid sequence analysis (14) of phosphoserine derivatives. In a micro one-step batch reaction, phosphoserine is converted to stable S-ethyl-cysteine starting with less than 100 pmol of peptide, and the reaction conditions are so gentle, that no further peptide purification is required.

We first present our results from experiments performed with a small model peptide and the development of the procedure. Next, we explain our understanding of the Edman chemistry concerning phosphoserine residues in peptides during degradation in the gas phase sequencer and the advantage of on-line PTH-amino acid analysis. Finally, we demonstrate the capability of the new method; we report the identification of autophosphorylation sites in rabbit skeletal muscle phosphorylase kinase.

## II. Materials and Methods

Chemicals and reagents. Kemptide ($H_2N$-Leu-Arg-Arg-Ala-Ser-Leu-Gly-OH) was from Serva, Heidelberg, W. Germany. Proteinase Lys-C was from Boehringer, Mannheim, W. Germany and proteinase Glu-C (S. aureus V8) was obtained from Miles, ICN Biomedicals, Eschwege, W. Germany . Chemicals and solvents

Table I.

| Reagent | Abbreviation | Source |
|---|---|---|
| Acetic Acid, HPLC grade | HAc | Baker |
| Adenosine Triphosphate | ATP | Serva |
| Acetonitrile, HPLC grade | MeCN | Roth |
| Calcium Chloride, p.a. | $CaCl_2$ | Baker |
| Dimethylsulfoxide, p.a. | DMSO | Merck |
| Dithiothreitol | DTT | Calbiochem |
| Ethanethiol | EtSH | Pierce |
| Ethanol, HPLC grade | EtOH | Baker |
| Ethylenediamine- -tetra Acetic Acid | EDTA | Roth |
| Ethyleneglycol-bis-(ß-Amino Ethyl Ether) N,N'- tetra acetic acid | EGTA | Sigma |
| S-Ethyl-Cysteine | S-Et-C | Sigma |
| Endoproteinase Lys C | Lys-C | Boehringer |
| ß-Glycerophosphate | GP | Merck |
| Hexafluoroacetone, pro synthesis | HFA | Merck |
| Hydrochloric Acid | HCl | Merck |
| N-(2-Hydroxyethyl)piperazine- -N'-2-Ethanesulfonic Acid | HEPES | Serva |
| Magnesium Chloride | $MgCl_2$ | Baker |
| Methylamine | $MeNH_2$ | Linde |
| ß-Methylamino-alanine | MAA | |
| Phenylisothiocyanate | PITC | Beckman |
| Polybrene | | Pierce |
| Sodium Dodecylsulfate | SDS | BioRad |
| Sodium Hydroxide | NaOH | Roth |
| Tetrahydrofuran | THF | Roth |
| Trishydroxyaminomethane | Tris | Merck |
| Triethylamine | TEA | Pierce |
| Trichloroacetic Acid | TCA | Merck |
| Trifluoroacetic Acid | TFA | Pierce |

for automated sequence analysis were from Applied Biosystems, Weiterstadt, W. Germany. Other solvents and reagents are listed in table 1. Chemicals or solvents not listed were of the highest purity available.

Further abbreviations are: cAMP: cyclic adenosine mono-phosphate, ATZ: anilinothiazolinone, AUFS: absorbtion units full scale, PTC: phenylthiocarbamyl, PTH: phenylthiohydantoin. HPLC columns and packings for peptide separations were from Vydac, Hesperia, CA, USA. Off-line PTH amino-acid analysis was performed on a 5 µ cyano-column from IBM. On-line PTH-amino-acid analysis was done with the Applied Biosystems PTH-column.

$^{32}$P-ATP was prepared according to Walseth and Johnson (15). The catalytic subunit of bovine heart cAMP-dependent protein kinase was prepared according to Beavo et al. (16). Rabbit skeletal muscle phosphorylase kinase was purified according to Cohen (17) with the modifications introduced by Jennissen and Heilmeyer (18) and by Hessova et al. (19).

Phosphorylation of kemptide (14,20). Kemptide (0.2 mg) was dissolved in 4.1 ml of 40 mM potassium phosphate pH 7.3 containing 1 mM ATP, 5 mM magnesium acetate and 7.5% glycerol. The reaction was started by the addition of 100 µg of catalytic subunit of cAMP-dependent protein kinase and proceeded for 30 minutes at 26°C. The phosphorylated kemptide was purified by reversed phase HPLC. The conditions used were those described in legend to figure 3.

Phosphorylase kinase autophosphorylation. Phosphorylase kinase (6 mg) was dialysed against 2x 250 ml 50 mM HEPES, 1.5 mM GP buffer pH 6.8 containing 1 mM DTT. Dialysis was at 4°C overnight plus an additional buffer change (250 ml) for 1h. The reaction mix consisted of 2.1 ml dialysate, 150 µl 22.45 mM $CaCl_2$, 150 µl 20 mM EGTA, 150 µl 60 mM ATP (1.5-3.0 mCi $^{32}P$-ATP) and 200 µl HEPES buffer (0.5 M HEPES, 15 mM GP, 10 mM DTT pH 6.8). The concentration of free $Ca^{2+}$ in this system was calculated to be 100 µM (21). The reaction was started with the addition of 150 µl of 180 mM $MgCl_2$ in 10 mM Tris-HCl pH 6.8. After 1 h at 24°C, the reaction was stopped by cooling on ice. Numbers for incorporated phosphate were calculated from radioactivity in TCA precipitated protein; 5 to 6 mol of phosphate per mol of phosphorylase kinase tetramer (306,000 D) were usually found under these conditions.

Purification of phosphorylase kinase subunits. The subunits of phosphorylase kinase were isolated by preparative reversed phase HPLC according to Crabb and Heilmeyer (22,23). The separations were performed on Vydac 10µ $C_4$ material in a 20x50 mm column. The solvents were 0.1% TFA and 84% MeCN, 0.08% TFA; the flow rate was 7.8 ml/min. Fractions from several runs were combined and 7.5 µl 10% SDS were added to a pool volume of about 60 ml. The volume of the pool was reduced in vacuo to a final SDS concentration of 0.025%. Precipitated protein went into solution after adjusting the pH to 8.6 at a buffer concentration of 25 mM Tris and 0.1 mM EDTA.

Incorporation of phosphate was again determined by precipitation of protein bound radioactivity.

Protease cleavage of the subunits. The alpha-subunit (~5 nmol) was digested with either 2x 10% (w/w) trypsin (Worthington) for 2x 2h at 37°C or with 2x 2.5% (w/w) Lys-C also 2x 2h at 37°C. For trypsin digestions, no SDS was added to the pools. The beta subunit (~5 nmol) was digested using the same conditions with either 2x 2.5% (w/w) Lys-C or with 2x 5% (w/w) endoproteinase V8. The pH of the V8 digestion buffer was 7.6.

Peptide purification. Peptides were purified by reversed-phase HPLC; details are given in the legends to the figures.

Conversion of phophoserine residues to ß-methylamino-alanine or S-ethyl-cysteine. The procedures for the conversion of phophoserine residues to ß-methylaminoalanine or to S-ethyl-cysteine have been described by Meyer et al. (13,14). The optimized protocol includes the following steps: The dry peptide sample is incubated for 1h at 50°C under nitrogen with 50 µl of a reaction mix containing 0.5 M NaOH, 33% DMSO, 13% EtOH and either 10% EtSH or 10 M methylamine. After cooling to room temperature, the reaction mix is acidified with acetic acid ( 10 µl) and directly applied to the glassfiber-filter disk for Edman degradation in the gas-phase sequencer. Samples for amino-acid analysis are prepared without DMSO.

Prior to hydrolysis, the acidified mix is vacuum dried at room temperature. Drying times for amino-acid analysis samples were kept as short as possible by using a vacuum system ($10^{-3}$ mbar) equipped with a small volume trap (800 ml) cooled to -80°C. Optimum results were achieved if not more than ten samples were processed at a time.

Amino-acid analysis. Peptides and proteins were gas-phase hydrolysed at 150°C for 1h. Amino-acids were separated by reversed phase HPLC on a 3 μ Spherisorb ODS-2 column (4.6 x 125 mm) after precolumn derivatization with PITC as described (13,14,24).

Sequence analysis. Edman degradation was performed by the G. Tarr manual microsequencing methods (25,26) or with the Applied Biosystems Model 470A gas-phase sequencer using the 03RPTH program. PTH-amino acids were analysed on an Applied Biosystems Model 120A on-line HPLC system.

## III. Results and Discussion

Studying the model phosphopeptide. PTH-amino-acid analyses from the manual Edman degradation of the unmodified phosphorylated kemptide are shown in figure 1. The chromatogram of cycle 5 shows some by-product peaks and overlapping alanine and arginine from cycles 4 and 3. PTH- Phosphoserine from position 5 does not yield an identifyable PTH-derivative.

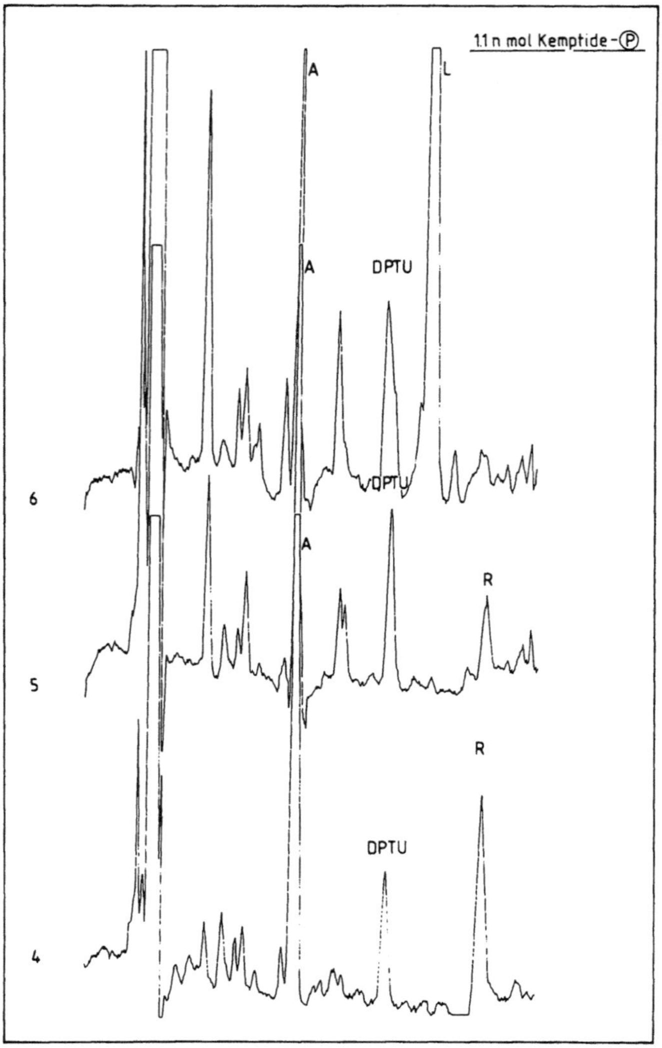


Figure 1: Sequence analysis of phosphorylated kemptide. PTH-amino-acid analyses of degradation steps 4, 5 and 6 are shown. The peptide was sequenced using the G. Tarr manual micro-sequencing methods. PTH-amino-acids were analysed on an IBM 5µ cyano column (4.6x250 mm) with 18.6 mM sodium acetate, 5% tetrahydrofuran pH 5.25 as solvent A and 100% acetonitrile as solvent B at a flow rate of 1 ml/min. The detector was a Kratos model 773 spectrophotometer set to 0.005 AUFS at 269 nm.

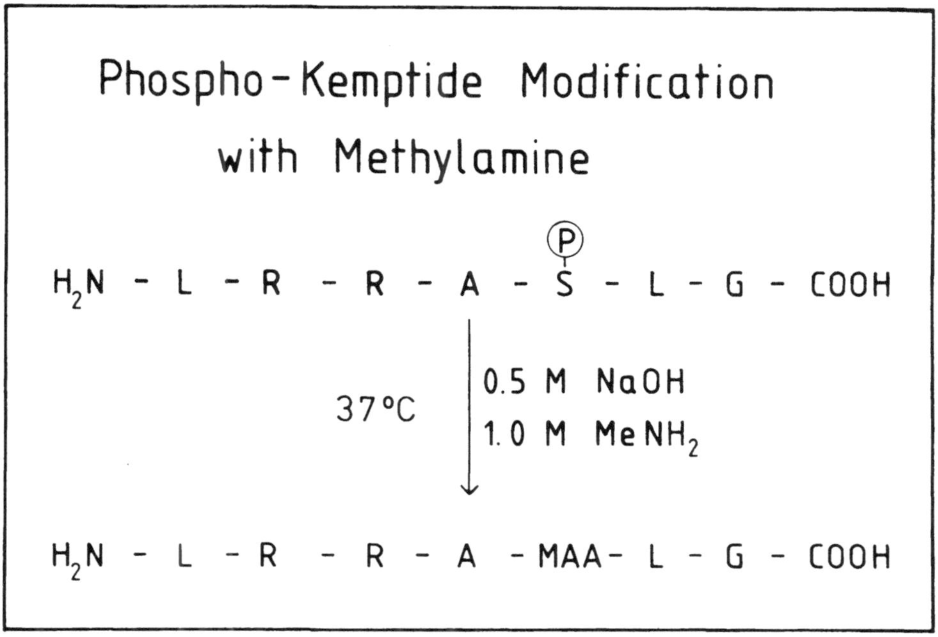


Figure 2: Conversion of phospho-kemptide to ß-methylamino-alanine kemptide.

However, the Edman degradation of the peptide chain continues through the following positions and the chromatogram of cycle 6 shows the expected PTH-leucine peak. In order to unambiguously identify phosphoserine residues in polypeptide chains, we tried to convert these residues into derivatives which are stable during Edman degradation and which can be identified by PTH-amino acid analysis.

A modification reaction which leads to the conversion of phosphoserine to ß-methylamino alanine has been suggested by Annan (8). Figure 2 shows how our model peptide is converted to the MAA-derivative. HPLC analysis of the reaction products after 30 min incubation time at 37°C shows that unreacted phosphokemptide, the two diastereomers of MAA-kemptide and the dehydroalanine form of kemptide elute with increasing retention times (Fig. 3). The effects of incubation time and base concentration were investigated (Fig. 4.). An incubation time of 1h and a sodium hydroxide concentration of 0.5 M were found to be suitable conditions for the conversion of phosphoserine to ß-methylamino alanine. HPLC analyses after 2 or 3h of incubation time at a base concentration of 0.5 or 0.9 M show additional peaks. Amino-acid analysis revealed that these products are not decomposed forms of the peptide. We assume that diastereomeric forms of the same peptide, caused by amino-acid racemization under basic conditions, elute with different retention times. In a related study, Mega et al. (11) observed peak broadening during HPLC of base

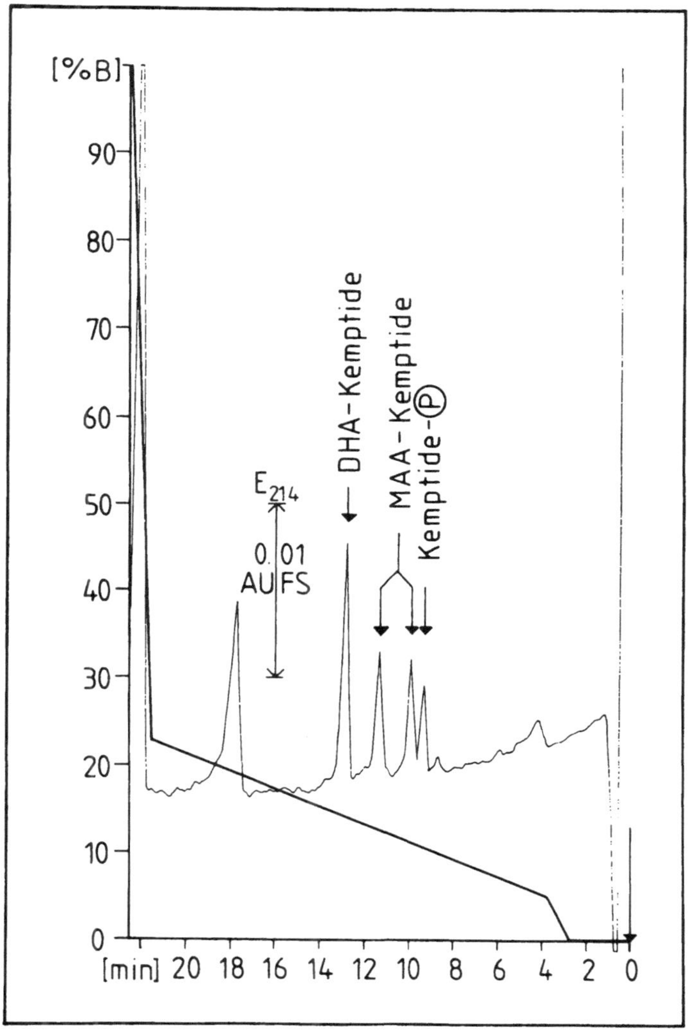


Figure 3: HPLC analysis of base/methylamine treated phospho-kemptide. Phospho-kemptide (11 nmol) was incubated for 30 min at 37°C with 0.5 M NaOH/1.0 M $MeNH_2$. The reaction mixture (20%) was injected on a Vydac 218TP5 5µ $C_{18}$ column (8x50 mm). Solvent A was 0.1% TFA in water and solvent B was 84% MeCN, 0.08% TFA. A gradient at a flow rate of 3 ml/min was applied as indicated. The detection wavelength was 214 nm.

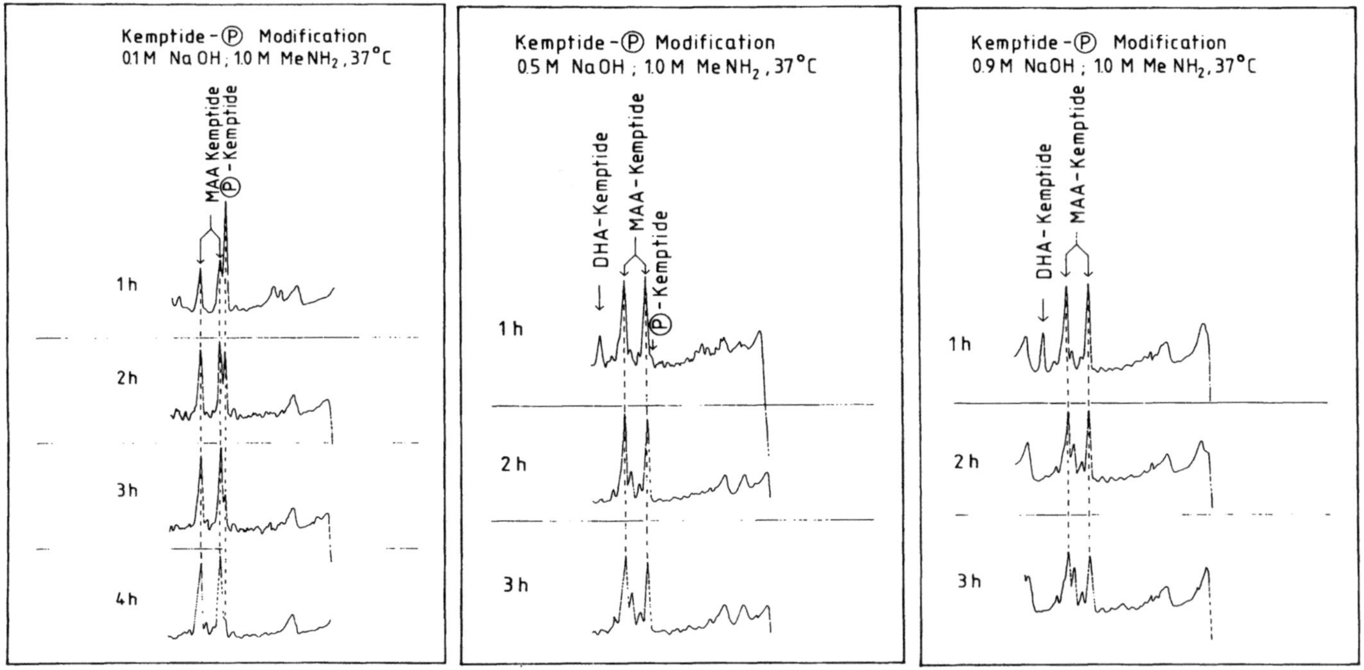


Figure 4: <u>HPLC analysis of base/methylamine treated phosphokemptide after increased incubation times at different base concentrations.</u> Phospho-kemptide (11 nmol) was incubated at 37°C at 0.1, 0.5 and 0.9 M NaOH. At times indicated, aliquots (20%) of the samples were injected on the column. HPLC conditions were those given in figure 2.

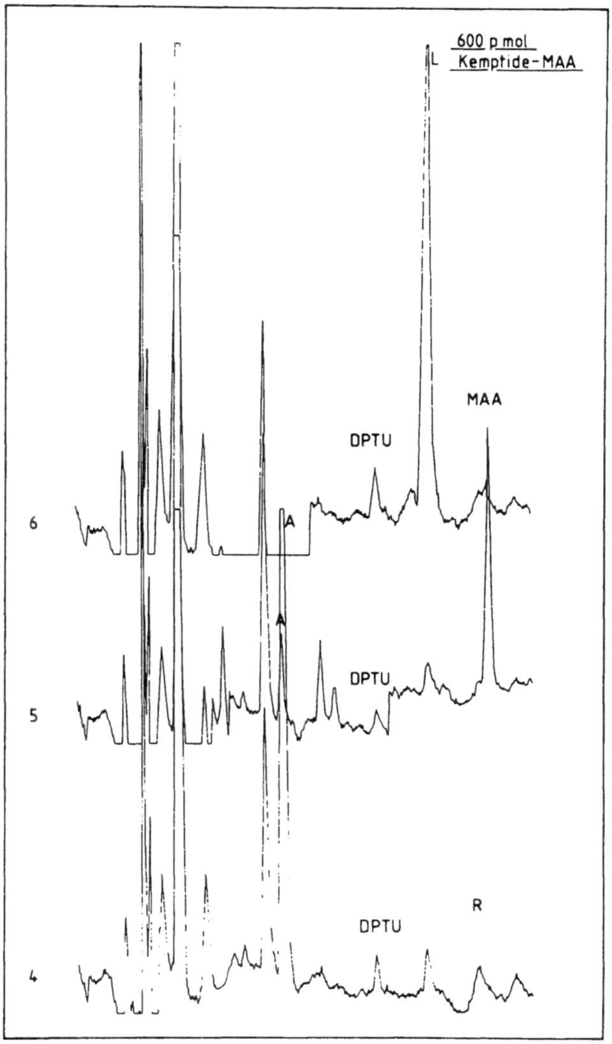


Figure 5: <u>Sequence analysis of base/methylamine treated phospho-kemptide.</u> PTH-amino-acid analyses of degradation steps 4, 5 and 6 are shown. The sample was sequenced with an Applied Biosystems gas-phase sequencer. PTH-amino acids were identified off-line using the same conditions as described in legend to figure 1.

treated peptides. What they explain as decomposition is probably also the amino-acid racemisation effect.
Phosphoserine can easily be detected during Edman degradation as the PTH-derivative of ß-methylamino-alanine. This is shown in fig. 5 for our model peptide. PTH-MAA elutes as a sharp peak slightly later than PTH-arginine. A PTH-MAA standard, kindly provided by Dr. Annan, eluted with identical retention time.

Phosphoserine during gas-phase sequencing. Our interpretation of how underivatized phosphoserine decomposes during automated Edman degradation is given in fig. 6. The example is a decapeptide isolated from the ß-subunit of phosphorylase kinase after autophosphorylation. The phosphoserine in position 4 remains intact during the first three Edman degradations. When, after 3 cycles, phosphoserine is in the amino-terminal position, it reacts with PITC to the PTC-phosphopeptide. Binding of PITC under alkaline conditions facilitates ß-elimination of the phosphate group and the PTC-dehydroalanine peptide is formed. Most of the liberated phosphate remains bound to the cationic groups of polybrene. If the sample is radioactive, an increased level of radioactivity can be detected in this cycle. The recovered radioactivity usually comprises only about 1% of the total radioactivity (4,5) and what is determined is washed out phosphate rather than phosphoserine. More phosphate will be washed out during subsequent Edman degradations, leading to the typical tailing

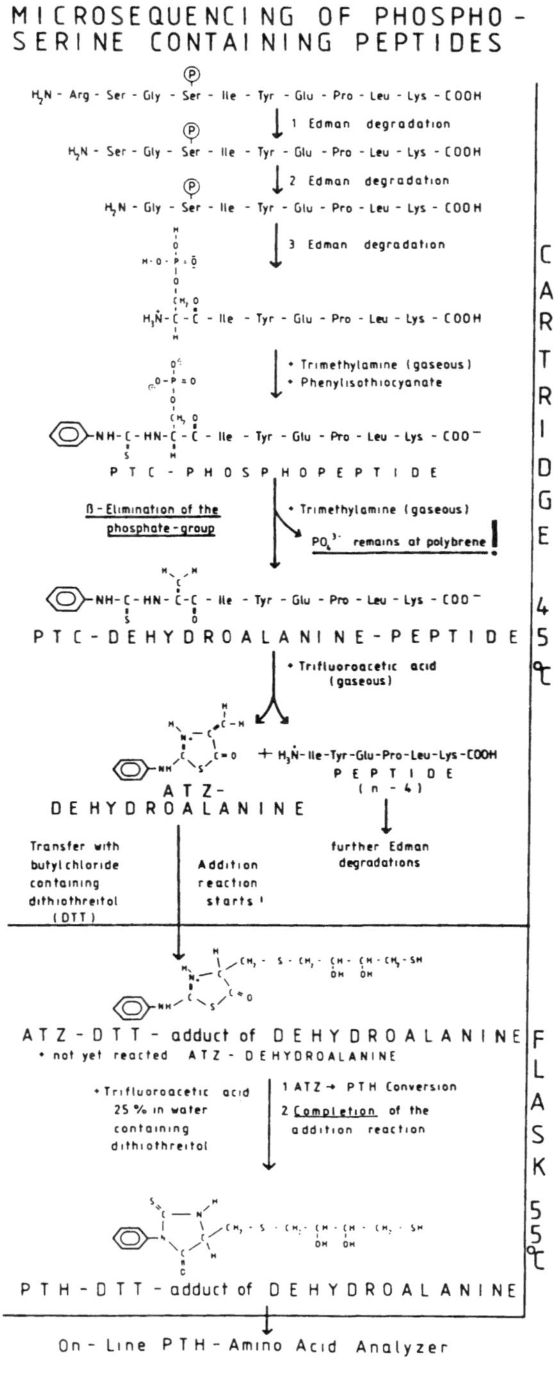
MICROSEQUENCING OF PHOSPHO-SERINE CONTAINING PEPTIDES
H2N - Arg - Ser - Gly - Ser - Ile - Tyr - Glu - Pro - Leu - Lys - COOH
1 Edman degradation
H2N - Ser - Gly - Ser - Ile - Tyr - Glu - Pro - Leu - Lys - COOH
2 Edman degradation
H2N - Gly - Ser - Ile - Tyr - Glu - Pro - Leu - Lys - COOH
3 Edman degradation
+ Trimethylamine (gaseous)
+ Phenylisothiocyanate
PTC - PHOSPHOPEPTIDE
ß - Elimination of the phosphate - group
+ Trimethylamine (gaseous)
PO4 3- remains at polybrene!
PTC - DEHYDROALANINE - PEPTIDE
+ Trifluoroacetic acid (gaseous)
ATZ-DEHYDROALANINE
PEPTIDE (n - 4)
further Edman degradations
Transfer with butylchloride containing dithiothreitol (DTT)
Addition reaction starts!
ATZ - DTT - adduct of DEHYDROALANINE
+ not yet reacted ATZ - DEHYDROALANINE
+ Trifluoroacetic acid 25 % in water containing dithiothreitol
1 ATZ → PTH Conversion
2 Completion of the addition reaction
PTH - DTT - adduct of DEHYDROALANINE
CARTRIDGE 45°C
FLASK 55°C
On - Line PTH - Amino Acid Analyzer

Figure 6

of the radioactivity vs. cylcle no. curves while an increased overlap of amino-acids is not observed.

During gas-phase sequencing, the PTC-dehydroalanine peptide is cleaved by gaseous TFA and the ATZ-dehydroalanine is extracted by butylchloride into the conversion flask. Both butylchloride as well as the conversion reagent (25% aqueous TFA) contain DTT which reacts with dehydroalanine side chains to yield the DTT-dehydroalanine adduct and this adduct is then converted to its PTH-derivative. The overall yield for the conversion of phosphoserine residues within the peptide chain to the PTH-DTT-dehydroalanine adduct is essentially quantitative. The adduct rapidly reacts with oxygen to yield a variety of products and this explains why phosphoserine-specific PTH-derivatives have not been detected using off-line PTH-amino-acid analysis.

The unstable PTH-DTT-dehydroalanine adduct can easily be detected quantitavely with an on-line PTH amino-acid analysis system. Fig. 7 illustrates these findings. The peptide containing an autophophorylation site was subjected to gas-phase sequencing without any modification. Serine in position 2 yields both PTH-serine as well as the PTH-DTT-dehydroalanine adduct. The formation of the DTT-dehydroalanine adduct is due to partial ß-elimination at serine side chains. On our system, we find for serine residues a ratio of PTH-serine to the PTH-DTT-dehydroalanine adduct of about 25 to 75. The almost exclusive presence of the PTH-DTT-dehydroalanine ad-

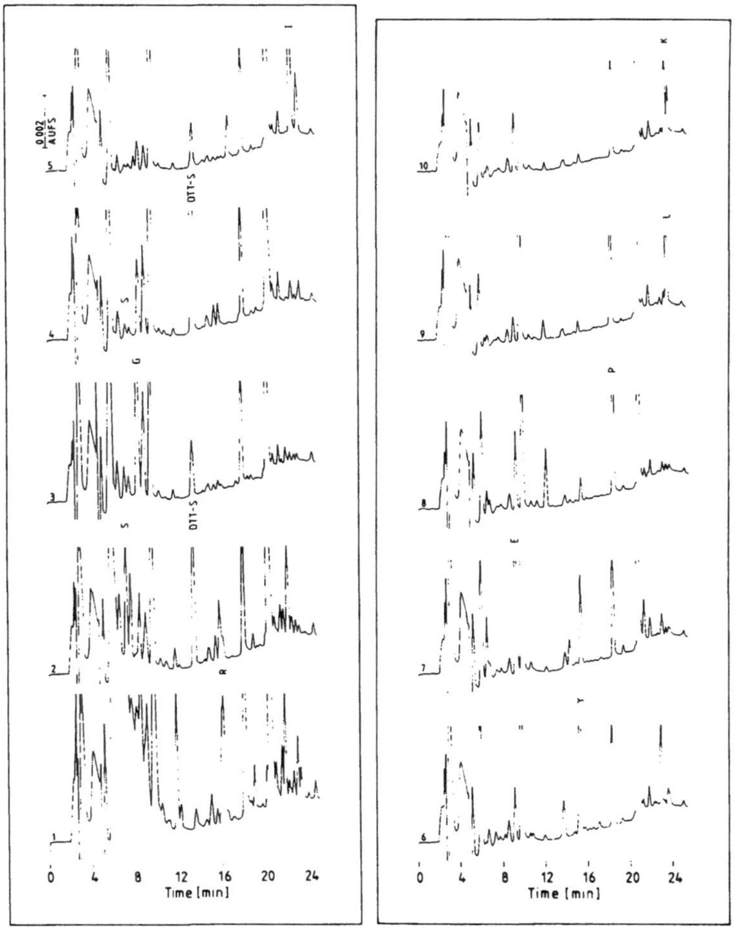

Figure 7: <u>Sequence analysis of a phospho-peptide from the ß-subunit of phosphorylase kinase.</u> The phospho-peptide was isolated from a Lys-C digest of the ß-subunit after autophosphorylation of the holoenzyme. The purified $^{32}$P-phospho-peptide (150 pmol) contained 1.08 mol of phosphate per mol of peptide. The sample was applied directly onto the glassfiber-filter and sequenced using the 03RPTH program. PTH-amino-acid analysis was performed with the Applied Biosystems model 120A on-line HPLC-system.

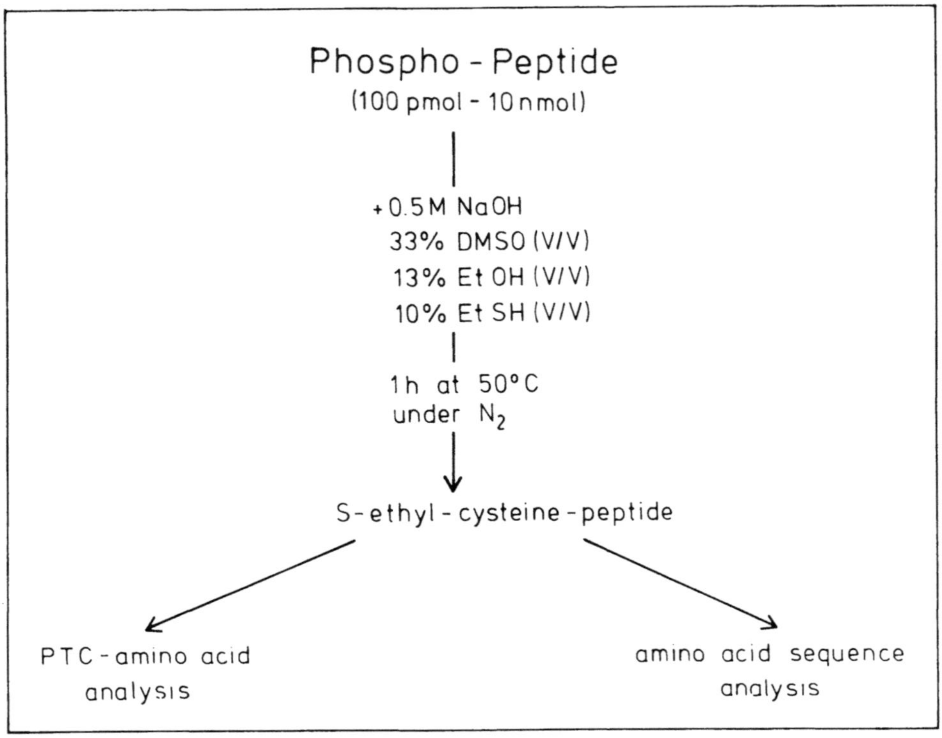


Figure 8: Conversion of phospho-serine containing peptides to the S-ethyl-cysteine derivatives.

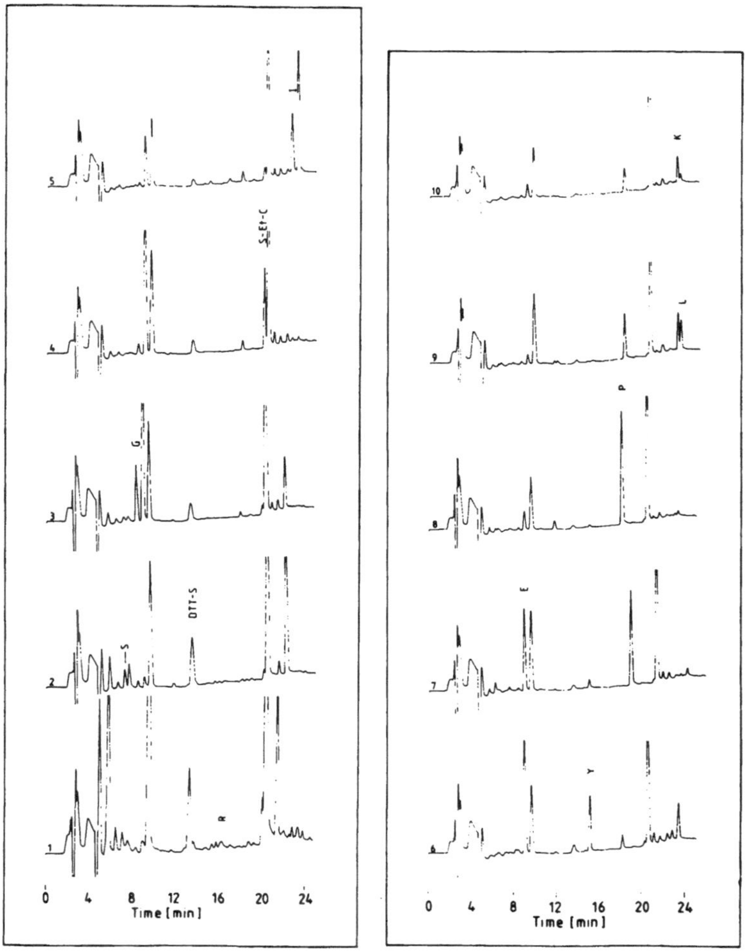


Figure 9: Sequence analysis of a phospho-peptide from the ß-subunit of phosphorylase kinase after modification with ethanethiol. The peptide decribed in legend to figure 5 was converted to the S-ethyl-cysteine derivative. The sample (50 pmol) was sequenced with the gas-phase sequencer using on-line PTH-amino-acid analysis.

duct in cycle 4 indicates that a phosphorylated serine was in this position. This was confirmed by sequencing the same peptide after modification with ethanethiol which converts phosphoserine to S-ethyl-cysteine (fig. 8). The PTH-amino-acid analyses are shown in fig. 9. PTH-S-ethyl-cysteine, eluting slightly earlier than diphenylthiourea, is present only in cycle 4, clearly demonstrating the position of phosphoserine in the peptide chain. Serine in position 2 yields PTH-serine and the PTH-derivative of the DTT-dehydro-alanine adduct in the usual 25 to 75 ratio.

Autophosphorylation sites in phosphorylase kinase. The ß-subunit of phosphorylase kinase is phosphorylated in three positions by autophosphorylation. Fig. 10a shows the sequences of phosphorylated peptides isolated after Lys-C cleavage. The RSGS... peptide was isolated from the first radioactive fraction. This fraction contained a second phosphopetide which was repeatedly lost during rechromatography. This phospho-peptide was then isolated from a V8 digest as shown in fig. 10b. The three autophosphorylation sites on the alpha-subunit were identified on two tryptic peptides (fig. 11a) and on one Lys-C peptide (fig. 11b). Two phosphorylation sites were found on the larger tryptic peptide. The peptide was derivatized with ethanethiol and 27 pmol of the peptide were sequenced with the Applied Biosystems gas-phase sequencer. Phosphoserine residues were identified as PTH S-ethyl-cysteine in positions 7 and 13 and were unambiguously distin-

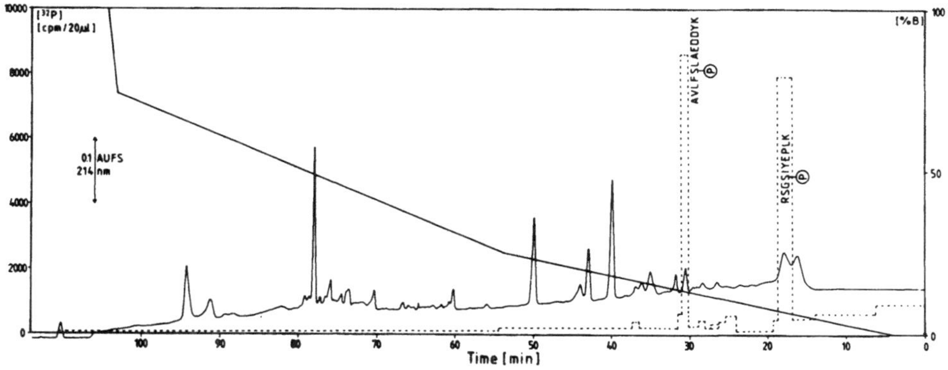


Figure 10a: Separation of endoproteinase Lys-C peptides from the ß-subunit of phosphorylase kinase. The digestion mixture (1.7 nmol in 3.0 ml) was chromatographed on a 10 µ Vydac 228TP104 pH-stable reversed-phase column (4.6x250 mm) using a hexafluoroacetone containing solvent system. Solvent A: 0.2% HFA/$NH_3$ pH 8.6, solvent B: 84% acetonitrile, 0.03% HFA pH 8.6. A gradient was applied as indicated at a flow rate of 1.0 ml/min. Radioactivity was determinted from 20 µl aliquots. Radioactive fractions were rechromatographed on the same column using the same solvent system with pH 7.2 instead of pH 8.6.

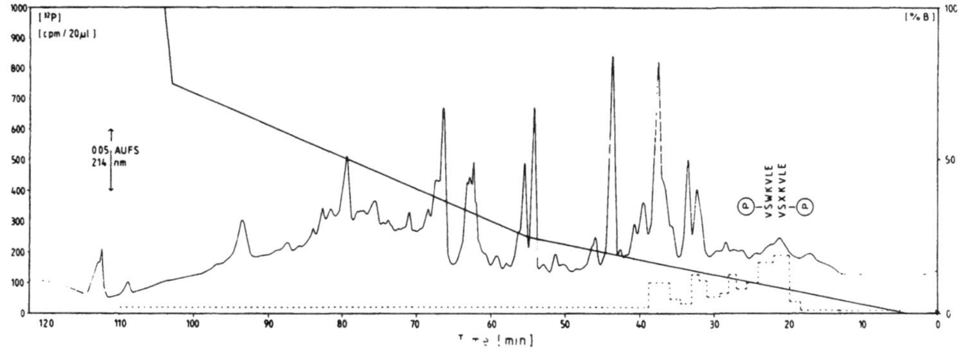


Figure 10b: Separation of endoproteinase Glu-C peptides from the ß-subunit of phosphorylase kinase. The digestion mixture (2.8 nmol in 4.1 ml) was injected on a 10 µ Vydac pH-stable column (4.6x250 mm). The chromatography conditions were the same as those used for the Lys-C peptides.

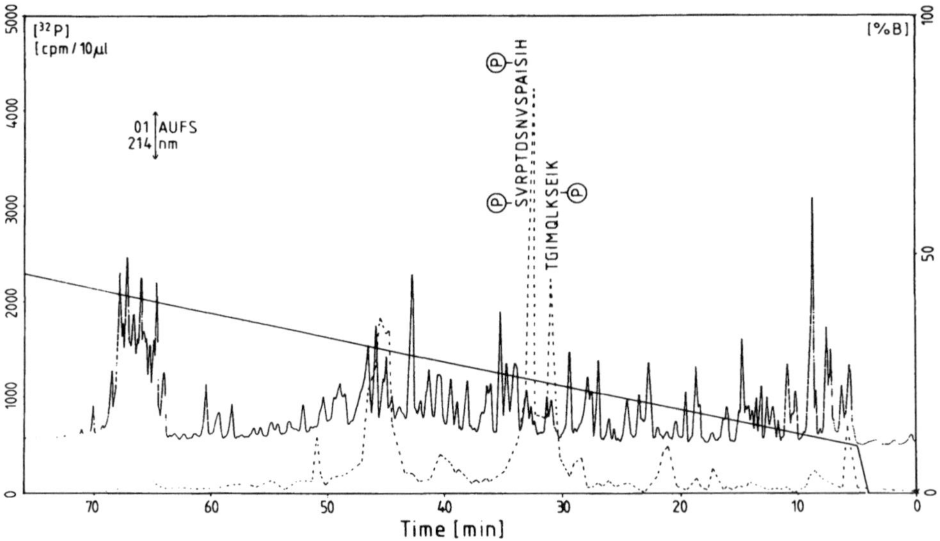


Figure 11a: <u>Separation of tryptic peptides from the alpha-subunit of phosphorylase kinase.</u> The digestion mixture (2.3 nmol in 2.0 ml) was chromatographed on a 5 µ Vydac $C_{18}$ columnn (4.6x150 mm) using 0.1% TFA as solvent A and 84% acetonitrile, 0.08% TFA as solvent B at a flow rate of 1.0 ml/min. Radioactivity was determined from aliquots of 10 µl. Radioactive fractions were rechromatographed on the same column using the same solvent system as described in legend to figure 10a with a pH of 7.8 instead of pH 8.6.

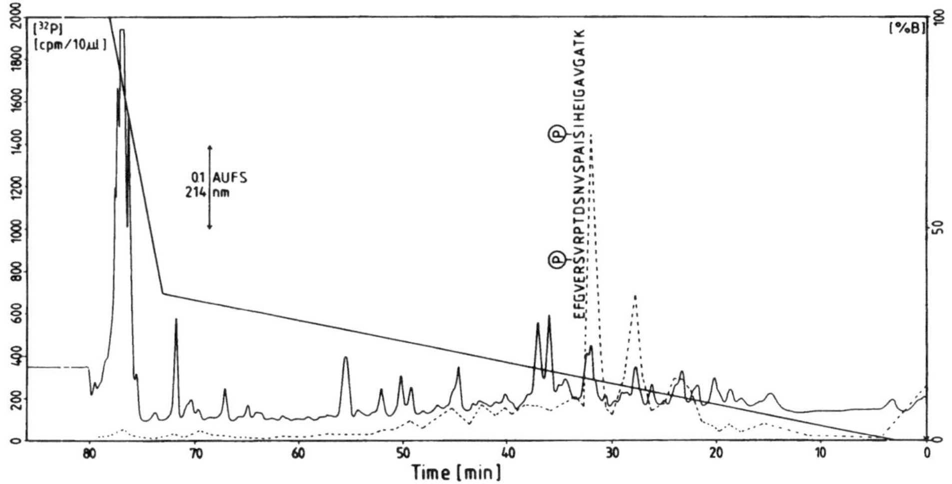


Figure 11b: <u>Separation of endoproteinase Lys-C peptides of the alpha-subunit of phosphorylase kinase.</u> The digestion mixture (700 pmol in 900 µl) was injected on the Vydac pH-stable column. Chromatography conditions were the same as described in legend to figure 10a.

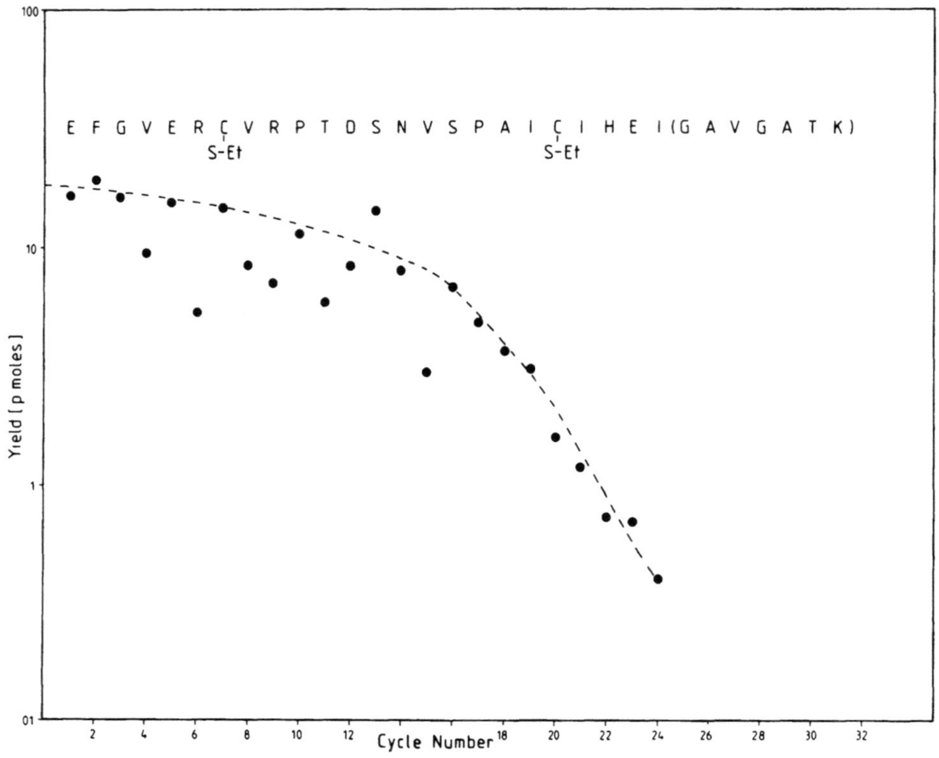


Figure 12: Sequence analysis of a Lys-C phosphopeptide from the alpha subunit of phosphorylase kinase. Yields of amino acids identified in given cycles are plotted versus cycle number. The peptide contains two serines and two phosphoserines. The sample (27 pmol) was sequenced with the Applied Biosystems gas-phase sequencer and on-line PTH-amino-acid-analyser.

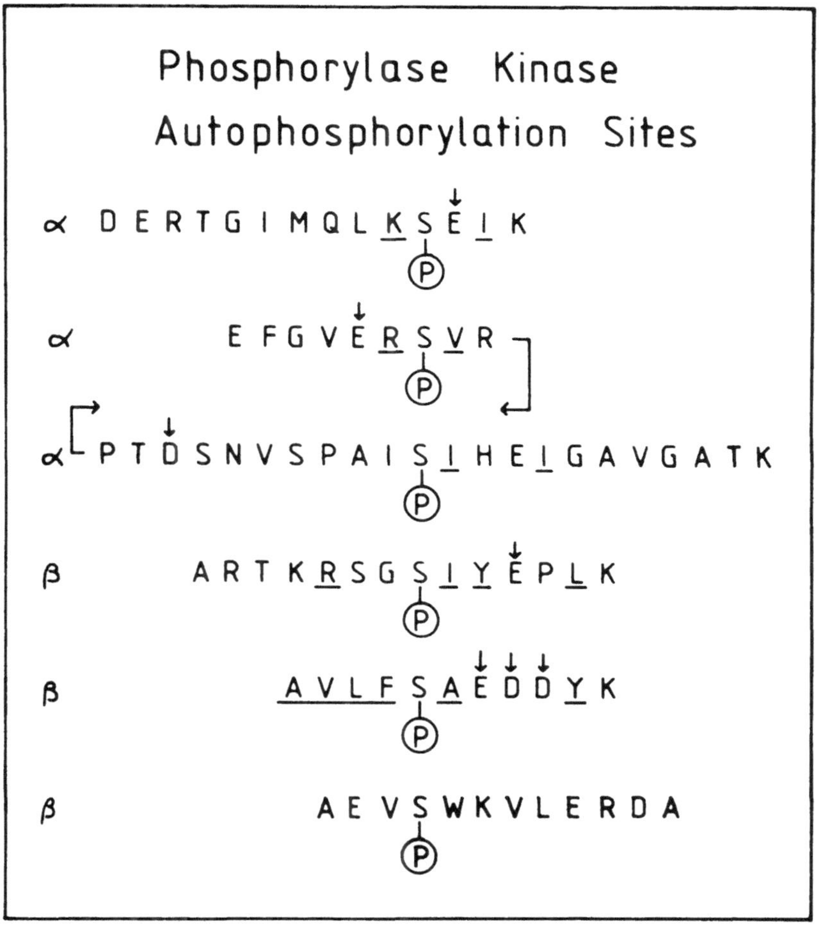
Phosphorylase Kinase
Autophosphorylation Sites
α D E R T G I M Q L K S E I K
α E F G V E R S V R
α P T D S N V S P A I S I H E I G A V G A T K
β A R T K R S G S I Y E P L K
β A V L F S A E D D Y K
β A E V S W K V L E R D A
P

Figure 13

guished from the serine residues in positions 13 and 16 (fig. 12). The autophosphorylation sites of rabbit muscle phosphorylase kinase are listed in figure 13. The first ß-subunit autophosphorylation site is identical with a site phosphorylated by cAMP-dependent protein kinase (27). All other phosphorylation sites are not related to sites recognized by cAMP-dependent protein kinase. All six autophosphorylation sites do not seem to be phosphorylated constitutively as concluded from the numbers of radioactive phosphate groups per mol of isolated peptide.

Acknowledgements: We gratefully acknowledge the expert assistance of H. Korte. This work was supported by Deutsche Forschungsgemeinschaft, SFB 168 and the Minister für Wissenschaft und Forschung des Landes Nordrhein-Westfalen.

# IV. References

1. Kemp, R.G., Foe, L.G., Latshaw, S.P., Poorman, R.A., Heinrikson, R.L.: J. Biol. Chem. **256**, 7282-7286 (1981)

2. Murray, K.J., El-Maghrabi, M.R., Kountz, P.D., Lukas, T.J., Soderling, T.R., Pilkis, S.J.: J. Biol. Chem. **259**, 7673-7681 (1984)

3. Pearson, R.B., Jakes, R., John, M., Kendrick-Jones, J., Kemp, B.E.: FEBS Letters **168**, 108-112 (1984)

4. Caudwell, F.B., Higara, A., Cohen, P.: FEBS Letters **194,1**, 85-90 (1986)

5. Holmes, C.F.B., Kuret, J., Chisholm, A.A.K., Cohen, P.: Biochem. Biophys. Acta **870**, 408-416 (1986)

6. Simpson, D.L., Hranisavljevic, J., Davidson, E.A.: Biochemistry **11**, 1849-1856 (1972)

7. Kolesnikova, V.Y., Sklyankina, V.A., Baratova, L.A., Nazarova, T.I., Avaeva, S.M.: Biokhimiya **39,1**, 235-240 (engl. translation) (1974)

8. Annan, W.D., Manson, W., Nimmo, J.A.: Anal. Biochem. **121**, 62-68 (1982)

9. Clark, R.C., Dijkstra, J.: Int. J. Biochem. **11**, 577-585 (1979)

11. Mega, T., Hamazume, Y., Nong, Y.-M., Ikenaka, T.: J. Biochem. **100**, 1109-1116 (1986)

12. Downs, F., Pigman, W. in: Methods in Carbohydrate Chemistry VIII, Whistler, R.L., BeMiller, J.N.: eds., Academic Press, New York, pp. 200-204 (1976)

13. Meyer, H.E., Swiderek, K., Hoffmann-Posorske, E., Korte, H., Heilmeyer, L.M.G. jr.: J. Chromatogr. in press (1987)

14. Meyer, H.E., Hoffmann-Posorske, E., Korte, H., Heilmeyer, L.M.G. jr.: FEBS Letters **204,1**, 61-66 (1986)

15. Walseth, T.F., Johnson, R.A.: Biochem. Biophys. Acta **562**, 11-31 (1979)

16. Beavo, J.A., Bechtel, P.J., Krebs, E.G. in: Methods in Enzymology, O'Malley, B.W., Hardman, J.G.: eds., Academic Press, New York, **Vol. 38**, Part C, p. 307ff. (1974)

18. Jennissen, H.P., Heilmeyer, L.M.G. jr.: Biochemistry **14**, 754-760 (1975)

19. Hessova, Z., Varsanyi, M., Heilmeyer, L.M.G. jr.: J. Biol. Chem. **260**, 10111-10117 (1985)

20. Zetterqvist, Ö., Ragnarsson, U., Humble, E., Berglund, L., Engström, L.: Biochem. Biophys. Res. Commun. **70**, 696-701 (1976)

21. Moisescu, D.G., Thieleczek, R.: J. Physiol. **275**, 241-262 (1978)

22. Crabb, J.W., Heilmeyer, L.M.G. jr.: J. Biol. Chem. **259**, 6346-6350, 14314 (1984)

23. Crabb, J.W., Heilmeyer, L.M.G. jr.: J. Chromatrogr. **296**, 129-141 (1984)

24. Heindrikson, R.L., Meredith, S.C.: Anal. Biochem. **136**, 65-74 (1984)

25. Tarr, G.E. in: Microcharacterization of Polypeptides: A Practical Manual, Shively, J.E., ed., Humana Press, New Jersey, pp. 155-194 (1986)

26. Kuhn, C.C., Crabb, J.W. in: Advanced Methods in Protein Microsequence Analysis, Wittmann-Liebold, B., Salnikow, J., Erdmann, V.A.: eds., Springer, Berlin, pp. 64-76 (1986)

27. Cohen, P., Watson, D.C., Dixon, G.H.: Eur. J. Biochem. **51**, 79-92 (1975)

# NEW MODULES FOR THE CONSTRUCTION OF A GASPHASE-SEQUENCER

**Frank Reimann*, Brigitte Wittmann-Liebold and Stefan Fischer**
Max-Planck-Institut für Molekulare Genetik,
Abteilung Wittmann, D-1000 Berlin 33

*present address: Fa. Dr. H. Knauer,
Hegauer Weg 38, D-1000 Berlin 37

A new sequencer of modular construction has been designed which is governed by a personal computer that allows to accomodate different types of automatic Edman degradation (solid-, gas-, liquid-phase and pulsed-liquid sequencing) of polypeptides. The basic modules of the new sequencer were developed for assembly into a common basic machine to which various types of reactors for the different protocols can be connected, e.g., a column, cartridge, or any of the new reactors which are presently under construction and investigation.

The machine consists of the following units which can be freely assembled into a flexible rack system:

1. Power supply;
2. Personal computer (IBM or IBM compatible) with printer;
3. Interface;
4. Gas distribution system with electronically controlled pressure regulators;
5. Flask battery;
6. Delivery and waste or collect valve system;
7. Reactor(s) and electronically controlled heating system;
8. Converter and electronically controlled heating system;

Modern Methods in Protein Chemistry, Vol. 3

9. On-line HPLC detection system;
10. Fraction collector (evacuated and pressurized with inert gas).

Prototype versions for most of the parts have been developed at this institute and tested through long test-periods, e.g., the delivery and collect valve system, valves for the delivery of inert gas, the flask system, the conversion device and the on-line detection system in the isocratic mode (1 – 2). Further, the appropriate software for programming of the different protocols in a "multipurpose sequencer" was elaborated (3–4) and new reactors designed (unpublished results).

For a final commercialization of the machine (through Dr. H. Knauer GmbH, Hegauer Weg 38, D-1000 Berlin 37, Zehlendorf) important modifications to some of the parts were made (5): further miniaturization of the delivery valve system which makes it more suitable for industrial production and new gas pressure regulators were constructed. The sequencer control program provided is flexible and programming is assisted by an editor-type menu. These new sequencer modules are briefly discussed below.

## Interface

The programmer governs 48 binary functions, which can be activated in any desired combination in a flexible manner and for the appropriate times. The interface board is connected to the computer by a V-24 connection. The activated functions give a positive impuls for switching electro-pneumatic valves. The interface can govern 8 electric functions and 40 different membran valves. The interface is mounted together with these valves, with electric elements and with the status control lamps in one unit.

**Sequencer program**

The program is especially suitable for repeated chemical processes, such as the stepwise degradation of polypeptides. In the program it is also possible to repeat loops within one degradation cycle by combining a series of individual **STEPs** to **BLOCKs** or **CYCLEs** for selected repetitions of individual events within the degradation cycle (1). The BLOCKs facilitate programming for the degradation of difficult sequence regions, where repetions of the cleavage (at hydrophobic residues) or of the coupling (accumulation of basic amino acids) is mandatory. The program is executed following the hierarchical order as verified by the section **RANGE** (which contains the STEPs, BLOCKs and CYCLEs in the order of the wanted execution).

The program consists of two main parts, a) the **EDIT** mode in which the program is monitored, changed, stored and printed, and b) the **RUN** mode for execution of a program. The programming is menu-assisted by the aid of function keys which are easy to learn:
HELP (F1); EXIT (F2); RUN (F3); EDIT (F4); NEXT (F5); LAST (F6); LOAD (F7); SAVE (F8); MODIFY (F9); PRINT (F10).

In the **HELP** mode explanations for editing or modifying a program are given; thereafter, the program stays at the same event as before. In the **EXIT** mode editing is closed and execution after security check is initialized. The execution can be started as dictated by the RANGE or at any appropriate STEP. In the RUN mode the STEPs and BLOCKs are executed as programmed in the RANGE manner. **EDIT** edits a new program or changes the old one by calling STEPs (CTRL+S), BLOCKs (CTRL+B), CYCLEs (CTRL+C) or the RANGE (CTRL+R). By means of the cursor functions the wanted characters or numbers are entered into a mask on the screen (e.g., the STEP name, time in seconds, and the function name of the valves). With **NEXT** the following STEP(s) are actuated; **LAST** actuates the previous STEP(s). **LOAD** finds a program which is stored on the disk for loading into the computer. The sequencer program is saved on a disk by **SAVE** (with appropriate warnings for overwriting of an existing program under the same name). All programs on the disk can be listed on the screen. With **MODIFY** STEPs can be changed (the time or the functions) in the executed sequencer program by stopping the execution and storing the new data into the memory of the computer. Thereafter, the program can be continued with **START**, **NEXT** or **LAST**, as required. With the command **PRINT** the program is listed by the printer. This is done in the STEP sequence of the RANGE, either in a short form (in the EDIT mode, listing all entries in the arrays STEP, BLOCK, CYCLE and RANGE) or as full data set (in the RUN mode, listing all STEPS with given names, times, and functions in the sequence to be executed). The program has an error routine which shows reading errors of the disk, etc. The program further demands the entry of a documentation file for storing actual data of the sample (e.g., the name, origin and preparation; date, number of cycles and run time) which is printed as hard copy before starting the sequencer run.

In conclusion, the control program for the sequencer is flexible and comfortable in its use, and especially useful for adaptation of new protocols for the degradation, conversion or on-line detection.

**Pressure regulators**

The pressures of the inert gas regulators are controlled electronically. The regulators maintain a constant pressure independent of flow rate and atmospheric pressure. This is advantageous in comparison with mechanically controlled regulators. The possible pressure range of these new regulators is 30 to 500 hPa (= 30 to 500 mbar or 0.003 MPa to 0.05 MPa).

**Membrane valve system**

The membrane valves open or close the nitrogen dosage lines, the delivery lines of all solvents and reagents, and the waste and collect lines of the sequencer. They are constructed from a block of KEL F, and provided with TEFLON adaptors for the inlet and outlet teflon tubings similar as described recently (1). They have been further miniaturized and updated to accomodate a straight central dosage line in the surface area of the KEL F valve block. One block contains eight or ten valve ports, respectively. Any combination of ports can be assembled as desired. The valve block is closed with a TEFLON PFA 1000 LP film (DUPONT, Wilmington, Delaware) by means of pressurized air or nitrogen of 0.3 MPa. The valves are opened by applying a vacuum (membrane vacuum pump).

**Acknowledgements**

The design of the sequencer for a commercialization was made in collaboration with Prof. Dr. H. Tschesche and Dr. L. Meinecke, Institut für Biochemie, Universität Bielefeld, Prof. Dr. W. Machleidt and A. Esterl, Institut für Physiologische Chemie, Physikalische Biochemie und Zellbiologie, Universität München, and Dr. H. Knauer, Wissenschaftliche Gerätebau GmbH, Berlin 37.

We are grateful to Mr. M. Steiffa, who helped us with the adaptation of the programm for the PC Commodore and to Mr. H. Graffunder and Mr H. Kohls for many fruitful technical suggestions and continuous help in constructing sequencer parts.

**REFERENCES**

1. Wittmann-Liebold, B.: in "Modern Methods in Protein Chemistry", H. Tschesche, ed., W. deGruyter Verlag, Berlin, New York (1983), pp. 229-266.

2. Wittmann-Liebold, B.: in "Modern Methods in Protein Chemistry", H. Tschesche, ed., W. deGruyter Verlag, Berlin, New York (1985), pp.303-327.

3. Reimann, F. and Wittmann-Liebold, B. in: Advanced Methods in Protein Microsequence Analysis, B. Wittmann-Liebold, J. Salnikow and V. A. Erdmann, eds., Springer Verlag, Berlin, Heidelberg, New York, London, Paris, Tokyo (1986), pp. 118-125.

4. Wittmann-Liebold, B. in: Methods in Protein Microcharacterization, J. E. Shively, ed., Humana Press, Inc. (1986), pp. 249-277.

5. Reimann, F., Graffunder, H. and Wittmann-Liebold, B. in: Methods in Protein Sequence Analysis, K.A. Walsh, ed., Humana Press, Clifton, N.J., in press.

# A NOVEL SOLID SUPPORT FOR GAS-LIQUID MICROSEQUENCING OF PROTEINS AND PEPTIDES - DERIVATIZED POROUS GLASS REPLACES POLYBRENE-COATED GLASFIBER FILTER

Lothar Meinecke and Harald Tschesche

University of Bielefeld, Faculty of Chemistry, Department of Biochemistry, D-4800 Bielefeld 1

## Introduction

Since the description of the first automated sequenator - that needed upper nanomole quantities of protein for sequencing in the liquid phase - 20 years ago by Edman and Begg (1) many developments and improvements of this technique led to different concepts for more sensitive sequencing. These findings together with the principle of gas-phase sequencing introduced 30 years ago by Fraenkel-Conrat (2) and Shelton and Schroeder (3) were the prerequisites for the construction of a gas-liquid phase sequenator (4) that allows microsequencing of proteins and peptides in the lower picomole range.

The wash-out of small proteins and peptides, one of the main disadvantages in liquid phase sequencing, is minimized during gas-phase sequencing by the immobilization of the proteins with polybrene, a polymeric quaternary ammonium salt, and also by the gaseous delivery of trimethylamine for the coupling step and trifluoroacetic acid (TFA) for the cleavage step. On the other hand polybrene has the effect of a "swamp" for byproducts of the Edman degradation as are thioureas, phenylthiocarbamyldimethylamine and others that appear during the subsequent HPLC-identification of the phenylthiohydantoin (PTH) amino acids. Therefore polybrene has to be purified by performing several "empty" degradation cycles without protein. Besides this procedure that is never quantitative for the removal of UV absorbing material proteins and peptides are not quantitatively immobilized by polybrene as in the method of solid phase sequencing. Basic proteins or pep-

Modern Methods in Protein Chemistry, Vol. 3

tides are not retained as good as acidic ones. In gas-liquid sequencing proteins and peptides are normally applied in solution into the reactor called cartridge to form a homogeneous film after drying on the glass fiber filter. This protein film shrinks into a small ring after some degradation steps and is not as easy permeable for solvents and reagents as the thin film. This fact and difficulties for a quantitative reaction of the reagents with proteins embedded in polybrene forming the interphase between the liquid, solid and gaseous materials are the reasons for the lower repetitive yields when compared to classical automated liquid or solid phase sequencing. There are several attempts to circumvent or solve the polybrene problem. Common to all these procedures is the covalent modification of the glass fiber support either by TFA etching or reaction with quaternary ammonium silanols (5,6). Taking these activated glass fiber sheets for direct electroblotting of proteins after separation by sodium dodecyl sulfate polyacrylamide gel electrophoresis as little as 10 - 20 pmol of protein material can be used for gas-phase sequencing (4).

The purpose of this paper is to present first results and experiences with a novel porous and cone shaped glass support for protein microsequencing. The new glass reactor is covalently modified on its surface and easily permeable for gases and liquids. The cone shape and the specific covalent activation makes it the ideal support for the immobilization of proteins and peptides during microsequencing. Proteins or peptides loaded on the top of the modified cone have space to slip during washing cycles and this "moving" takes place in the directing of the liquid or gas flow in contrast to the cartridge reactor.

The results were obtained with the new reactor used in the modular sequenator that was constructed in cooperation with MPI Berlin, Dept. B. Wittmann-Liebold, University Munich, Dept. W. Machleidt and the manufacturing company Dr. Knauer KG Berlin. The design, construction principle and the sequencer control program have been reported in some articles recently (7,8,9).

## The glass material

The new material is called Siran and described as open-pore sintered glass in the product information 6179e, Schott information no. 1/1983, Schott Glaswerke, Mainz, West Germany. It is made from borosilicate glass, also called Duran 50, has a maximum water-absorbing capacity of 0.8 $ml/cm^3$, is chemically resistant, can be produced with pore sizes from 10 μm to about 400 μm, exhibits high flow rates and can be mechanically tooled by sawing, drilling, turning, milling, etc. Cones (6 mm diameter, 4 mm height) are produced with the turning machine of our laboratory workshop from rods of 6 mm diameter and 200 mm length; they have a water absorption capacity of 30 μl; the porous glass we used for our first experiments had a maximum pore size of 55 μm.

## The glass reactor

In Fig. 1 a diagrammatic view of the glass reactor is shown. It consists of two teflon rods of 35 mm length and with inner bore hole. At the end of the upper rod the hole is expanded for the exact taking up of the porous glass cone. The glass rods are guided in a hole of two screwable aluminium blocks and are sealed with teflon "pull-through" sealings in hollow screws. The aluminium block is heated to 50 - 60$^{o}$C in a suitable aluminium device (90 mm diameter, 100 mm height) on a cook-plate, regulated by a contact thermometer. An electronically regulating heating-device is currently built in our electronic workshop.

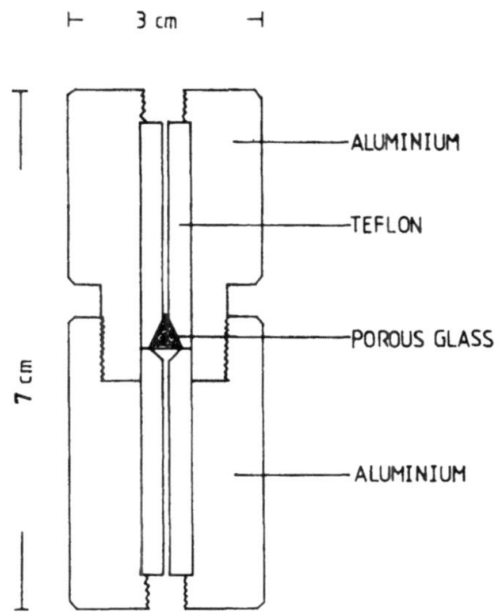


FIG. 1 Diagrammatic view of the glass reactor

The chemical modification

The great surfaces of the porous glass cones are chemically derivatized for an optimal immobilization of proteins and peptides during sequencing. In Fig. 2 this kind of modification is shown.

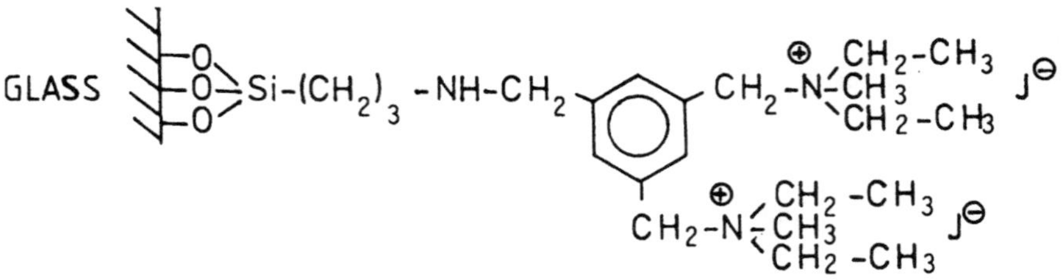


FIG. 2

It contains two quaternary ammonium groups for ionic and an aromatic system for π-π-interaction with the side chains of proteins and peptides. Furthermore the modifying group carries a secondary amine capable to act as a "scavenger" to neutralize aminoreactive compounds that may be generated within highly purified reagents during a sequencer run (10,11). The modification starts from aminopropylglass (AP-glass) that is synthesized from underivatized glass cones and 3-aminopropyl thiethoxysilane (Merck) according to the prescription of Machleidt (12), Fig. 3.

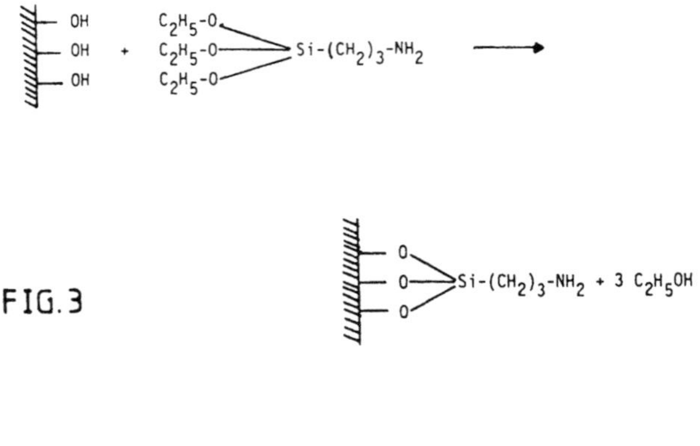


FIG. 3

The next step is the coupling of trimesoylchloride (that is either synthesized from 1,3,5-benzene-tricarboxylic acid according to (13) or purchased (Janssen Chemical)) with AP-glass to get the acid chloride modified glass (AC-glass), (14), Fig. 4.

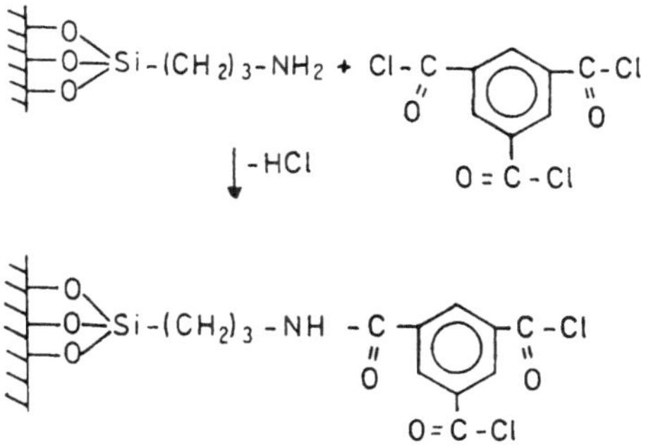


FIG. 4

In all these reactions with solid porous glass care has to be taken that all reagents are quantitatively absorbed and that reaction products are carefully washed out. This is achieved by repeated degassing and careful moving of the cones so that they are not mechanically destroyed.
Diethylamine is reacted with AC-glass to yield trimesoylamide-glass (TMA-glass) according to (14), Fig. 5.

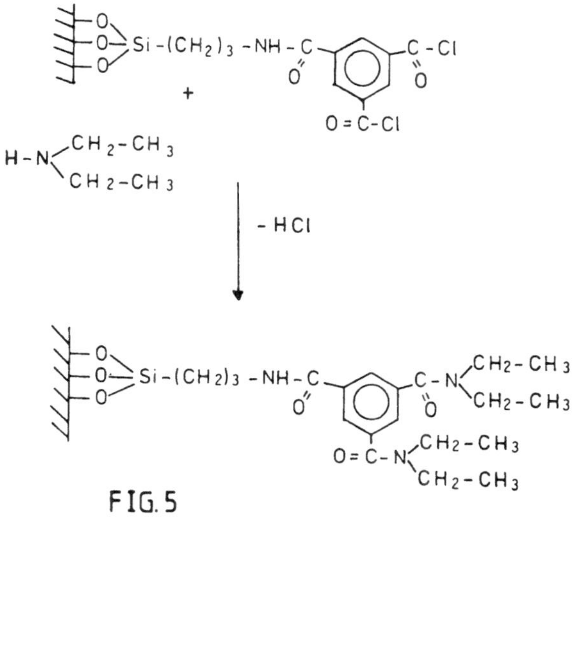


FIG. 5

After reduction with $LiAlH_4$ (15) the diquaternary amino benzene derivative of glass (DQB-glass) is synthesized by reaction with methyliodide (16), Fig. 6.

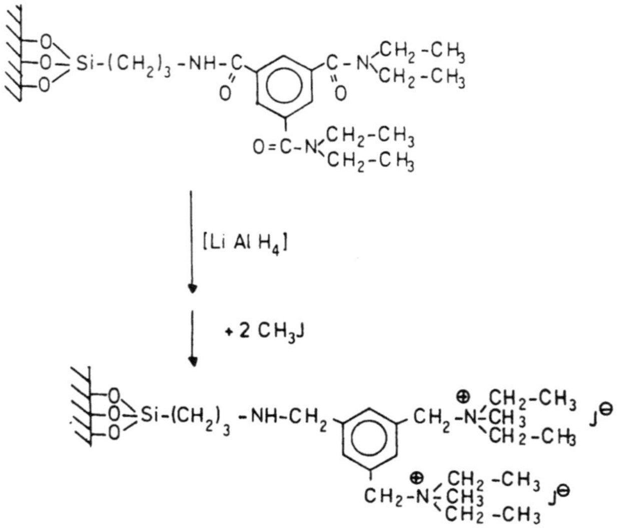


FIG.6

The success of this reaction can be checked qualitatively with silver nitrate: Dipping of the modified cone in a 1 - 5 % silver nitrate solution turns it yellow from silveriodide that after some minutes becomes black from precipitated and reduced silver. Titration of the iodide with silver nitrate would allow quantitation of the derivatized group on the porous glass support. The modified cones prepared for use in the reactor are stored cool in an exsiccator under vacuum.

Reagents and solvents

All reagents and solvents used for microsequencing are purified according to our recommendations published recently (17); in some cases reagent 2, 12.5 % trimethylamine in water, from Applied Biosystems was used, when it was not purified as in (18). We are planning to replace it by alternative bases, that are easier to handle and to purify as described by Braunitzer (19).

## Degradation program

The degradation temperature is 50°C. The protein is dried with nitrogen; no vacuum pump is used. The program consists of 3 steps of 400 seconds with $R_2$, 12.5 % trimethylamine in water, volume 2.5 - 2.8 ml/min and 2 - 3 deliveries with 2 - 5 % phenylisothiocyanate in heptane. A 3 ml ethylacetate wash is followed and after drying a 1200 seconds $R_3$, trifluoroacetic acid, step with 2.5 - 2.8 ml/min. The thiazolinone is extracted with 900 µl $S_3$, butylchloride, dried down quickly and converted with 20 % TFA in water at 65°C in 15 min; after drying the resultant PTH-derivatives are dissolved in 250 µl HPLC-buffer without salt (20) and 100 µl are injected automatically.

## Sample handling

The test protein for the sequencer is β-lactoglobulin A from bovine milk. It is dissolved with a concentration of 1 nmol in 5 µl TFA and stored frozen at -25°C. When lactoglobulin A is dissolved in water instead of TFA and stored frozen at -25°C it precipitates when thawed again. For sequencing 2.5 - 25 µl are applied onto the top of the glass cone and dried. Besides the modification described above the glass cones are also activated by TFA etching, aminopropyl-modification and polybrene-absorption.

## Acknowledgement

This work was supported by a BMFT-grant no. 03 8598 0 from the Federal Minister of Research and Technology, Bonn.

The sequenator used for these investigations is an instrument developed in cooperation with Prof. Dr. B. Wittmann-Liebold, MPI für Molekulare Genetik, Berlin; Prof. Dr. W. Machleidt, Institut für Physiologische Chemie, Physikalische Biochemie und Zellbiologie, Universität München; and Dr. H. Knauer, Wissenschaftliche Gerätebau GmbH, Berlin.

References


1. Edman, P. and Begg, G. (1967) Eur. J. Biochem. 1, 80 - 91.
2. Fraenkel-Conrat, H. (1954) J. Am. Chem. Soc. 76, 3606 - 3607.
3. Schroeder, W.A., Shelton, J.R., Shelton, J.B., Cormick, J. and Stones, R.T. (1963) Biochemistry 2, 992 - 1008.
4. Hewick, R.M., Hunkapiller, M.W., Hood, L.E. and Dreyer, W.J. (1981) J. Biol. Chem. 256, 7990 - 7997.
5. Aebersold, R.H., Teplow, D.B., Hood, L.E. and Kent, S.B.H. (1986) J. Biol. Chem. 261, 4229 - 4238.
6. Frank, R. and Ashman, K. (1986) Biol. Chem. Hoppe-Seyler 367, 573
7. Reimann, F. and Wittmann-Liebold, B. (1987) Biol. Chem. Hoppe-Seyler 368, 441 - 442.
8. Fischer, S. and Wittmann-Liebold, B. (1987) 18th FEBS Meeting, Ljubljana, Yugoslavia, Poster 1659.
9. Reimann, F., Wittmann-Liebold, B. and Fischer, S. (1987), this volume.
10. Frank, G. (1982) in: Elzinga, M. (ed.) Methods in protein sequence analysis, Humana Press, Clifton, NJ, 91 - 97.
11. Bhown, A.S., Mole, J.E. and Bennet, J.C. (1981) Anal. Biochem. 110, 355-360.
12. Machleidt, W., Borchart, U. and Ritonja, A. (1986) in: Wittmann-Liebold, B., Salnikow, J. and Erdmann, V.A. (eds) Advanced Methods in Protein Microsequence Analysis, Springer-Verlag, Berlin, New York, 91 - 107.
13. Organikum (1977) VEB Deutscher Verlag der Wissenschaften, Berlin, 527 - 528.
14. Organikum (1977) VEB Deutscher Verlag der Wissenschaften, Berlin, 513.
15. Organikum (1977) VEB Deutscher Verlag der Wissenschaften, Berlin, 614 - 615.
16. Organikum (1977) VEB Deutscher Verlag der Wissenschaften, Berlin, 260.
17. Meinecke, L. and Tschesche, H. (1986) in: Wittmann-Liebold, B., Salnikow, J. and Erdmann, V.A. (eds) Advanced Methods in Protein Microsequence Analysis, Springer-Verlag, Berlin, New York, 126 - 148.
18. Reimann, F. and Wittmann-Liebold, B. (1986) in: Wittmann-Liebold, B., Salnikow, J. and Erdmann, V.A. (eds) Advanced Methods in Protein Microsequence Analysis, Springer-Verlag, Berlin, New York, 118 - 125.
19. Braunitzer, G. and Huber, F. (1987) Biol. Chem. Hoppe-Seyler 368, 195 - 198.
20. Ashman, K. (1986) in: Wittmann-Liebold, B., Salnikow, J. and Erdmann, V.A. (eds) Advanced Methods in Protein Microsequence Analysis, Springer-Verlag, Berlin, New York, 219 - 225.

# NEW AUTOMATED AMINO ACID ANALYSIS BY HPLC PRECOLUMNS DERIVATISATION WITH FLUORENYLMETHYLOXYCARBONYLCHLORIDE

I. Betnér / P. Földi*
LKB-Produkter AB, S-161 26 Bromma, Sweden

## Introduction

In the last few years, precolumn derivatisation of amino acids with orthophthalaldehyde (OPA), phenylisothiocyanate (PITC) or fluorenylmethyloxycarbonylchloride (FMOC) followed by reversed phase HPLC, have become accepted methods of separation and it is likely that the popularity of these methods will further increase because of their higher sensitivity, shorter analysis time and versatility compared to other methods.
However, each of these methods has some specific drawbacks [1]. For instance, whilst precolumn derivatisation with OPA offers a considerable increase of sensitivity and sample output over the conventional postcolumn ninhydrin approach, the method suffers from the major disadvantage that OPA does not react with secondary amines. On the other hand the PITC method, which reacts with both primary and secondary amines, is less sensitive than the OPA procedure and in addition it cannot be fully automated.
The application of precolumn derivatisation with FMOC has been hampered by the need to extract the excess FMOC and the object of this study was to overcome this drawback by a chemical modification of the derivatisation technique.

Modern Methods in Protein Chemistry, Vol. 3

## Materials and Methods

Apparatus: The HPLC systems (LKB, Bromma, Sweden) consisted of a two pump gradient delivery system (Model 2150 pumps and Model 2152 controller for gradient programming), however, test control experiments have also been performed successfully with one pump gradient mixing devices.
The samples (5-20 $\mu$l) were injected by an autoloader (Model 2157).
The fluorescence was monitored with a Model F1000 Hitachi spectrofluorimeter ($\lambda_{exc}$ 330 nm, $\lambda_{em}$ 450 nm for OPA derivatives and $\lambda_{exc}$ 260 nm, $\lambda_{em}$ 310 nm for FMOC derivatives) with a 2 $\mu$l flow cell and a Xenon lamp. UV-detection was carried out with an LKB-VW monitor (LKB Model 2151) or a multidiode array monitor (LKB Model 2140). Three different recording systems have been used. Simple chromatograms were recorded with a two channel recorder (Model 2210); for the evaluation and quantitation based on the Nelson Analytica 3000 Series Chromatography Data System a Canon printer plotter PJ-1080A was used; continuous on-line quantitation was obtained either with a recording integrator (LKB Model 2220, Shimadzu C-R3A) or with an Olivetti PC M24 computer and a Nelson Analytica interface (760 series).
HPLC grade chemicals and solvents were used throughout. All the solutions, eluents, buffers, etc. were prepared as already described [1, 5, 6]. Amino acid standards were used for the optimization of the chromatographic conditions but real samples such as protein hydrolysates were used for the evaluation of the analytical technique.

## Discussion and Results

When discussing a chromatographic method such as an amino acid analysis done either by postcolumn derivatisation with ninhydrin or by precolumn derivatisation with OPA, PITC, FMOC and others, it is often convenient to characterize the method according to three criteria: resolution, sensitivity and speed, which may be referred to as the "magic triangle of chromatography". According to this "magic triangle", the best chromatogram of each method must always be a compromise [2, 3] (Fig.1).

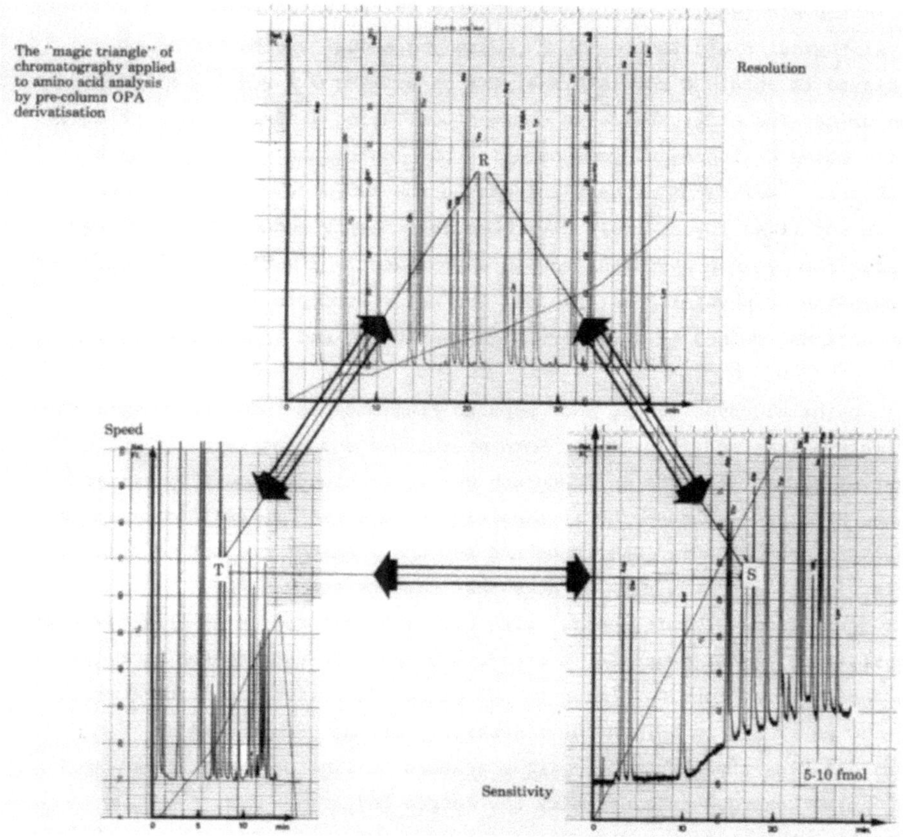


Fig. 1

Chromatographic conditions (see ref. [6, 12])
(a) Physiological standard (43 compounds)
(b) Rat plasma (taken by biopsy)
(c) Microbore chromatography - Protein hydrolysate

Up to the mid-1960, a complete separation of the amino acids in a physiological fluid took up to 24 hours and a dual system had to be employed to obtain a complete analysis on between 0.1 and 1.0 ml of plasma and urine. The highly reliable conventional amino acid analysers of today, still based on ion-exchange separation and postcolumn separation with ninhydrin, operate using sample loads in the 100 picomole to 10 nmole range and times for analysis have been drastically reduced to less than 1 hour for protein hydrolysates and approximately 2 hours for the complete separation of physiological fluids. The major strength of this conventional method is still its high resolution and high reliability [4].

Orthophthalaldehyde is the most popular fluorescence-inducing reagent for primary amines and is used for both postcolumn and precolumn derivatisation. Its use in precolumn derivatisation followed by reversed phase HPLC becomes superior if sensitivity (low fmol range) and speed of analysis are of prime importance and secondary amines are not of interest (Fig. 1) [5]. It is worthy of note that adducts such as the OPA-derivatives of amino-acids, also have a UV-absorption so that, if high sensitivity is not required, a standard UV-monitor can be used to detect the amino-acids routinely down to 200 pmole (Detection limit 20-50 pmole) at 230 nm or 330 nm. Often the instability of the OPA-adducts is considered as a weakness but with automated on-line derivatisation, this cannot be regarded as a drawback any longer [6].
Recently, the precolumn derivatisation of amino-acids with PITC has become more and more accepted by protein chemists. The reaction is well understood, as it is identical to the first step of the Edman degradation as used in protein sequence analysis. The method is easily manageable because of its simple chromatographic procedure and has good sensitivity with UV-detection (detection limit 5 pmol) [7]. However, it suffers from two drawbacks, first because of an evaporation step during the derivatisation of the amino acids it cannot be fully automated and secondly the carbamoyl derivatives are only chromophors. Since they do not fluoresce, the method cannot give the high sensitivity (fmol range) which will be demanded by protein chemists in the near future.

In the search for a superior alternative precolumn reagent and in order to compensate for the disadvantages inherent in the methods described above, Josefsson et al. [8, 9] used FMOC to determine amino-acids in protein hydrolysates and seawater. As indicated by its name fluorenylmethyloxycarbonylchloride, FMOC has a high fluorescence. Its excitation and emission spectra are nearly identical to that of the amino-acid derivatives [1] and therefore it is essential to remove excess FMOC, for instance by pentane extraction, so that its elution peak does not interfere with the amino acid derivatives. The advantages of this reagent are as follows: the amino acid derivatives are formed very quickly (~30sec.). They are stable (more than 30 hours) (Fig. 2), they can be detected down to the lower fmole range and lastly FMOC reacts with primary and secondary amines such as proline or hydroxyproline. Because of these advantages the FMOC-procedure seemed to be well suited for fully automated "on line" derivatisation, for amino acid analysis. Even with the disadvantage of pentane extraction for the removal of the excessive FMOC it has been used extensively and this new method seems to fulfil all the demands of resolution (Fig. 3), speed of analysis (Fig. 4) and high sensitivity. Narrow or microbore columns give a further increase in sensitivity especially when baseline subtraction is used (Fig.5) [11]. From Figure 5 one can conclude that the sensitivity limits are not due to the instrument but are dependent on contamination even with HPLC-grade chemicals.

After the reproducibility of the method had been checked [10] - compare Fig. 6 - the only point of the procedure, which could be regarded as a weakness, was the lack of a fully automated precolumn derivatisation of the amino acids.

The new idea to overcome this weakness is as follows: As can be seen in Figures 3 to 5, the lysine FMOC-complex is eluted at the end of the chromatogram, which means that under the given conditions lysine is the most hydrophobic amino acid. Now as usual the amino acids are derivatisized with an excess of FMOC and after some seconds an excess of an amine, which is even more hydrophobic than lysine may be added. This FMOC-hydrophobic amine complex should be eluted at the end of a chromatogram without interfering with any of the other amino acid derivatives and nearly all the disturbing excess FMOC is derivatisized.

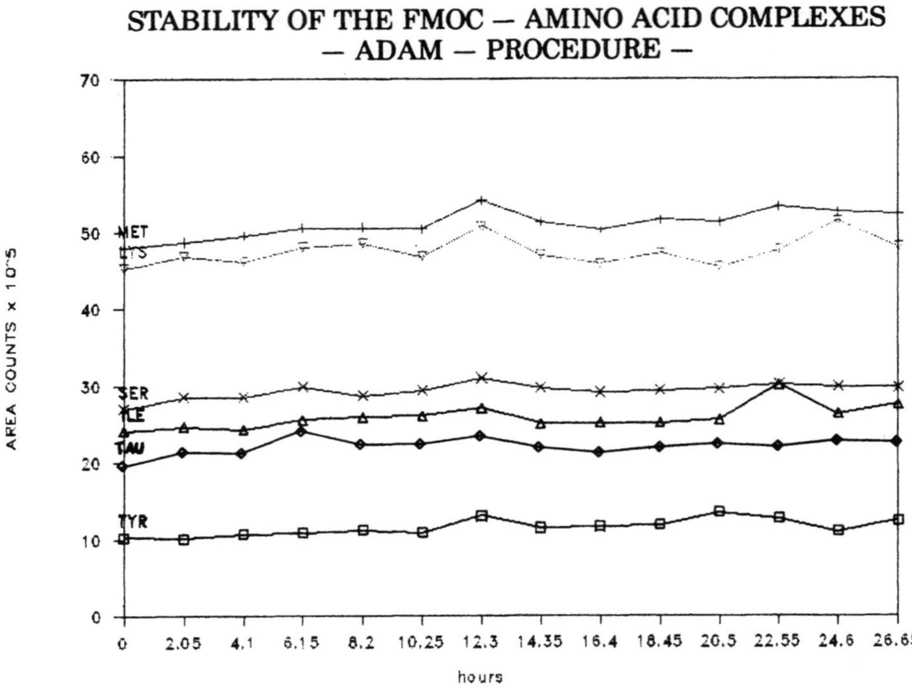


Fig. 2

The stability of the FMOC derivatives over a period of more than 30 hours can be proven independently how the excess of FMOC is treated (pentane extraction etc., see also ref. [10]).

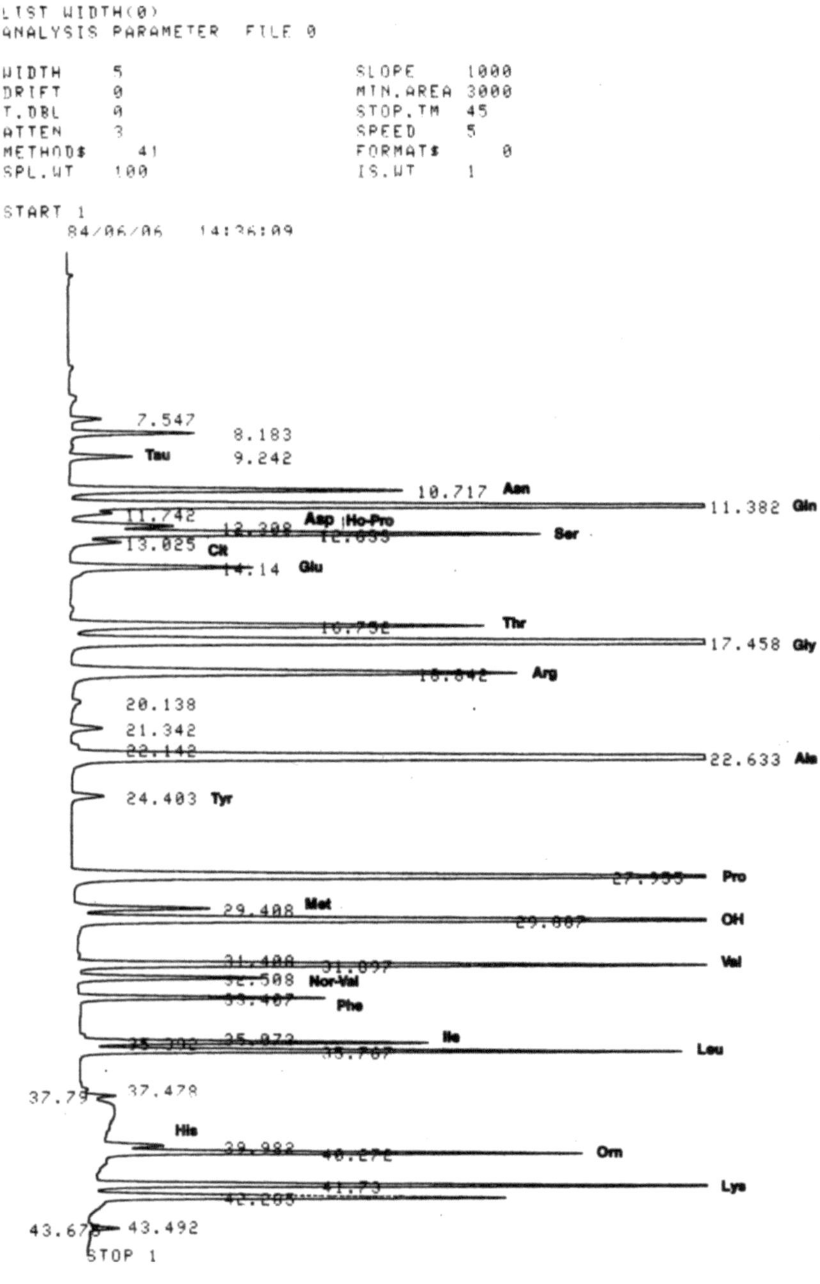


Fig. 3

FMOC derivatised physiol. amino acid standard (s. ref. [11, 12])

| | |
|---|---|
| Injection volume: | 20 µl, approx. 25 pmol/amino acid; |
| Analytical system: | LKB 2 pump gradient HPLC; |
| Column: | 250 x 4.0 mm, LiChrosphere CH8, 4 µm; |
| Temperature: | 35° C; |
| Eluent A: | 20 % ACN in 50 mM Na-acetate, pH 4.2; |
| Eluent B: | 70 % ACN in 50 mM Na-acetate, pH 4.2; |
| Flow rate: | 1.5 ml/min; |
| Detection: | Fluorescence, 260 nm excitation, 310 nm emission |

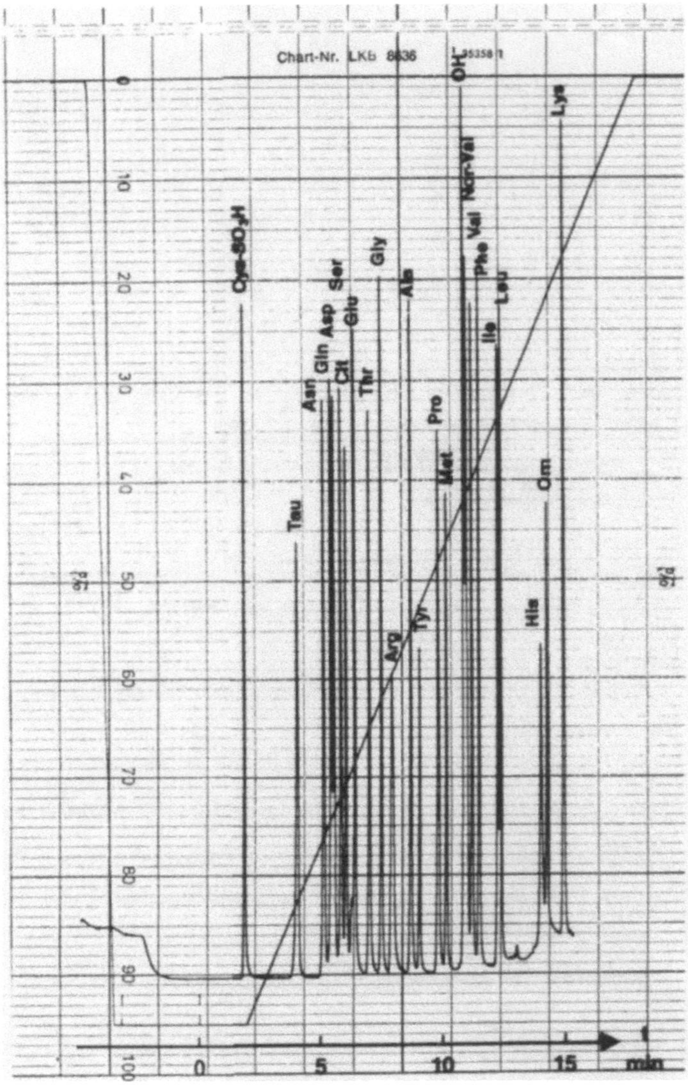


Fig. 4

Physiological standard (5 pmol/AA, Orn=2.5 pmol, His=15 pmol) (ref.[11, 12])

| | |
|---|---|
| Column: | LiChroCART Superspher CH 8, 4 µm, 125 mm x 4 mm, Merck, Darmstadt |
| Eluent A: | 20 % ACN in 50 mM Na-acetate Buffer, pH 4.2 |
| Eluent B: | 70 % ACN in 50 mM Na-acetate Buffer, pH 4.2 |
| Flow: | 1.75 ml/min |
| Detection: | Fluorescence $\lambda_{exc}$ 260 nm, $\lambda_{em}$ 310 nm |
| Temperature: | 35° C |

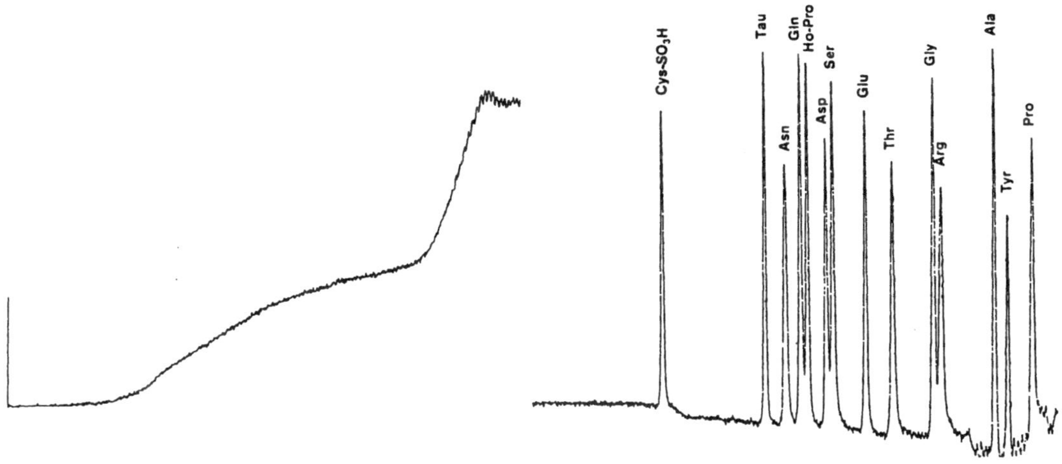


Fig. 5

Baseline subtraction applying an artificial mixture of amino acids - 300fmol each - precolumn derivatized with FMOC (ref.[11])

| | |
|---|---|
| Column: | LiChroCart Superspher CH8 - 4/um, 250 mm x 2 mm (Merck, Darmstadt) |
| Eluent A: | 20 % ACN in Na-acetate buffer, 50 mM; pH 4.2 |
| Eluent B: | 70 % ACN in Na-acetate buffer, 50 mM; pH 4.2 |
| Flow: | 350/ul/min |
| Pressure: | 240 bar |
| Temperature: | 22° C |
| Detection: | Fluorescence; $_{exc}$ 260 nm, $_{em}$ 310 nm |
| Injection volume: | 1/ul |

LINEARITY FMOC 1-AMINOADAMANTANE
FL FLUORESCENCE MONITOR
260/310 nm 0.1—10 pmol OF EACH AMINO ACID

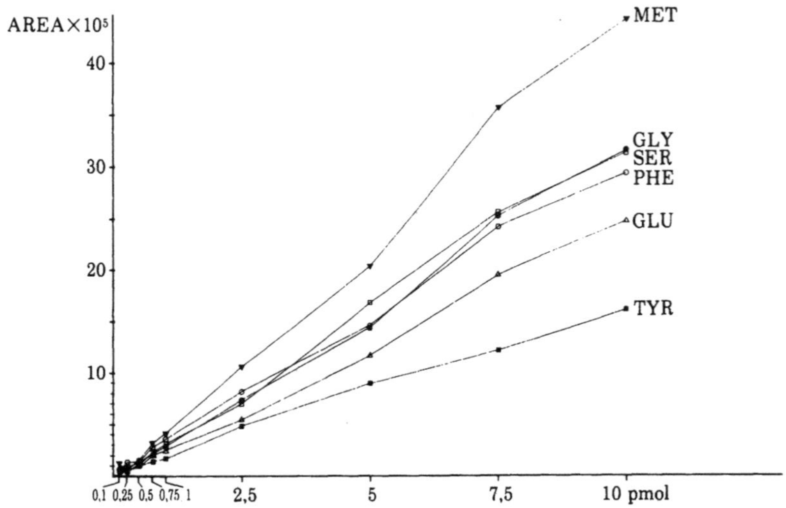


LINEARITY FMOC 1-AMINOADAMANTANE
UV MONITOR
UV 260 nm 25—500 pmol OF EACH AMINO ACID

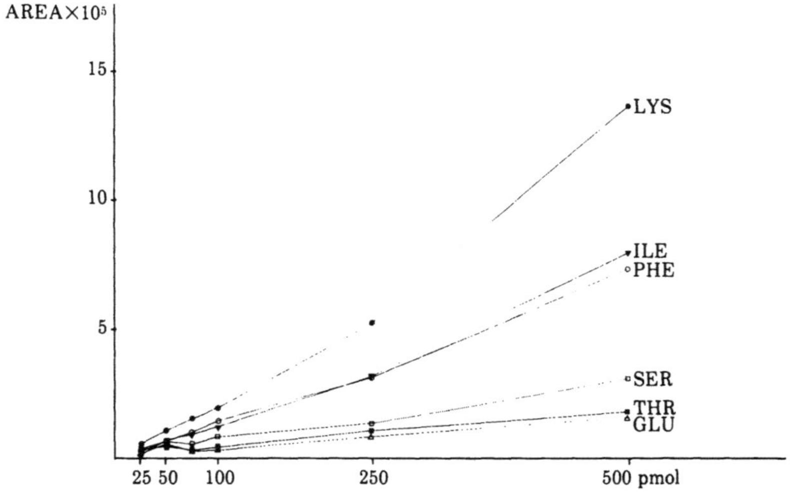


Fig. 6

Table I shows a list of some of the investigated hydrophobic amines, among which for several reasons like purity of the available compounds, stability, solubility in the borate buffer, etc., 1-aminoadamantane (ADAM) is the preferred material. Figure 7 shows a standard chromatogram of the FMOC derivatisized amino acids, for which the excess FMOC has been removed by derivatisation with an excess of 1-aminoadamantane (ADAM) in a second step.

This simple method can be managed quite easily in a manual mode; for instance to a given volume (5-10 $\mu$l) of amino acid samples in a carousel of an autoloader 10 $\mu$l borate buffer (0.5 M - pH 7.7) are added followed by 10 $\mu$l FMOC (1 mM) and, after 30-45 seconds, 15 $\mu$l ADAM (12 mM). In cases where this degree of automation is not sufficient, the addition of the two latter chemicals can be performed by autoloaders.

This method at the given concentrations is linear up to 25 pmol per amino acid injected on 4 mm ID columns. - For larger samples which have been outside of our interest the derivatisation has to be modified accordingly. - Further it is worth to be noted, that an excess of FMOC ($\geq$ 20 fold) for amino acids is needed in order to achieve nearly complete reaction. A too large FMOC excess, however, will disturb the chromatography because of the high FMOC-OH peak then formed. The ratio FMOC:ADAM should be approximately 1:20 during derivatisation to minimize that OH-peak. If a too high ADAM concentration is choosen, problems with its solubility may occur [2, English version].

Special precautions have to be taken for achieving column life time of ~ 1000 runs in routine. There are several "conditiones sine quae non" such as only HPLC grade solvents and suprapure chemicals have to be used. It is also advisable for high sensitivity ($\leq$ 1 pmol), i.e. high sophisticated applications, specially when Microbore Columns are chosen, to purify even the HPLC grade eluents by reversed phase chromatography, prior to use. In addition it is an absolute "must" to use precolumns, not only for filtering eluents and samples but also to avoid exbleeding of the stationary phase, which otherwise usually occurs after 150 to 200 runs only. Recommended stationary phases for the described application are Spherisorb C8 (LKB or Phase Separations), Mino RPC (Pharmacia) or Lichrospher 60-CH8 (Merck). After approximately each 100 runs, depending on the samples applied, it is recommended to rinse the column in 100 % ACN.

**Table I.** FMOC Derivatisation of amino acids. Comparison of the retention times from ten different amines.

| No. | Sample | FMOC removing chemical | Structure | R.T. FMOC | R.T. Lysine | R.T. Amine |
|---|---|---|---|---|---|---|
| 1 | STD IV | Pentane extraction | | 13.20 | 29.59 | |
| 2 | STD IV | 1-aminoadamantane | 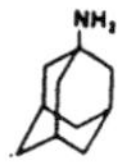 | 13.34 | 29.54 | 33.07 |
| 3 | Lysine | | | 12.79 | 29.05 | |
| 4 | Lysine | Pentane extraction | | 13.20 | 2948 | |
| 5 | Lysine | ε-aminocaproic acid | $H_2N-(CH_2)_5-COOH$ | 13.21 | 29.49 | 19.98 |
| 6 | Lysine | ω-aminocaprylic acid | $H_2N-(CH_2)_7-COOH$ | 13.12 | 29.47 | 25.63 |
| 7 | Lysine | DL-α-aminocaprylic acid | $H_3C-(CH_2)_5-CH(NH_2)-COOH$ | 13.10 | 29.47 | 25.91 |
| 8 | Lysine | 2-amino-3-picoline | 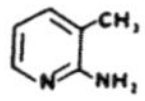 | 12.62 | 28.91 | 27.67 |
| 9 | Lysine | 2-amino-5-picoline | 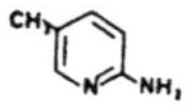 | 13.01 | 29.42 | –* |
| 10 | Lysine | 2-amino-6-picoline | 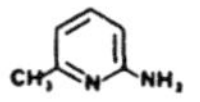 | 12.58 | 28.92 | 28.35 |
| 11 | Lysine | 1-aminoadamantane | 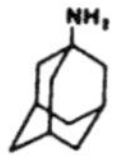 | 12.65 | 28.97 | 32.86 |
| 12 | Lysine | 2-aminooctane | $H_3C-(CH_2)_5-CH(NH_2)-CH_3$ | 13.13 | 29.47 | 33.19 |
| 13 | Lysine | Octylamine | $H_2N-(CH_2)_7-CH_3$ | 13.08 | 29.43 | 33.25 |
| 14 | Lysine | Dodecylamine | $H_2N-(CH_2)_{11}-CH_3$ | 12.66 | 29.96 | 34.76 |
| 15 | STD IV | 1-aminoadamantane | 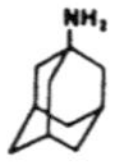 | 12.69 | 29.42 | 32.86 |

* No peak found in chromatogram

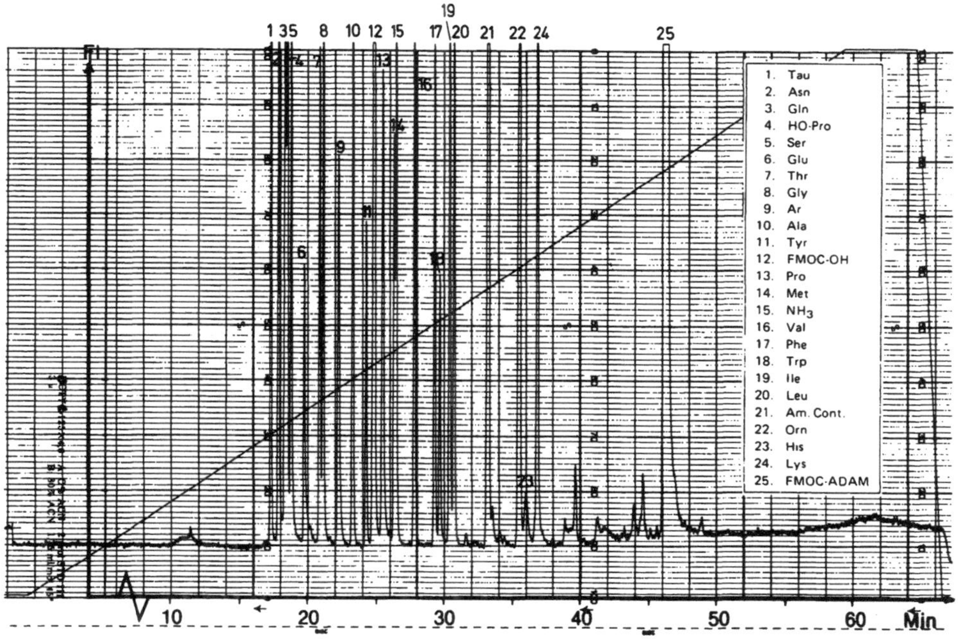


Fig. 7

Elution profile of an artificial mixture of amino acids precolumn derivatised with FMOC.

Excess of FMOC reacted with ADAM. 2 pmol of each amino acid injection.

Column: Octyl-Si-Spherisorb-3 µ, 125 x 4 mm, with guard column;

Buffer: see Fig. 4

Even as it was the intention to keep the elution system as simple as possible (only two components: acetate buffer and ACN). Sometimes depending on the complexity of the amino acid composition of the sample it may help up to add Tetramethylamine, Tetrahydrofuran or Methanol (1-5 %) as modifiers to Buffer A. For increasing the massexchange between the stationary phase and the FMOC amino acid complexes higher temperatures are often recommended. As this, however, can reduce the lifetime of a Column, 35 - 45° C are regarded as an acceptable compromise. But it is essential for reproducable retention times to keep the temperature of injectionport and column constant.

The performance of this simple and automatized method was tested (Table II and Fig. 8) with HCl-hydrolysates of previously sequenced proteins.

This new method appears to be superior to the earlier mentioned ones. However, the bottle neck for making use of the high sensitivity inherent in it is still the sample pretreatment, i.e. the HCl-hydrolysis or the deproteinisation, that is why in the future an interface (may be just a simple carousel or rack) could be developed to avoid contamination. This will be used to collect the samples with a fraction collector (for instance the peptides from a tryptical digestion of a protein), to hydrolyze these samples with HCl followed by the evaporation and to fit into the autoloader to do the precolumn derivatisation prior to the injection on the HPLC column.

**Table II.** Comparison of analysis data by IEX-chromatography and FMOC/ADAM procedure. Hydrolysate of RCM-Notechis II-1*

| Substance | Conventional Ninhydr. Meth. | | | Theory | HPLC — FMOC/ADAM | | | |
|---|---|---|---|---|---|---|---|---|
| | n moles per tube | corrected | 68.2 = 1 | | p moles+ per inj. | 1.8 = 1 | p moles++ per inj. | ×5.5 |
| CM-Cysteine | 951 | | 13.9 | 14 | | | 2.29 | 12.6 |
| Aspartic Acid | 1021 | | 15.0 | 15 | 34 | 18.3 | 3.06 | 16.8 |
| Methionine Sulfone | | | | | | | | |
| Threonine 0.96 | 589 | 614 | 9.00 | 9 | 15 | 8.2 | 1.72 | 9.4 |
| Serine 0.89 | 533 | 599 | 8.78 | 9 | 16.3 | 8.9 | 1.43 | 8.8 |
| Glutamic Acid | 538 | | 7.89 | 8 | 14.5 | 7.9 | 1.13 | 6.2 |
| Proline | 432 | | 6.33 | 6 | not analysed | | 1.29 | 7.09 |
| Glycine | 413 | | 6.06 | 6 | 11 | 6 | 0.99 | 5.4 |
| Alanine | 727 | | 11.0 | 11 | 20 | 10.9 | 2.04 | 11.2 |
| Half-cystine | | | | | | | | |
| Valine | 215 | | 3.15 | 3 | 5.5 | 3 | 0,52 | 2.86 |
| Methionine | 39.4 | | 0.58 | 1 | not analysed | | 0.11 | 0.6 |
| Isoleucine | 202 | | 2.96 | 3 | 5.5 | 3 | 0.47 | 2.6 |
| Leucine | 273 | | 4.00 | 4 | 19 | 10.3 | 1.27 | 6.9 (contam.) |
| Norleucine | 476 | | | | | | | |
| Tyrosine | 607 | | 8.90 | 9 | 15.5 | 8.5 | 1.69 | 9.2 |
| Phenylalanine | 204 | | 2.99 | 3 | 5.5 | 3 | 0.55 | 3.02 |
| Glucosamine | | | | | | | | |
| Galactosamine | | | | | | | | |
| Histidine | 136 | | 1.99 | 2 | not analysed | | 1.65 | 1.8 |
| Lysine | 610 | | 8.94 | 9 | 16.2 | 9.1 | (0.61) | ( 3.3 ) |
| Ammonia | 1426 | | | | | | | |
| Tryptophan 0 | | | 2 | | | | | |
| Arginine | 341 | | 5.00 | 5 | 9.5 | 5.1 | 0.66 | 3.63 |

* The sample was a gift of D. Eaker, Biomedical Center, Uppsala, Sweden. Further we thank D. Eaker for providing the data of conventional amino acid and sequence analysis.

+ Found on Spherisorb C8,3 μ, 4 × 125 mm Column

++ Found of Lichrospher 60—CH8,4 μ, 4 × 250 mm Column

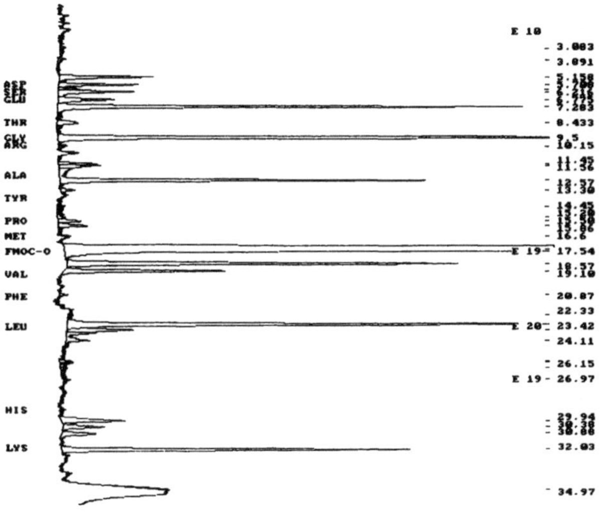


Analysis of tryptical Peptide from β-Tubulin (res. 3—19)*

| | Ninhydrin | Sequence | FMOC/ADAM |
|---|---|---|---|
| CM-Cys | 0.85 | 1 | (not anal.) |
| Asp | 1.08 | 1 | 1.07 |
| Thr | — | — | 0.19 |
| Ser | — | — | 0.34 |
| Glu | 3.92 | 4 | 4.2 |
| Pro | — | — | 0.08 |
| Gly | 2.97 | 3 | 3.12 |
| Ala | 1.91 | 2 | 2.07 |
| Half-Cys | — | — | — |
| Val | 1.27 | 1 | 0.87 |
| Met | — | — | — |
| Ile | 2.61 | 3 | 2.6 |
| Leu | — | — | 0.42 |
| Tyr | — | — | — |
| His | 1.2 | 1 | (not anal.) |
| Lys | 1 | 1 | 1.05 |

* This hydrolysed sample was a gift of H. Pontstingl. — DKFZ Heidelberg, Germany — who also provided the ninhydrin and sequence data. Chrom.gr.cond. s. Fig 4, except 250 x 4 mm Superspher Column (CH 8,1 Merck)

Fig. 8

Analysis of a hydrolysate of a peptide from ß-tubulin consisting of 17 amino acid residues (pos. 3-19 - 0.2-4.0 pmol injected per amino acid).

References

[1] H.J. Schneider, H.R. Hoepker, P. Földi, "HPLC-Aminosäuren-Analyse - Precolumn Derivatisierung", printed by LKB GmbH, Gräfelfing (1986). New English version in process.

[2] H. Engelhardt, "High Performance Liquid Chromatography" (Chem. Laboratory Practice), Springer, Berlin-Heidelberg-New York (1979).

[3] H.J. Schneider, P. Földi, GIT Fachz. Lab. 8, 783-786 (1986); Part II 9, 873-889 (1986).

[4] G.D. Condon, "Amino Acid Analysis" - Theory & Laboratory Techniques by LKB-Biochrom, Cambridge (1985).

[5] Th. Graser, H. Godel, P. Földi, P. Fürst, Anal. Biochem. 151, 142-152 (1985).

[6] H. Godel, Th. Graser, P. Földi, P. Fürst, J. Chromatogr. 297, 49-61 (1984).

[7] R.R. Granberg, L.C. Magazine 2, 776-781 (1984).

[8] S. Einarsson, B. Josefsson, S. Lagerkvist, J. Chromatogr. 282, 609-618 (1983).

[9] S. Einarsson, J. Chromatogr., 348, 213-220 (1985).

[10] H. Godel, Thesis, Inst. Biol. Chem. and Nutrition, Univ. of Hohenheim (FRG) (1986).

[11] H. Godel, Th. Graser, P. Földi, P. Fürst, presented at "International HPLC-Symposion" Kyoto (Japan), Jan. 1985.

[12] H. Godel, Th. Graser, P. Földi, P. Fürst, applications from "Aminosäuren-Mappe", printed by LKB GmbH, Gräfelfing.

We thank Friedr. Vieweg & Sohn Verlagsgesellschaft mbH for permission to reprint major parts of Chromatographia Vol. 22, No. 7-12, December 1986.

# MICROCALORIMETRY OF PROTEIN - LIGAND INTERACTION AND PROTEIN UNFOLDING

H.-J. Hinz

Institut für Biophysik und Physikalische Biochemie der Universität Regensburg, Universitätsstraße 31, D-8400 Regensburg, FRG.

**INTRODUCTION:**

Biological systems are subject to the laws of physics and chemistry. They obey the rules of mechanics and energy conversion just as much as a steam engine does. Therefore thermodynamics provides the appropriate means for characterizing them. Though this sentence appears to be a truism from the chemist's point of view, in biology the logical consequence i.e. precise determination of the thermodynamic parameters of biochemical reactions has not been fully taken. The rising interest in a thermodynamic description of biological macromolecules and biological phenomena in general could only be met after the development of highly sensitive, easy to work with microcalorimetric equipment that permits precise determination of reaction- and transition enthalpies and heat capacities.

A widespread misunderstanding of the inherent cognitive value of thermodynamic quantities in improving insight into the energetic basis of structure and function of biopolymers resulted from the somewhat unqualified feeling that for such systems as proteins a full quantum mechanics treatment would be required. Since that is, however, beyond present time capabilities, a characterization by classical statistical thermodynamics would be of little value.

This argument had been refuted indirectly in the various treatments of thermodynamics of small systems (1,2). Recently it has been particularly clearly shown incorrect in a lucid article by A. Cooper (3). His reasoning is summarized in Table 1.

Modern Methods in Protein Chemistry, Vol. 3

Table 1: Treatment by Classical Statistical Thermodynamics is Adequate for Proteins

| |
|---|
| $\Delta\varepsilon \cdot \Delta t \approx h$ |
| $\Delta\varepsilon = k_B.T$ |
| $\Rightarrow \Delta t \approx 2.10^{-14}$ s |
| assuming a typical velocity of |
| $v = 10^5$ cm $s^{-1}$ |
| one obtains a limiting dimension |
| $\Delta x \approx 0.2$ A |
| => classical treatment adequate for |
| $\Delta t > 10^{-13}$ s |
| $\Delta x > 1$ A |

From the quantum mechanical uncertainty relation of Heisenberg it is known that quantum effects become significant for processes involving energy changes, $\Delta\varepsilon$, over time intervals, $\Delta t$, for which $\Delta\varepsilon.\Delta t \approx h$. Typical thermal energies, $\Delta\varepsilon \approx kT$, give $\Delta t = 2.10^{-14}$ s which, for typical atomic velocities of order $10^5$ $cm.s^{-1}$ at room temperature corresponds to a limiting dimension of order 0.2 A. Thus if we restrict our attention to thermal processes slower than about $10^{-13}$ s in systems larger than 1 A the consequences of quantized energy changes may be studied by classical statistical thermodynamics.

Apart from their value for improving the understanding of fundamental properties of biopolymers, thermodynamic parameters have become of increasing significance in applied biochemistry and biotechnology. This rising interest has several causes, important ones are the following:

Thermodynamic quantities such as reaction enthalpies and their temperature coefficients are required for quantitative predictions of the course of reactions under a different set of environmental conditions.

Rational protein design could be envisaged to involve stability considerations already on the basis of the nucleic acid sequence of a protein. For this to become realistic, thermody-

namic parameters would be required that allowed estimates of the changes in overall stability of the macromolecule as a function of amino acid exchanges.
Finally, thermodynamic studies and in particular microcalorimetric measurements have an immense analytical potential for characterization of proteins under extreme conditions such as high concentration or very low water content.
This brief summary emphasizes the need for and the suitability of thermodynamic parameters to provide an adequate characterization of the energetics of protein systems.

**Thermodynamic Quantities Relevant to Biological Systems**

The thermodynamic quantities that are most appropriate for a characterization of ligand-binding equilibria and protein studies in general are the apparent standard Gibbs energy change, $\Delta G^{0\prime}$, the apparent standard enthalpy change, $\Delta H^{0\prime}$, the apparent standard entropy change, $\Delta S^{0\prime}$, and the apparent heat capacity change, $\Delta C_p{}'$. They have been summarized in Table 2.
The apparent standard Gibbs energy change can be obtained from relationship 1 if both $\Delta H^{0\prime}$, and $\Delta S^{0\prime}$ are known at temperature T, or, more frequently, from the apparent equilibrium constant, $K_{eq}{}'$, using relation 2, in which R is the gas constant (8.34 $J.mol.^{-1}.degree^{-1}$) and T the absolute temperature. Emphasis is laid upon the term "apparent" because usually activity coefficients are ignored in the formulation of the equilibrium constant, simply because they are unknown for the majority of reactions. In practice, however, many equilibrium constants are determined at low enough protein concentration, where the assumption of a value of unity for the activity coefficient will be justified. The enthalpy changes of ligand binding reactions can be either calculated from the temperature dependence of the equilibrium constant (so called van't Hoff enthalpies $\Delta H_{vH}$) according to the van't Hoff equation 3 or, preferably, by direct microcalorimetric measurements of

the heat effects involved in the binding process (calorimetric enthalpies $\Delta H_{cal}$).

Table 2: Thermodynamic Quantities

$$\Delta G^{o\prime} = \Delta H^{o\prime} - T.\Delta S^{o\prime} \quad (1)$$

$$\Delta G^{o\prime} = -R.T.\ln K_{eq}' \quad (2)$$

$$d \ln K_{eq}'/dT = \Delta H^{o}{}_{vH}'/RT^2 \quad (3)$$

$$\Delta C_p' = R\ (\partial/\partial T).[-\partial \ln K_{eq}'/\partial(1/T)] \quad (4)$$

$$\Delta C_p' = C_{p,f}' - C_{p,b}' = d\Delta H^{o\prime}/dT \quad (5)$$

$$\Delta S^{o\prime} = (\Delta H^{o\prime} - \Delta G^{o\prime})/T \quad (6)$$

$$S^{o}(T) = \int_0^{T_1} C_p{}^{o}\ dT/T + \frac{\Delta H(T_1)}{T_1} + \int_{T_1}^{T_2} C_p{}^{o}\ dT/T + \frac{\Delta H(T_2)}{T_2} + \ldots \quad (7)$$

$$\Delta G_d = \Delta H_d(1 - T/T_m) + \Delta C_p.T.\ln(T_m/T) - \Delta C_p(T_m - T) \quad (8)$$

These two quantities need not be identical for complex equilibria, since calorimetric enthalpies are the algebraic sum of all heat effects occuring in the system, while the numerical value of the overall equilibrium constant and therefore its variation with temperature is dependent on the way in which the reaction is formulated and possibly also on the experimental technique used to follow the process (ultraviolet absorption, fluorescence, circular dichroism, etc.)
Such a different answer does usually not occur when simple equilibria are involved; however, for cooperative or multiligand binding reactions these differences may become apparent and can be of discriminative value.
The standard entropy of a pure substance can be calculated from relation 7, if the variation with temperature of the heat capacity $C_p{}^{o}$ and the enthalpy changes $\Delta H(T_i)$ at temperatures

$T_i$ are known. Algebraic summation of $S^0$ values for all reactants would then yield the standard entropy change, $\Delta S^0$.
For complex biochemical reactions such an evaluation is, however, not possible, and the usual procedure is to determine $\Delta G^{0\prime}$ and $\Delta H^{0\prime}$ at temperature T and to calculate $\Delta S^{0\prime}$ from equation 6.
The fourth thermodynamic quantity that is particularly characteristic for protein-ligand interaction is the variation with temperature of the reaction enthalpy given by eq. 5. The apparent heat capacity change $\Delta C_p{}^0$ can in principle also be estimated from the degree of curvature of the van't Hoff plot, eq. 4, however, the precision required for the values of the equilibrium constant to permit a meaningful second derivative with respect to temperature is rarely attained. Therefore direct microcalorimetric measurements of the apparent reaction enthalpy $\Delta H^{0\prime}$ at different temperatures are the most reliable source of heat capacity data.

## CALORIMETRIC TECHNIQUES

### Isothermal Microcalorimetry

The majority of measurements of thermodynamic parameters has been performed using isothermal microcalorimeters either of batch or flow type (4).
The measuring principle is the determination of a heat effect $\Delta Q$ that is proportional to the number of moles of the reactant that has been converted to product

$$\Delta Q = \Delta H^0 \cdot \Delta n \qquad (9)$$

Obviously in the simplest case involving a very large equilibrium constant, such as that of repressors binding to operator DNA, practically all the reactant is converted to product. This allows direct determination of the molar $\Delta H^0$ by the ratio of the measured heat effect and the number of moles of reac-

tant $\Delta n$ given to the mixture.
Frequently, however, equilibrium constants are in the order of $10^{-2}$ to $10^{-5}$ M leaving significant amounts of the substrates unreacted in the equilibrium mixture. In such cases the equilibrium constant must be known for calculation of the actual number of moles of reactant to be inserted in eq. 9.
Calorimetry offers a very convenient procedure to determine both, the equilibrium constant K and the reaction enthalpy $\Delta H^0$ simultaneously by performing a thermal titration. Assuming i titrations steps, the criterion for calculation of the correct equilibrium constant is that $\Delta H^0 = \Delta Q_i/\Delta n_i$ must hold for each titration step.
For a protein-ligand binding reaction of type

$$[E] + [L] = [EL] \tag{10}$$

the association constant is given by

$$K_a = [EL] / [L] [E] \tag{11}$$

The free ligand and enzyme concentration can be obtained from the conservation equations

$$[E] = [E_0] - [EL] \tag{12}$$

$$[L] = [L_0] - [EL] \tag{13}$$

and the number of moles reacted, $\Delta n_i$, is given by $\Delta n_i = [EL].V_i$, $V_i$ being the volume in the calorimetric reaction cell after the $i^{th}$ titration step. Introducing these expressions into the mass law, we obtain for each titration step i a quadratic equation

$$K_a[EL]_i^2 - \{([L_0]_i + [E_0]_i).K_a + 1\}.[EL]_i + K_a[E_0]_i.[L_0]_i = 0 \tag{14}$$

that can be solved for $[EL]_i$ and $K_a$ using conventional numerical methods. The procedure can of course be extended to more complex reaction schemes. The greater number of adjustable parameters requires then a greater number of titration steps to obtain acceptable precision. Preferentially in such a situation different techniques will be employed in addition to calorimetry to collect independent information on the equilibria.

The potential of microcalorimetry as an analytical tool that provides quantitative thermodynamic parameters can only fully be exploited for biochemistry, if the sensitivity of the method is high enough as to allow measurements on small samples. The next table shows an order of magnitude calculation that illustrates the possibilities of present time isothermal microcalorimetry.

Table 3: Order of Magnitude Calculation of Sensitivity of Isothermal Microcalorimetry

$$M = 100\ 000\ g/mol$$
$$\Delta H_0 = 100\ 000\ J/mol$$
$$\Delta Q = \Delta H_0\ .\ \Delta n_p$$
$$10.10^{-9}\ moles = 1\ mg$$
$$\rightarrow \Delta Q = 10^8\ mJ/mol\ .\ 10^{-8}\ mol$$
$$\Delta Q = 1\ mJ$$

$$\text{Reaction time} = 3\ min = 180\ s$$
$$\Delta Q = 1000\ \mu W\ .\ s$$

Recorded Signal:

$$\Delta P = \frac{\Delta Q}{\Delta t} = \frac{1000}{180}\ \mu W \approx 5.6\ \mu W$$

Sensitivity of modern instrumentation:

$0.5 \mu W$

Assuming a molecular weight of 100 kD for the protein and a molar reaction enthalpy of 100 kJ a sample of 10 nano moles will produce a heat effect of 1 mJ. Provided the reaction time extends over 3 min, the average power signal during this time would be 5.6 $\mu W$. This signal is approximately ten times larger than the sensitivity of modern instrumentation. In reality the reaction signal will of course not assume a rectangular shape but that of a peak having the identical area. An example of a high sensitivity microcalorimetric curve is shown in Fig. 1.

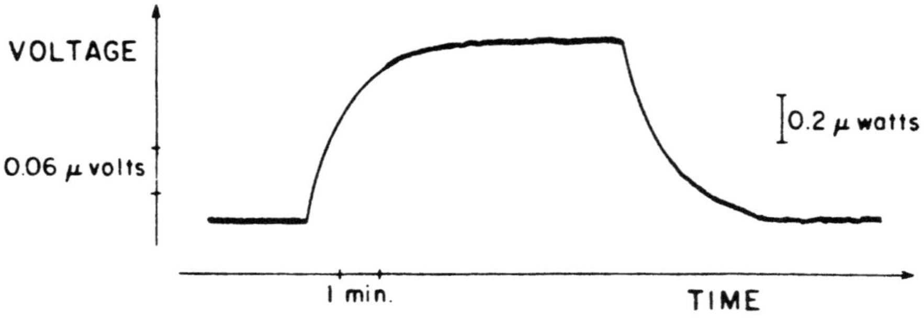


Figure 1: Time response of calorimetric signal for an applied power of 0.763μW (After (5)).

**SCANNING MICROCALORIMETRY**

The second approach which has become rather popular recently is scanning microcalorimetry. Various reviews covering technical and analytical aspects of the method have been published (6 and the references therein).

In this survey we confine the description of the method to some basic details that facilitate understanding of application of scanning microcalorimetry to biochemical problems.

The quantity primarily determined is the variation with temperature of the apparent heat capacity of the sample e.g., the protein. Table 4 summarizes some quantities relevant to application of scanning calorimetry.

To get an idea of the shape of the reaction cell of a modern solution scanning microcalorimeter, Figure 2 shows the helical reaction vessels of the DASM-4 microcalorimeter constructed by P.L. Privalov.

Table 4: Quantities relevant to scanning microcalorimetry of biochemical systems

| | |
|---|---|
| $$\frac{dH}{dt} = m_P \cdot c_{p,P}(T) \cdot r \qquad (15)$$ | |
| $$c_{p,P}(T) = c_{p,S} \cdot \frac{V_P(T)}{V_S(T)} - \frac{\Delta(T)}{m_P} \qquad (16)$$ | |
| dH/dT [J/deg] | = recorded power signal |
| $m_P$ [g] | = mass of protein in reaction cell |
| $c_{p,P}$ [J/(g.deg)] | = partial specific heat capacity of sample P |
| $c_{p,S}$ [J/(g.deg)] | = partial specific heat capacity of equilibrium buffer S |
| $V_P$, $V_S$ [$cm^3$/g] | = partial specific volumes of sample P and Buffer S, respectively |
| $\Delta(T)$ [J/deg] | = signal difference between sample and reference curve (Fig. 3) |

The magnitude of the experimental power signal [dH/dt] resulting from reactions such as conformational changes of the protein, ligand- or subunit dissociation, and protein association depends, according to eq. 15, on the mass, $m_P$, of the sample in the reaction cell, on the heating rate, r, [deg/s], and on the apparent specific heat capacity, $c_{p,P}$ [J/(g.deg)]. The variation with temperature of $c_{p,P}$ of a small globular protein exhibits in general three phases, as illustrated in Fig. 3 by a schematic transition curve.

Figure 2: Cell arrangement of the DASM-4 twin adiabatic scanning microcalorimeter constructed by P.L. Privalov and collaborators.

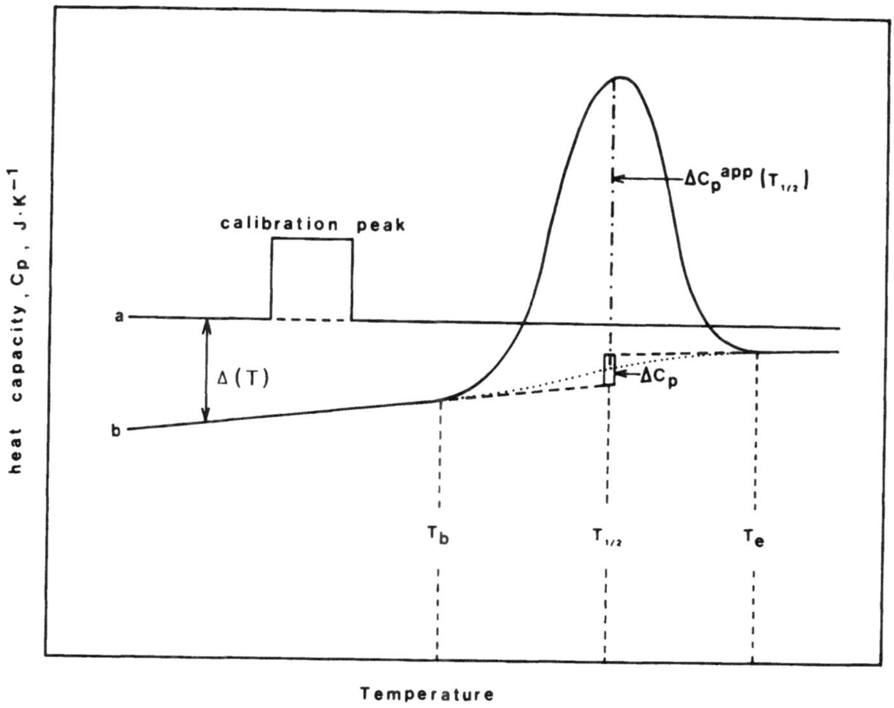


Figure 3: Schematic diagrams of a) base line with calibration peak and b) heat capacity vs. temperature curve of a typical protein unfolding reaction. (After (6)).

Prior to the so called transition region which is defined by two temperatures $T_b$ and $T_e$, the variation with temperature of the apparent heat capacity is determined by the small temperature coefficient of the specific heat capacity of proteins (approximately $4.10^{-3}$ $J.g^{-1}.K^{-2}$), since all other effects are compensated for by a physically identical reference cell, containing all components except for the protein.
The value of $c_{p,P}(T)$ can be obtained from eq. 16. Derivation of eq. 16 assumes constant volume of the calorimetric cells. Replacement of buffer by protein results in a decrease of heat capacity due to the lower heat capacity of protein (approximately 1.34 $J.g^{-1}.deg^{-1}$) as compared to that of buffer (approximately 4.18 $J.g^{-1}.deg^{-1}$ for the diluted buffers employed in biochemical research).

Apparent heat capacity in the transition region is determined by the enthalpy changes associated with the temperature induced conformational changes or shifts in the ligand binding equilibria of the protein. Thus the magnitude of the recorded power signal will depend on an intrinsic factor, the characteristic reaction enthalpy of the sample, which is a property of the system, and two extrinsic factors, which can be varied by the experimenter: the mass, $m_p$, in the calorimetric cell and the heating rate of the scan. For adiabatic instruments having reaction cells of 0.5 to 1 ml solution volume, the heating rate is usually limited to 1 or 2 deg/min, to avoid temperature gradients in the cells. Instruments constructed on the basis of the heat flow principle have often much smaller reaction vessels; they have been successfully employed using higher heating rates (7) to discriminate kinetically between various stages of a reaction pathway.
With present day high sensitivity instrumentation quantitative thermodynamic transition parameters of proteins can be obtained employing concentrations as low as 0.5 mg/ml. The great advantage of the calorimeteric technique that it shares with only a few other methods is its non-invasiveness and the fact that samples need not be optically transparent or even dissolved. There is practically no upper concentration limit to the sample solutions, thus suspensions and dry systems can be equally well studied as normal solutions.

To illustrate the potential of the method three examples will be considered.

**A**: Isothermal microcalorimetric studies on the bienzyme complex tryptophan synthase to exemplify the significance of linkage phenomena between ligand binding, intersubunit forces and conformational changes.

**B**: Shape analysis of transition curves of proteins to elucidate the path of unfolding and to identify domain structure.

C: Scanning microcalorimetric studies on BPTI that provide energy parameters for the contributions of modified amino acids to the overall stability of proteins.

## A: Linkage Phenomena of Proteins Illustrated by Studies on Tryptophan Synthase

Tryptophan synthase is a tetrameric bienzyme complex that comprises all features of the biochemically interesting class of oligomeric multifunctional enzyme complexes.
Some of its characteristics are summarized in Table 5.

Table 5: Some Properties of Tryptophan Synthase

| | |
|---|---|
| Holoenzyme: | $\alpha_2\beta_2PLP_2$ |
| Apoenzyme: | $\alpha_2\beta_2$ |
| $M_\alpha = 29\ 000$; $M_\beta = 89\ 000$ | |
| Coenzym: | Pyridoxal-5'-phosphate (PLP) binds to the β subunit |
| Catalysed Reactions: | |
| Overall Reaction: | Indoleglycerol phosphate (IGP) + L-serine → L-tryptophan + D-glyceraldehyde 3-phosphate (D-GAP) + $H_2O$ |
| α-Reaction: | IGP = indole + D-GAP |
| β-Reaction: | Indole + L-serine → L-tryptophan + $H_2O$ |

It has been reasoned that one of the advantages of a multifunctional enzyme complex as compared to separated subunits resides in shorter diffusion times of substrates between different reaction centers, thereby improving the kinetics of the reaction. The better catalytic properties of the $\alpha_2\beta_2$ complex relative to those of the $\beta_2$ subunit can be nicely demonstrated by comparing the time course of the heat output associated with the synthesis of tryptophan from indole and L-serine (Fig. 4). When using the $\alpha_2\beta_2PLP_2$ holoenzyme as catalyst, essentially all material has reacted after about 12 s; for the

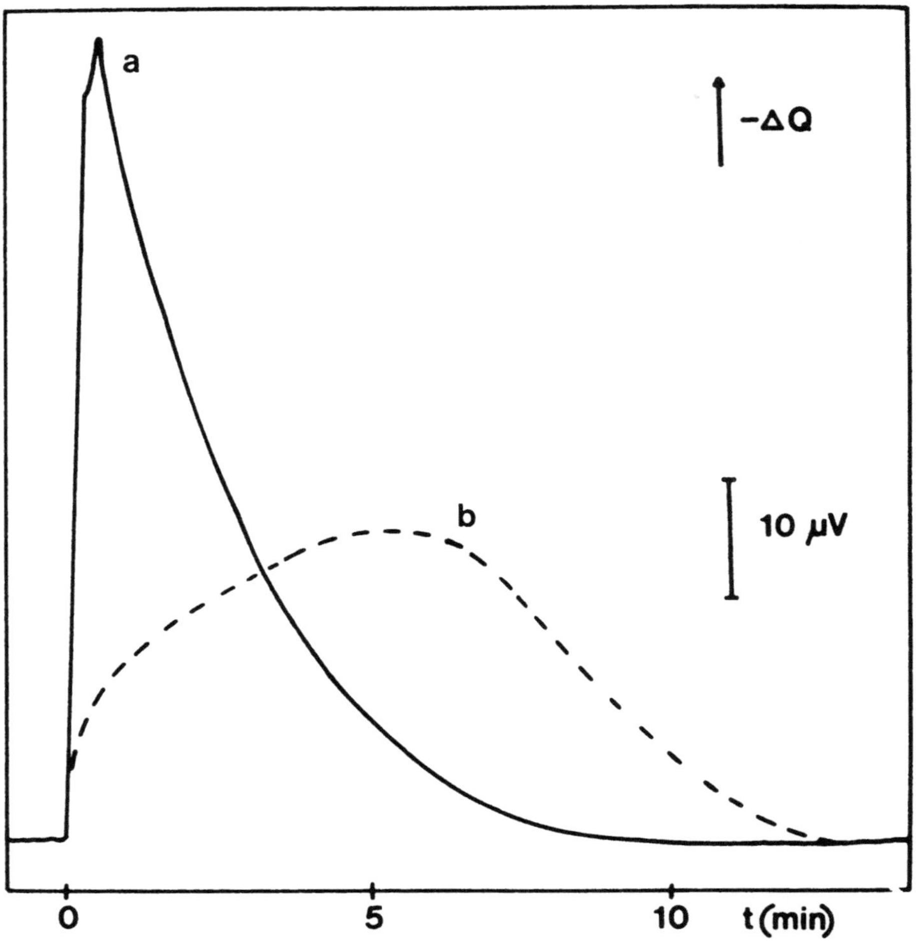


Figure 4: Time course of the β-reaction (Table 5) of tryptophan synthase catalysed by a) $\alpha_2\beta_2PLP_2$ and b) $\beta_2PLP_2$ (After (8)).

reaction to take place in the presence of $\beta_2PLP_2$, it takes approximately 6 min. The ratio of these values reflects the approximately 30 times higher specific activity of the $\alpha_2\beta_2$

tetramer (8) relative to the $\beta_2$ dimer. In addition to the kinetic information, the microcalorimetric studies provide also the thermodynamic reaction parameters of the condensation reaction, which would be difficult to obtain otherwise (8).

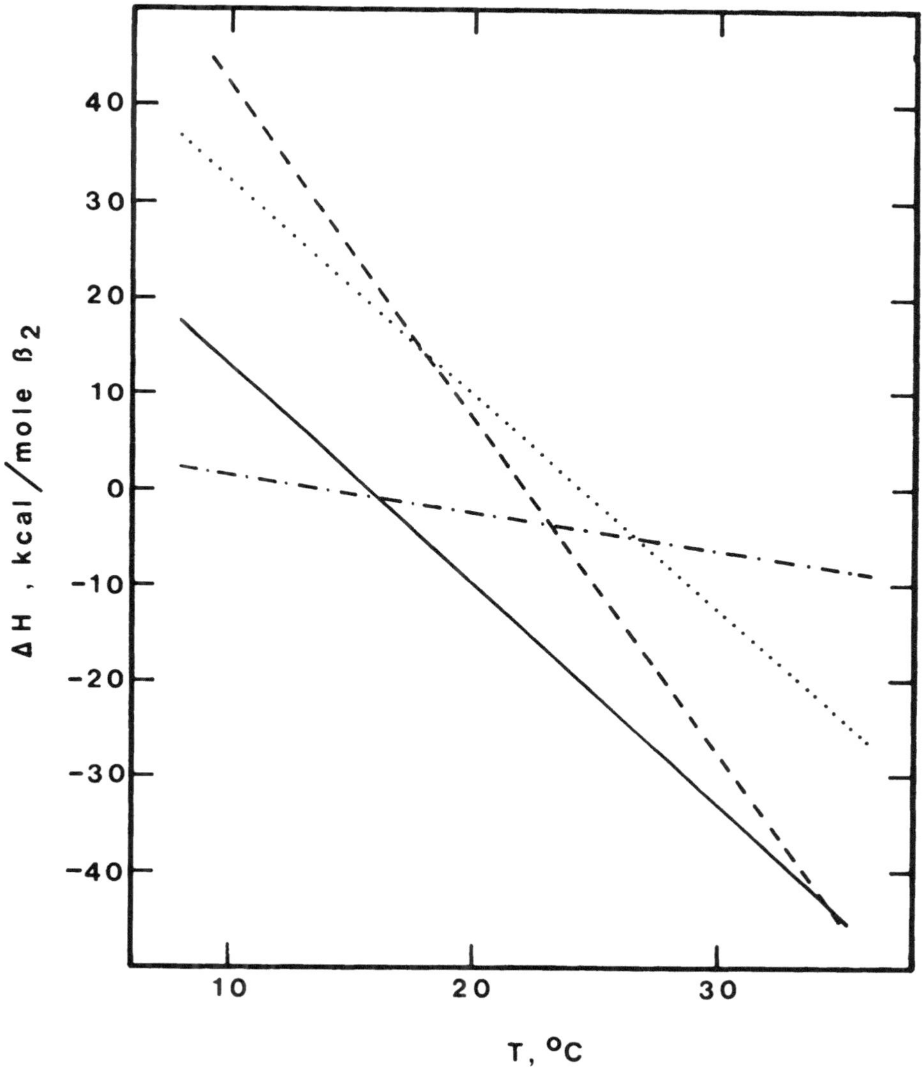


Figure 5: Apparent heat capacity changes associated with reactions of tryptophan synthase. The heat capacity changes are the slopes of the ΔH vs. T curves.
------- $2\alpha+\beta_2$ ——— $2\alpha+\beta_2PLP_2$
....... $\beta_2+2PLP$ -.-.-.- $\alpha_2\beta_2+2PLP$

Fig. 5 shows the variation with temperature of the association enthalpies of four binding reactions (9,10). The solid and dashed lines represent the temperature dependence of the reaction enthalpy of apo- and holoenzyme formation, while the dotted and dash-dotted lines show the temperature dependence of PLP binding to the $\beta_2$ and $\alpha_2\beta_2$ complex, respectively. The feature that is particulary pronounced in these reactions, but characteristic of all protein-ligand interactions studied so far, is the strong dependence on temperature of the binding enthalpy.

To put it in proper perspective you have to compare this phenomenon to reactions between small ligands in aqueous solution or in the gas phase. There, typically, both $\Delta H$ and $\Delta S$ do not vary by more than 1 to 3 % over a temperature range of 100 degrees. The change in $\Delta H$ for enzyme formation from tryptophan synthase subunits is larger than 100 % within 10 degrees, i.e. the change is thousand times larger than that in reactions between small ligands. Variations of this order of magnitude are not unusual for protein-protein interactions. They indicate that major energy redistributions occur on complex formation which may, and often will be, associated with detectable structural changes.

It is worth noting that for the vast majority of ligand-binding reactions $\Delta C_p$ has been found to be negative. This finding has been interpreted as being indicative of processes such as reversal of hydrohobic hydration or partial quenching of soft vibrational modes occuring on complex formation. Both changes would very likely involve "tightening" of the macromolecular structure. Support for this molecular interpretation of a macroscopic thermodynamic parameter has been provided by the determination of a positive $\Delta C_p$ for indole binding to the $\alpha$-subunit of tryptophan synthase (10). If we accept that a negative $\Delta C_p$ is suggestive of a tightening of the protein structure, a positve $\Delta C_p$ should indicate "loosening" of the structure. Indeed biochemical evidence showing better accessi-

bility to cysteines in the α-subunit (11) and higher sensitivity against thermal inactivation (12) is in good agreement with the interpretation of the heat capacity change.
Thermodynamic studies are very useful for discriminating between different models of ligand binding. An example for such an application is given in Fig. 6. It has been taken from reference (10). According to the reaction mechanism of tryptophan synthase indole must interact with both the α and the $\beta_2$ subunits. There have been different assumptions about the number of binding sites of indole to the α subunit.

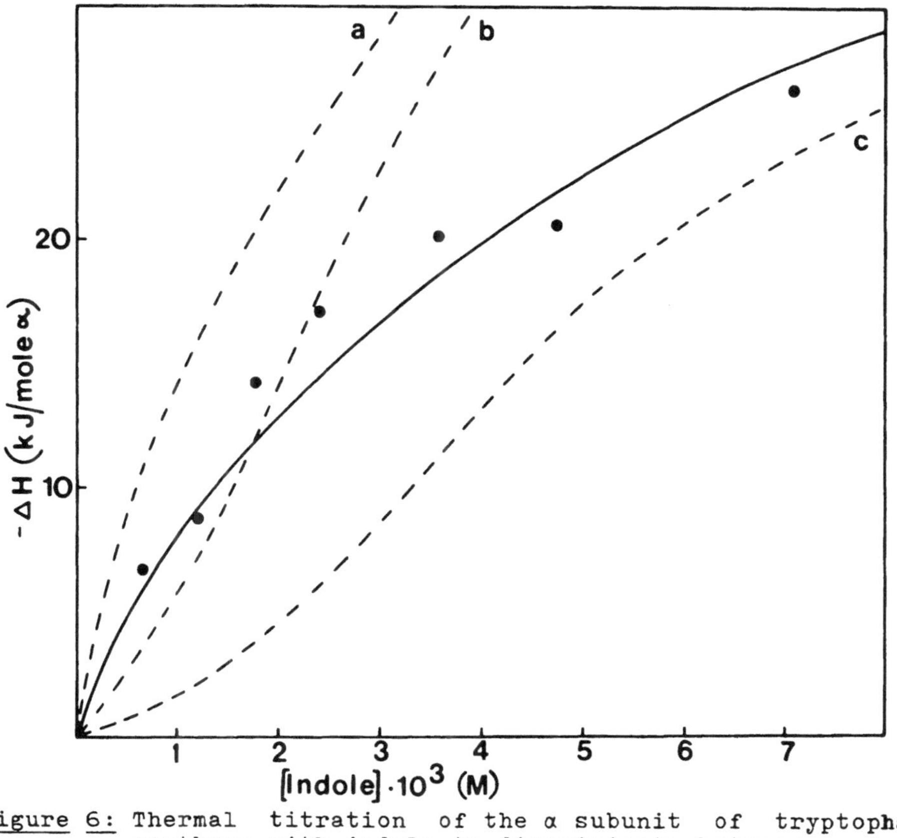


Figure 6: Thermal titration of the α subunit of tryptophan synthase with indole to discriminate between various binding models. (After (10)).

As demonstrated in Fig. 6 the thermal titration can clearly

differentiate between the various models. Only the assumption of a single binding site having a dissociation constant of 5 mM and an association enthalpy of -46 kJ/mol α is consistent with the experimental results. Other parameters cannot be reconciliated with the titration data.
These few examples give an impression of the great potential of microcalorimetric studies in providing insight into the physical basis of protein ligand interactions. Practically every information that can be obtained by other techniques such a fluorescence, CD, UV can also be derived from calorimetric studies. However, one obtains in addition the enthalpy and heat capacity changes of the equilibria, that cannot be determined otherwise with such precision.

**B: Shape Analysis of Transition Curves of Proteins**

The typical transition curve of a small globular protein as shown in Fig. 3 permits not only the calculation of the true overall enthalpy by integration of the area under the curve according to the formula

$$\Delta H = \int_{T_1}^{T_2} C_p \, dT \qquad (17)$$

but also determination of the equilibrium constant of unfolding as a function of temperature.
Application of the van't Hoff relation results in a value for the van't Hoff enthalpy. Thus both enthalpy values, the model free calorimetric enthalpy, $\Delta H_{cal}$, and the van't Hoff enthalpy, $\Delta H_{vH}$, that relies on the assumption of an all or none transition model, can be derived from the same measurement. This is a unique property of scanning calorimetric measurements.
The equality of the van't Hoff and the calorimetric enthalpy is an indication of the two state nature of a transition. Deviations from the ratio of 1 occur to both sides, i.e.,

van't Hoff enthalpies can be smaller or larger than the calorimetric ΔH values indicating independent domains or intermolecular cooperation, respectively.

A very obvious manifestation of domain structure of a protein is shown in Fig. 7 (13,14). λ-repressor exhibits on unfolding two clearly separated heat capacity peaks that unequivocally

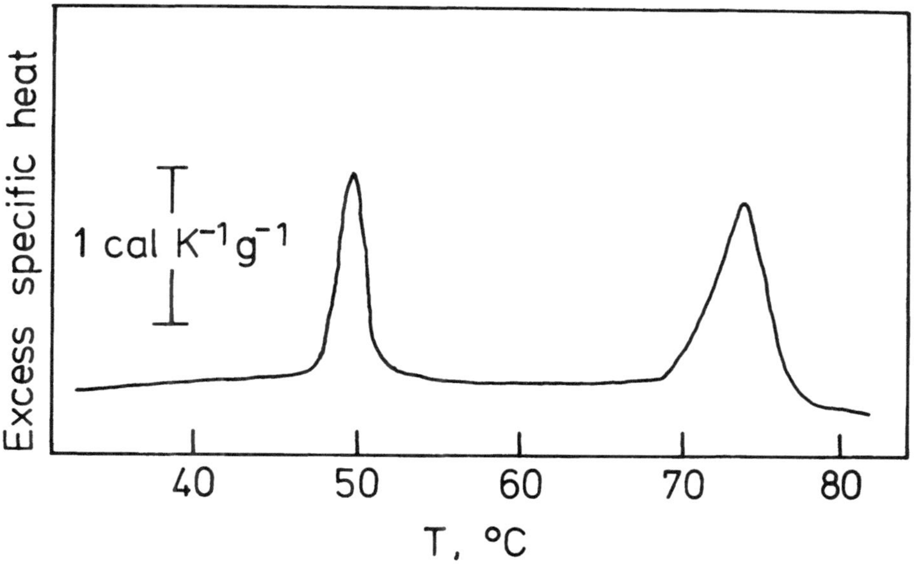


Figure 7: Manifestation of independent domains of λ-repressor in scanning microcalorimetry. (After (13)).

proove the existence of at least two domains of different stability. This interpretation can be substantiated by selective papain cleavage of the interdomain amino acid sequence and subsequent unfolding studies of the single λ-repressor fragments (13). The low temperature peak results from unfolding of the amino-terminal fragment whereas the peak near 70°C is due to denaturation of the carboxyl-terminal fragment. λ-repressor unfolding furnishes a very favourable example of a protein with two independent domains differing vastly in stability.

However, in general the analysis is not that straightforward. In these cases the ratio of $\Delta H_{vH}/\Delta H_{cal}$ is a reliable indication of the occurrence of thermodynamically relevant interme-

diates in unfolding.
Ratios of $\Delta H_{vH}/\Delta H_{cal}$ smaller than 1 will be encountered, if independent domains of similar stability occur. Unfolding of these domains will proceed in the same temperature range and usually on inspection the corresponding transition curve cannot easily be differentiated from the heat capacity curve of a fully cooperative system. However, evaluation of the two enthalpies by deconvolution algorithms will reveal the existence of these domains.
A sequential reaction scheme of the type

$$I_0 \overset{\Delta h_1}{\longleftrightarrow} I_1 \overset{\Delta h_2}{\longleftrightarrow} I_2 \longleftrightarrow \ldots\ldots I_{n-1} \overset{\Delta h_n}{\longleftrightarrow} I_n$$

can be analysed according to an algorithm developed by Freire and Biltonen (15,16).
The deconvolution can be quantitatively performed in an unique manner. The pertinent observation is that the experimentally observed enthalpy $\langle\Delta H\rangle$ at temperature T

$$\langle\Delta H\rangle = \int_{T_b}^{T} [C_p(T) - C_p(\text{baseline})]\, dT \qquad (18)$$

allows one to calculate the partition function Q(T) according to the relation

$$\ln Q = \int_{T_b}^{T} \frac{\langle\Delta H\rangle}{RT^2}\, dT \qquad (19)$$

$T_b$ is the smallest finite temperature at which a physically detectable deviation of excess heat capacity from baseline occurs. On the basis of the partition function Q molecular averages and distributions can be calculated.
Various examples of the deconvolution procedure are summarized in reference (17). A rather complex one is shown in Fig. 8. The deconvolution refers to the excess heat capacity function of a) carboxymethylated α-tropomyosin b) α-tropomyosin con-

taining a disulfide bridge. In panel b) only the first three stages of melting are shown.

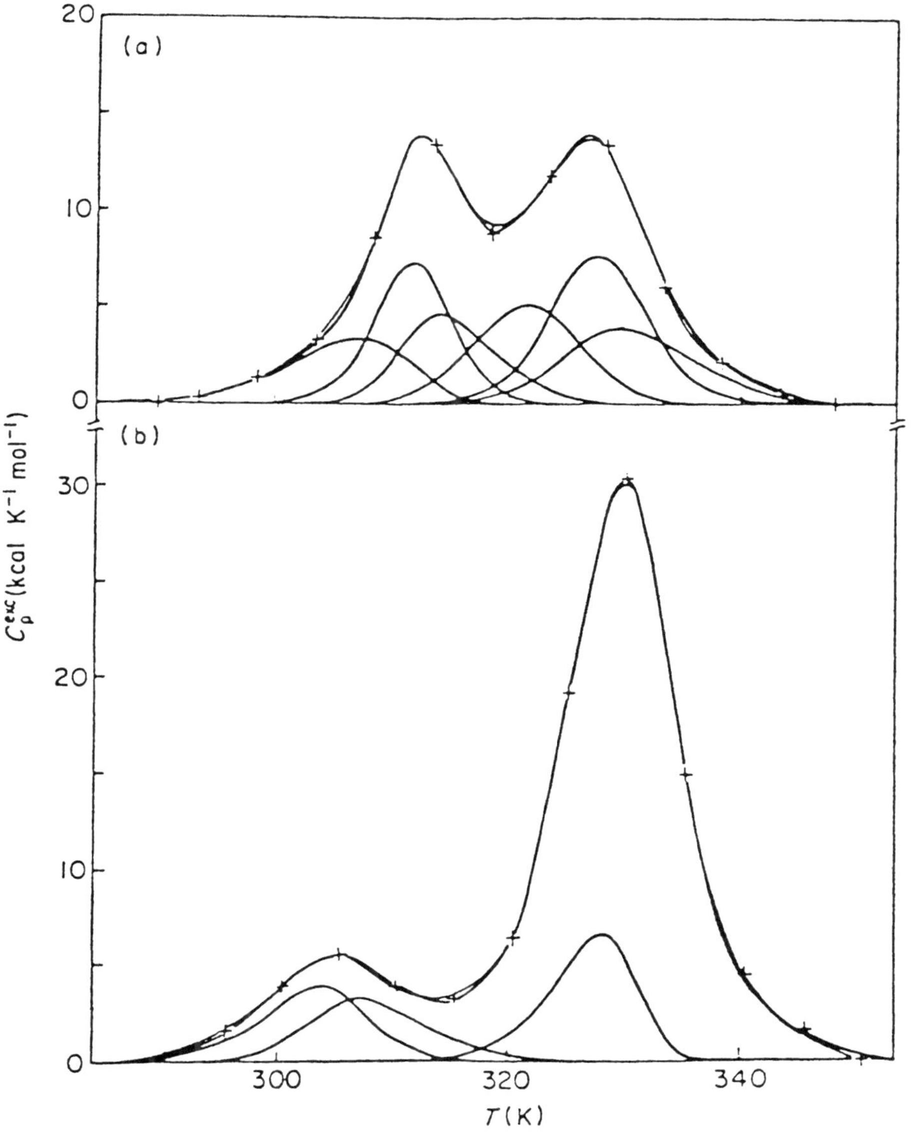


Figure 8: Example of a deconvolution procedure of a scanning microcalorimetric curve. (After (20)).
a) carboxymethylated α-tropomyosin
b) α-tropomyosin containing a disulfide bridge

Recently a more general treatment has been published which includes besides temperature also pressure and ligand concentration as variables (18), as well as an extension of the Freire-Biltonen treatment (19).
All of them, however, profit from the unique property of calorimetric transition curves to provide information of the partition function of the protein.

**C: Studies on BPTI:**
**Sequence Modification and Protein Stability**

The last topic that can be treated in an appropriate manner by microcalorimetry is the interdependence between sequence modification and stability of proteins.
This linkage becomes particularly important in view of the immense possibilities of genetic engineering in providing DNA sequences having well defined base pair exchanges. The progress in these techniques is contrasted by the theoretical inability of predicting quantitatively the influence of such well defined amino acid exchanges on the structure and function of a protein. One of the reasons for that unbalanced progress resides in the absence of appropriate energy parameters which could be used in the stability calculations.
Recently we started studies on Bovine Pancreatic Trypsin Inhibitor (BPTI) in collaboration with Prof. Tschesche and Dr. Wenzel that might help to improve the situation.
BPTI is a small very well characterized protein of 58 amino acids which is stabilized by 3 disulfide bonds (21). The protein has an enormous specific stability in comparison to other proteins as the selection of specific Gibbs free energy values shows (Table 6).

Table 6: Stability of Proteins (Specific Gibbs Free Energy Values, Δg, of Proteins have been taken from (22)

| Protein | Δg [J/g] |
|---|---|
| Ribonuclease | -1.6 |
| Lysozyme | -1.7 |
| α-Chymotrypsin | -1.9 |
| Trypsin | -2.8 |
| α-Subunit of Tryptophan Synthase | -0.8 |
| BPTI | -8.0 |

The question is of course: "What is the physical origin of that stability?"

The immediate answer will be that in all probability BPTI has a larger number of hydrophobic contacts and/or a larger number of hydrogen bonds.

Well, both these suggestions do not hold true as has already been shown by Privalov (23). Since the amino acid composition is not unusual either relative to that of other proteins, one is lead to suspect the 3 disulfide bonds to be the source of the high stability of BPTI.

Fig. 9 summarizes the modifications which can be introduced into BPTI employing chemical and enzymatic methods.

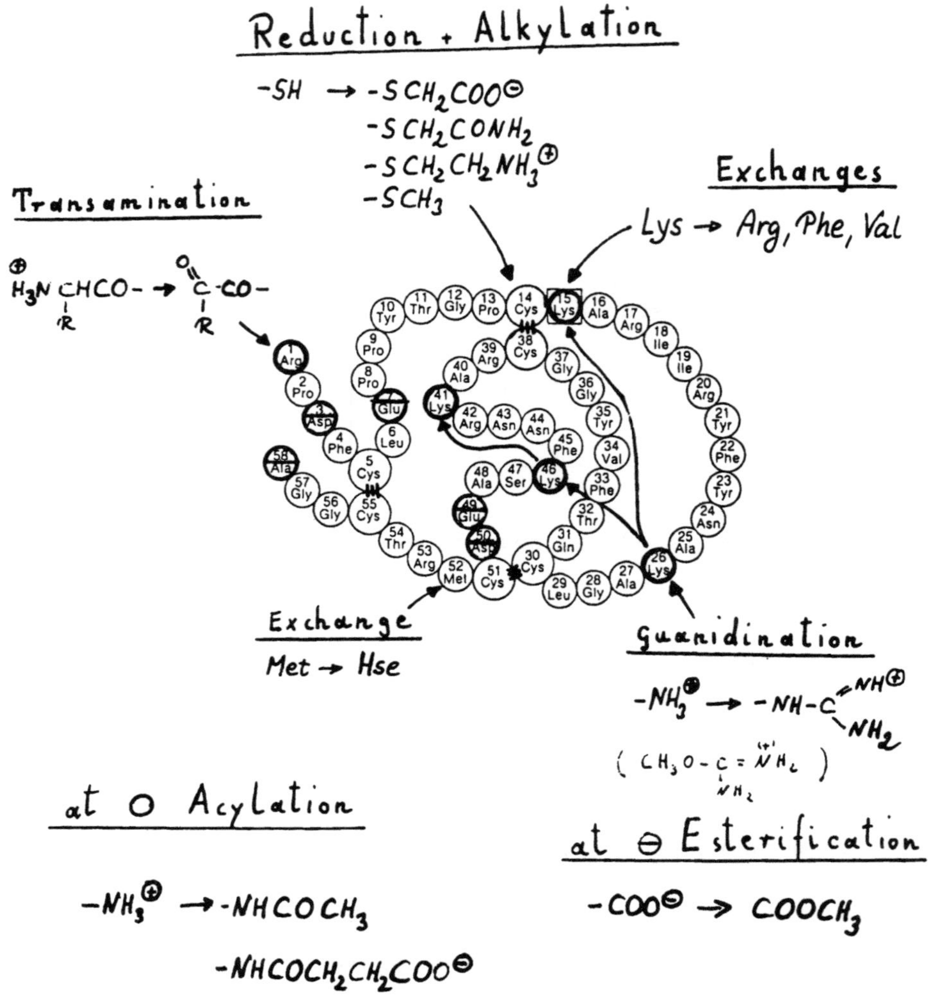


Figure 9: BPTI Analogs

These modifications permit us to investigate a wide spectrum of changes in molecular interactions including variations in hydrophobicity, charge, and degree of disulfide bonding. Particularly, changes in the disulfide bonds offer a large variety of well defined alterations of the native macromolecule as Fig. 10 summarizes.

According to the thermodynamic relationship

$$\Delta G = \Delta H - T.\Delta S \tag{20}$$

stability may result from a favourable negative ΔH and/or from a positive ΔS.

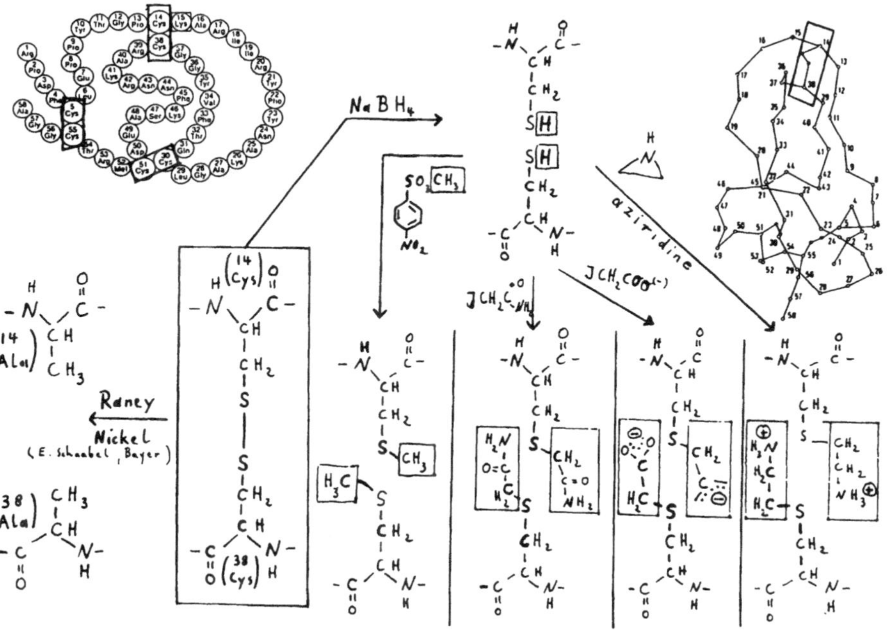


Figure 10: Modifications of the 14-38 Disulfide Bond of BPTI

We have studied so far the contribution to stability of the 14-38 disulfide bond in BPTI by investigating two analogs (24). BPTI-RCOM is obtained by reductive cleavage of the 14-38 disulfide bridge followed by carboxymethylation; BPTI-RCAM is analogous, only the cysteines in 14 and 38 positions are carboxamidomethylated after reduction. Table 7 shows the dramatic effects those modifications exert on the thermodynamic stability parameters.

Table 7: Thermodynamic Stability Parameters of BPTI Analogs (24)

| | pH | T [$_0$C] | BPTI | BPTI-RCAM | BPTI-RCOM |
|---|---|---|---|---|---|
| ΔG [kJ/mol] | 2 | 25 | -49.3 | -12.6 | -16.6 |
| ΔG [kJ/mol] | 5 | 25 | -59.7 | -23.5 | -26.2 |
| ΔH [kJ/mol] | | 25 | -284 | -57 | -131 |
| ΔS [J/K.mol] | 2 | 25 | -792 | -148 | -385 |

The overall stability as reflected in the Gibbs free energy of stabilization (calculated according to eq. 8 in Table 2 on the basis of calorimetric ΔH and $\Delta c_p$ values, and transition temperatures, $T_m$) is drastically reduced, if the 14-38 disulfide bridge has been opened. It is obvious that even the two additional negative charges BPTI-RCOM has at pH 5 in comparison to BPTI-RCAM influence the stability.
The most prominent result is, however, the strong decrease in energetic interactions that is associated with the reduction in the number of disulfide bonds from 3 to 2. This decrease overcompensates the less unfavourable entropy change so that, as a result, ΔG is also much less negative at 25$^0$C for both BPTI analogs.
This result is worth emphasizing: although visibly, e.g. by NMR techniques, there are only minor local perturbations detectable without any major changes in the overall structure (25), microcalorimetry reveals the drastic influence of the absent disulfide bond on stability. These studies will have to be carried further.
The information most urgently needed for predictions of secondary and tertiary structure is a set of quantitative energy parameters that describe the influence of single amino acid exchanges on stability of proteins. The microcalorimetric studies on BPTI analogs and other proteins will help filling that gap.
However, the possibilities of microcalorimetry go much

further; there are at least three major areas, where microcalorimetric studies are of paramount significance.
Firstly they are by far the best method to provide precise thermodynamic information on any biochemical system.
Secondly calorimetry is an excellent method to provide information on the domain structure in solution of very large proteins that are no longer amenable to X-ray studies or crystallization.
Thirdly the very fact that practically all reactions involve energetic changes renders calorimetry a powerful analytical tool to monitor biochemical reactions. It is almost unique in providing values for stability differences of proteins which result from interaction with macromolecular supports that are frequently used in biotechnological applications. Present day microcalorimetry provides instrumental sensitivity that makes studies of small samples feasable.

**Acknowledgement**

Financial support by the Deutsche Forschungsgemeinschaft is gratefully acknowledged.

**REFERENCES**

1. Hill, T.L.. 1960. An Introduction to Statistical Thermodynamics. Addison Wesley, Reading M.A.

2. Hill, T.L.. 1963. Thermodynamics of Small Systems. W.A. Benjamin Inc. Publishers. New York.

3. Cooper, A.. 1984. Protein Fluctuations and the Thermodynamic Uncertainty Principle. Progr. Biophys. Molec. Biol. 44, 181-214.

4. Hinz, H.-J.. 1983. Calorimetric Approaches to Protein-Ligand Interactions in: Topics in Molecular Pharmacology. (A.S.V. Burgen and G.C.K. Roberts, eds.). Elsevier Science Publishers B.V. pp.71-122.

5. McKinnon, I.R., L. Fall, A. Parody-Morreale and S.J. Gill. 1984. A twin titration microcalorimeter dor the study of biochemical reactions. Analytical Biochemistry 139, 134-139.

6. Hinz, H.-J.. 1986. Thermodynamic Parameters for Protein-Protein and Protein-Ligand Interactions. Methods in Enzymology 130, 59-79.

7. Sochava, I.V., T.V. Belopolskaya and O.I. Smirnova. (1985). DSC Study of reversible and irreversible thermal denaturation of concentrated globular protein solutions. Biophys. Chem. 22, 323-336.

8. Wiesinger, H. and H.-J. Hinz. 1985. Thermodynamics of the reactions catalysed by the multifunctional enzyme complex tryptophan synthase. Arch Biochem. Biophys. 242, 440-446.

9. Wiesinger, H., P. Bartholmes and H.-J. Hinz. 1979. Subunit interaction in tryptophan synthase of E.coli: Calorimetric studies on association of $\alpha$ and $\beta_2$ subunits. Biochemistry 10, 1979-1984.

10. Wiesinger, H. and H.-J. Hinz. 1985. Energetic adaptation of ligand binding to subunit structure of tryptophan synthase from E.coli. Biochemistry 23, 4928-4934.

11. Freedberg, W.P. and J.K. Hardman 1971. Structural and functional roles of the cysteine residues in the $\alpha$-subunit of the *Escherichia coli* tryptophan synthase. J. Biol. Chem. 246, 1449-1456.

12. Kirschner, K., R.L. Wiskocil, M. Foehn and L. Rezeau. 1975. An improved purification procedure for the $\alpha$-subunit and binding studies with substrate analogues. Eur. J. Biochem. 60, 513-523.

13. Pabo, C.O., R.T. Sauer, J.M. Sturtevant and M. Ptashne. 1979. The $\lambda$ repressor contains two domains. Proc. Natl. Acad. Sci. USA 76, 1608-1612.

14. Hecht, M.H., J.M. Sturtevant R.T. and Sauer. 1984. Effect of single amino acid replacements on the thermal stability of the $NH_2$-terminal domain of phage $\lambda$ repressor. Proc. Natl. Acad. Sci. 81, 5685-5689.

15. Freire, E. and R. Biltonen. 1978. Statistical mechanical deconvolution of thermal transitions in macromolecules I. Theory and application to homogeneous systems. Biopolymers 17, 463-479.

16. Biltonen, R. and E. Freire. 1978. Thermodynamic characterization of conformational states of biological macromolecules using differential scanning calorimetry. Crit. Rev. Biochem. 5, 85-124.

17. P.L. Privalov. 1982. Stability of Proteins. Adv. Prot. Chem. 35, 1-104.

18. Gill, S.J., B. Richey, G. Bishop J. and Wyman. 1985. Generalized binding phenomena in an allosteric macromolecule. Biophys. Chem. 21, 1-14.

19. Kidokoro, S. and A. Wada. 1987. Determination of thermodynamic functions from scanning calorimetry data. Biopolymers 26, 213-229.

20. Potekhin, S.A. and P.L. Privalov. 1982. Co-operative blocks in topomyosin. J. Mol. Biol. 159, 519-535.

21. Kassel, B. and M. Laskowski Jr.. 1965. The basic trypsin inhibitor of bovine pancreas. V. The disulfid linkages. Biochem. Biophys. Res. Commun. 20, 463-468.

22. Pfeil, W.. 1986. In: Thermodynamic Data for Biochemistry and Biotechnology. (H.-J. Hinz, ed.). pp.349-376.

23. Privalov, P.L.. 1979. Stability of Proteins. Adv. Protein Chem. 33, 167-241.

24. Schwarz, H., H.-J. Hinz, A. Mehlich, H. Tschesche, and H. Wenzel. 1987. Stability studies on derivatives of the bovine pancreatic trypsin inhibitor. Biochemistry 26, June edition.

25. Wagner, G., C.I. Stassinopoulou and K. Wüthrich. 1984. Amide-proton exchange studies by two dimensional correlated $^{1}$H NMR in two chemically modified analogs of the basic pancreatic trypsin inhibitor. Eur. J. Biochem. 145, 431-436.

# APPLICATION OF BIFUNCTIONAL REAGENTS FOR TOPOLOGICAL INVESTIGATION

**Roza Maria Kamp**

**Max-Planck-Institut fuer molekulare Genetik, Abteilung Wittmann, Ihnestrasse 63, D-1000 Berlin 33 (Dahlem)**

Chemical crosslinking of protein-protein neighbourhoods in ribosomes has found wide application in the study of protein topography of ribosomes derived from eubacterial and eukaryotic cells (1-12).

Ribosomes occur in all organisms and are considered well suited for studies on evolution of all living cells.

The structure of the Escherichia coli ribosomes is the best investigated and the topography of the protein constituents was investigated by immune electron microscopy (13) and neutron scattering (14,15).

However, these investigations only identified the neighbouring proteins or determined distances of the centre of the masses of some of them, but these approaches did not permit identification of individual neighbouring amino acids covalently bound by crosslinking reagents, with the exception of E.coli r-proteins S8-S5 (12). Such data could allow a more detailed understanding of the function and structure of ribosomes on a molecular level.

Determination of the crosslinked amino acids in a protein-protein crosslink requires considerable amounts of material.

Modern Methods in Protein Chemistry, Vol. 3

Therefore, in our studies only reagents that generate protein-protein pairs in relatively high yields could find application which allow further investigation at the peptide level.
Different reagents were tested for their ability to generate ribosomal protein pairs in sufficient amounts. Diepoxybutane (16-17) and cystamine (10) were found most suitable for crosslinking of ribosomal proteins.
Only reagents that crosslink proteins in reasonably high yields without causing conformational changes of the ribosome can be used.This means mild reactions at physiological conditions with a slight excess of the crosslinker can be applied.

Futher, sensitive methods for isolation of crosslinked protein and peptide pairs are required. Classical separation methods cannot be used as the amounts of crosslinks obtainable are rather limited, e.g.,below 10%,due to the mild reaction conditions used.The purification of the crosslinks is therefore only possible if sensitive HPLC methods are applied such as used for purification of ribosomal proteins from different organisms (18-19).

The present paper describes the criteria for chosing the most suitable crosslinking reagents and exemplifies the strategies used for preparative isolation and identification of crosslinked ribosomal protein pairs derived from B. stearothermophilus.

## CROSSLINKING REAGENTS

Several reagents were investigated for their use in generation of ribosomal protein-protein pairs.
Table 1 lists all tested cleavable reagents and shows their structure and properties.

| REAGENT | FORMULA | LENGTH | SOLUBILITY | CLEAVABILITY |
|---|---|---|---|---|
| 2'-di-sulfo-4,4'-diisothiocyanato-azo-benzene | 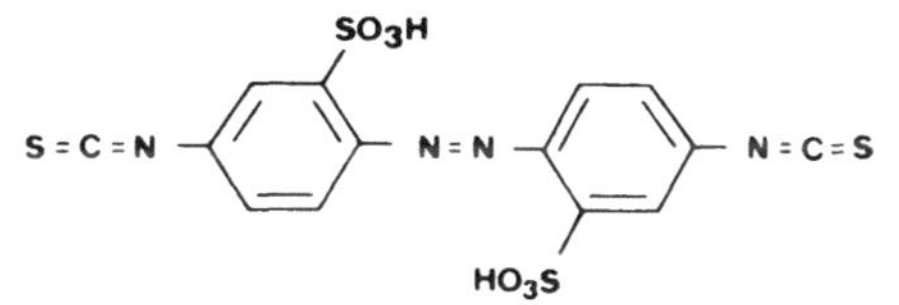 | 18Å | $H_2O$ | $Na_2S_2O_4$ |
| DST Disuccinimidyl tartarate | 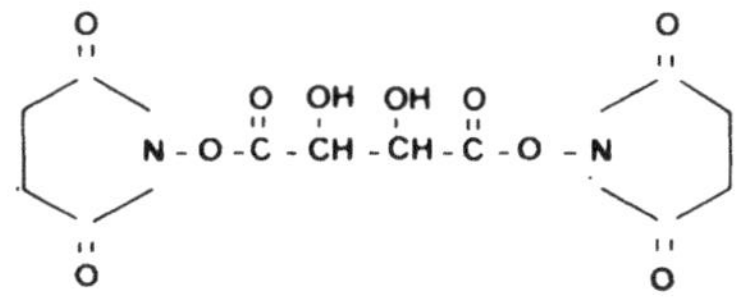 | 6Å | DMSO | $NaJO_4$ |
| EGS Ethylene glycol-bis(succinimidylsuccinate) | 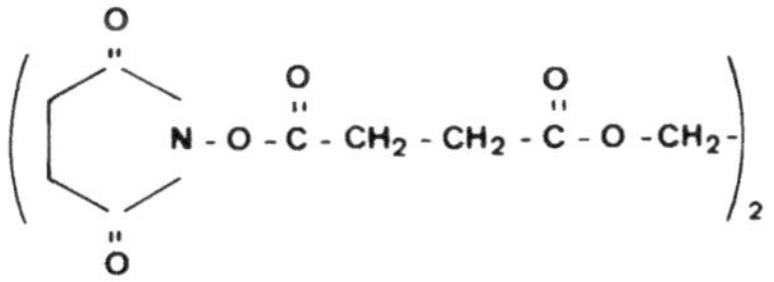 | 16Å | DMSO | hydroxylamine |
| DMS dimethyl suberimidate | 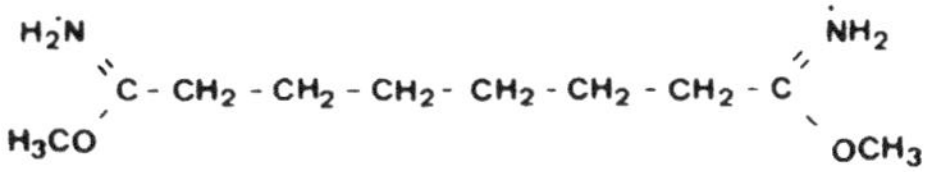 | 11A | $H_2O$ | $NH_4OH$ |
| TDA Tartryl diazide | 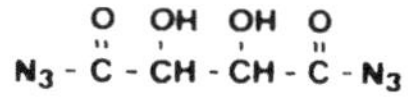 | 6A | $H_2O$ | $NaJO_4$ |

Table 1: *Crosslinking reagents*

| | | | | |
|---|---|---|---|---|
| **TDGA** Tartryl - di - (glycylazide) | $(N_3-\overset{O}{\overset{\parallel}{C}}-CH_2-NH-\overset{O}{\overset{\parallel}{C}}-\overset{OH}{\overset{\vert}{CH}}-)_2$ | 13Å | $H_2O$ | $NaJO_4$ |
| **TDCA** tartryl - di - (ε - aminocaproylazide) | $(N_3-\overset{O}{\overset{\parallel}{C}}-(CH_2)_5-NH-\overset{O}{\overset{\parallel}{C}}-\overset{OH}{\overset{\vert}{CH}}-)_2$ | 23Å | $H_2O$ | $NaJO_4$ |
| diepoxybutane | 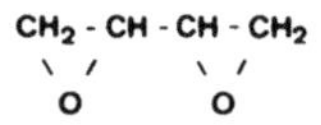 | 5Å | $H_2O$ | $NaJO_4$ |
| 2 - iminothiolane | 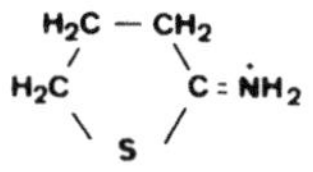 | 7Å | $H_2O$ | β-ME |
| **DTBP** Dimethyl - 3,3' - dithio - bis - propionimidate | $(H_3CO)(H_2\dot{N}{=})C-CH_2-CH_2-S-S-CH_2-CH_2-C({=}\dot{N}H_2)(OCH_3)$ | 12Å | $H_2O$ | β-ME |
| cystamine | $H_2N-(CH_2)_2-S-S-(CH_2)_2-NH_2$ | 15Å | $H_2O$ | β-ME |

In general, the reagents can be classified into two groups. One group reacts under mild conditions specifically with free N-terminal and the side chain amino groups of proteins.
Also sulfhydryl groups of cysteine can be attached by these reagents. A rather large group of this type of reagents is known, including disuccinimidyl tartrate (DST), ethylene glycol bis(succinimidylsuccinate)(EGS), dimethylsuberimidate (DMS), tartryl-diazide (TDA), tartryldi(glycylazide) (TDGA), tartryl-di(ε-amino-caproylazide) (TDCA), methyl-4-mercapto-butyrimidate (2-iminothiolane) and 3,3'-dithiobispropionimidate (DTBP).
Other types of reagents bind to free carboxyl groups of proteins after activation with a carbodiimide. One representative of this group is cystamine.
In our studies we used only cleavable reagents which facilitate the identification of crosslinked proteins and their peptides during purification. Some reagents such as 2-iminothiolane, DTBP or cystamine introduce a disulfide bond as the crosslink between different proteins. Cleavage is done by reduction to obtain monomeric proteins. On the other hand, this type of crosslinking reagents precludes the use of reducing agents during the isolation of crosslinked complexes. The crosslinks formed by reagents with vicinal hydroxyl groups or bisimidoesters are stable under reducing conditions and can be cleaved quantitativaly by mild treatment with sodium periodate or ammonolysis.

Reagents for crosslinking of ribosomal proteins should be chosen according to following criteria:
- water solubility of the bifunctional reagent; this allows crosslinking of proteins under physiological conditions without causing any conformational changes to the ribosome or its subunits;
- high yields for generation of protein complexes;
- stability of the crosslink bond to the amino acids at all

purification steps;

-cleavability of the crosslink under mild conditions for analysis of the proteins or peptides involved in the crosslink.

The length of reagents used usually for crosslinking ribosomal proteins varied between 4-20Å but more suitable are short reagents up to 10Å. Most of the reagents contain several highly flexible bonds and therefore can bind to longer side chains, e.g.,the amino groups of lysines.

Reagents longer than 10Å may cause generation of non specific inter-ribosomal crosslinks between two different subunits, e.g.,4,4'-diisothiocyanato-stilbene-disulfonic acid (DIDS). Such reagents are not desirable if highly specific intra-crosslinks between proteins whithin one subunit should be formed.

Crosslinking reagents based on bis(imidoesters) such as dimethylsuberimidate (DMS) or ethylene glycol bis(succinimidyl succinate) (EGS), have been used for studies of the structure of ribosomes; they react under mild conditions in a one step reaction to yield amidine derivatives.

Disadvantageous of both type of reagents is their non specific reactivity.

EGS attaches to all proteins in low yields and does not form one or a few protein pairs in yields, that would allow isolation of the crosslinked pairs in sufficient amounts.

Higher yields of crosslinked protein pairs were obtained with DMS and other tartryl derivatives. The reaction of proteins with DMS and the homologous bis(imidoesters) is shown in fig.1. The proteins react with suberimidate to yield amidine derivatives, which can be cleaved by ammonolysis or with higher efficiency by methylamine buffers in 75% acetonitrile (21). Tartryl derivates were synthetised as described in (22) as azides of three different lengths, namely 6Å (TDA) tartryl-diazide; 13Å (TDGA) tartryldi(glycylazide); and 16Å

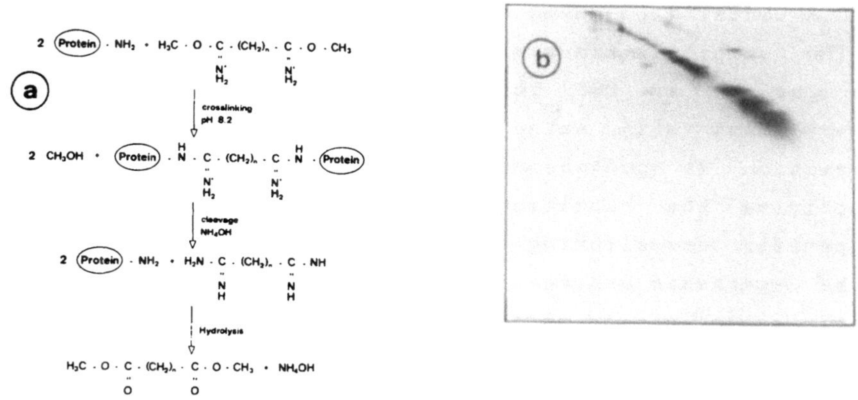


*Fig. 1:*

a) *Crosslinking reaction with bis(imidoesters)*
*n=2 succinimidate; n=4 adipimidate; n=6 suberimidate; n=8 sebacimidate*

b) *Diagonal electropherogram of the DMS-reacted proteins of the 70S ribosomes from B. stearothermophilus.*

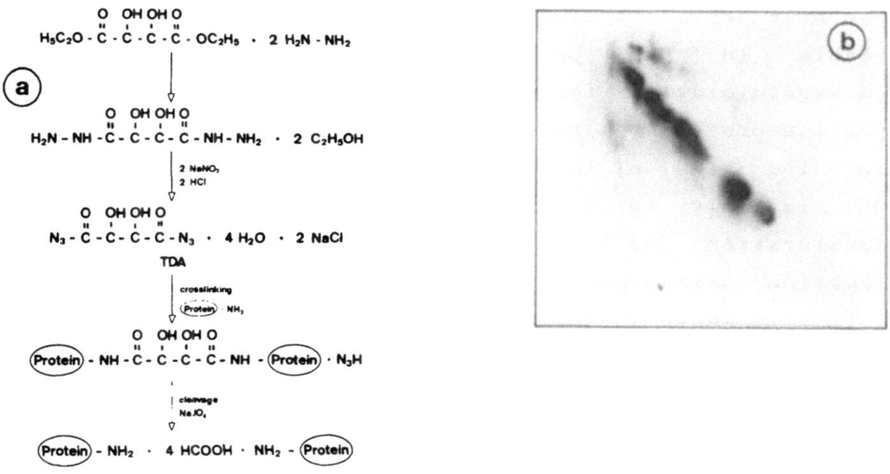


*Fig. 2:*

a) *Synthesis and reactions of the tartryl-diazide with proteins.*

b) *Diagonal electropherogram of the TDGA-reacted proteins of the 30S ribosomal subunits from B. stearothermophilus.*

TDCA tartryldi(ε-amino-caproylazide).

The stability and yields of tartryl crosslinks are higher in comparison to DMS. The azide-activated carbonyl groups react very fast with amino groups to produce crosslinks. This reaction is spontaneous and therefore it is more difficult to optimise the conditions for obtaining high yields and a specific crosslinking reaction. Fig.2 presents a scheme for the synthesis and reaction of the shortest representative of this group, namely of TDA.

The yields of generating protein complexes increase with the length of reagent, and TDCA clearly produces the highest amounts of crosslinks.

The crosslinks formed by tartryl derivates can be quantitatively cleaved by dilute sodium periodate treatment. Disadvantagous of these azides is the troublesome two-step synthesis of these reagents; the azides must be prepared freshly before use.

The next periodate cleavable bifunctional crosslinking reagent stable in the presence of reducing agents is DST (disuccinimidyl tartarate), which generates specific protein-protein pairs in relative high yields. Disadvantageous for the study of the protein topography of ribosomes is that DST is only soluble in organic solvents, and this causes denaturation of the native ribosomes. The crosslinking reaction using DST as a representative of succinimide active esters is shown in fig.3.

The most suitable reagent in our hands for crosslinking of ribosomal proteins was water-soluble diepoxybutane, which is commercially available and was first applied for crosslinking of E. coli ribosomal proteins to RNA (23).

Diepoxybutane reacts with free amino groups of the proteins to form covalently bound bridges, which contain vicinal hydroxyl groups that are easy cleavable with sodium periodate (fig.4). Diepoxybutane was found to form crosslinks in yields of about

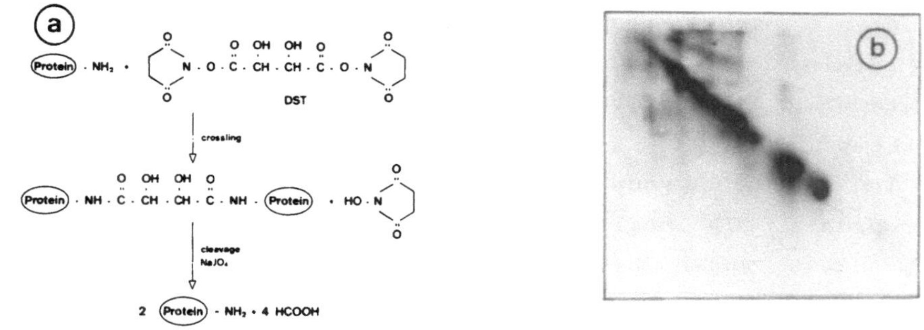


*Fig. 3:*

*a) Crosslinking reaction of proteins with disuccinimidyl tartarate (DST).*
*b) Diagonal electropherogram of proteins derived from DST crosslinked 30S subunits of B. stearothermophilus.*

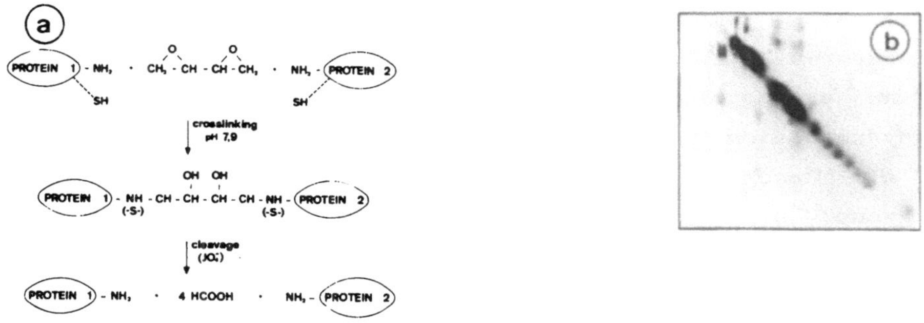


*Fig. 4:*

*a) The reaction of diepoxybutane with proteins.*
*b) Diagonal electropherogram of diepoxybutane crosslinked proteins derived from 50S ribosomal subunits of B. stearothermophilus.*

10%.

Proteins crosslinked with diepoxybutane are stable to high pH and high temperature and this makes their separation much easier.

The second group of cleavable reagents embraces disulfide reagents. The analysis of the composition of such crosslinked products would be performed after cleavage by reduction with ß-mercaptoethanol or dithiothreitol to regenerate monomeric proteins. 2-iminothiolane (methyl-4-mercaptobutyrimidate) has been extensively used by Traut et al. to study the protein arrangement in ribosomal subunits of E.coli (1) on the protein level. 2-Iminothiolane reacts with proteins in two steps: firstly, free amino groups are modified to form amidine linkages and secondly, by oxidation these derivatives can form disulfide bridges (fig.5). Another representative of the disulfide-linked reagents is DTBP (11.9Å) which has been applied for identification of neighbourhoods in rat liver 60S ribosomal subunits (25) and E.coli (24). The reaction is comparable to the reaction with other bis(imidoesters) with the exception, that the crosslink bond can be cleaved easily by reduction.

All described reagents specifically attach to N-terminal amino groups or ε-amino groups of lysines. Ribosomal proteins are also rich in glutamic and aspartic acids and for this reason also reagents which react with carboxyl groups can be employed for crosslinking studies.

One representative of this group is cystamine (15.5Å) which reacts with carboxyl groups after activation with water soluble carbodiimides. These crosslinks can be easily cleaved by reduction. The two-step reaction scheme of cystamine with proteins is shown in fig.6.

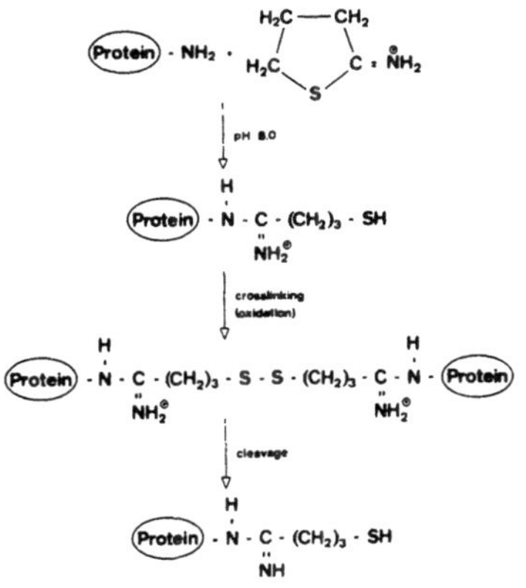


*Fig. 5:*

*The reaction of 2-iminothiolane with proteins.*

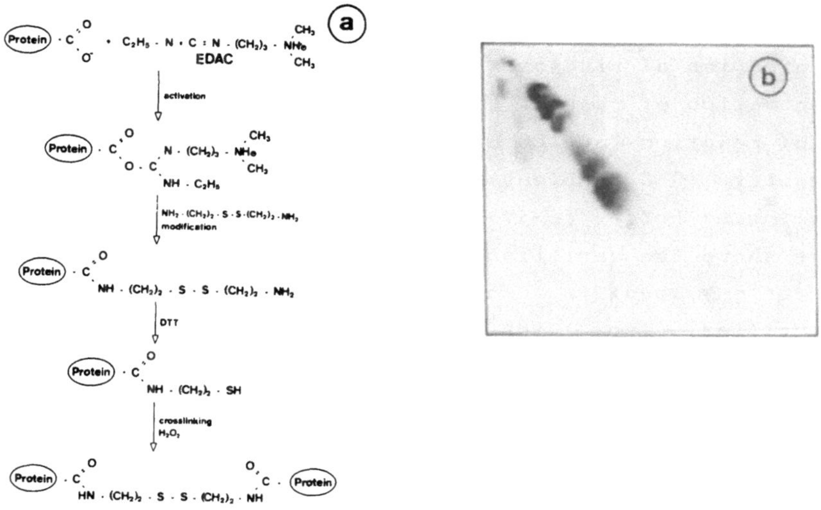


*Fig. 6:*

*a) Reaction of cystamine with proteins.*

*b) Diagonal electropherogram of proteins extracted from cystamine cross-linked 50S subunits of B. stearothermophilus.*

## OPTIMIZATION OF CROSSLINKING REACTION

Chemical crosslinking provides useful information on the topography of proteins in complexes, but this is only true if the crosslinking reaction is very specific between two proteins within the same complex or within one subunit (intra-crosslinks). On the other hand, the formation of crosslinks between proteins of different subunits (inter-crosslinks) must be avoided.

The specifity of reaction in case of crosslinking ribosomal proteins can be tested by sucrose gradient centrifugation of the derivatized subunits as shows fig.7. Peaks with a sedimentation coefficient higher than that of the 50S or 30S subunits indicates the formation of inter-crosslinks.

For optimization of the crosslinking reaction the following parameters can be varied:

- concentration of ribosomes (less than 50 OD/ml)
- concentration of reagent (5-200 mM)
- time of reaction (30-120 min)
- temperature (0°C, ambient or 37°C)
- crosslinking buffer (pH 7-8).

In case where the variation of all these paramaters does not allow for any specific reaction, the intra-crosslinks have to be purified from inter-crosslinks by sucrose gradient centrifugation which makes their futher investigation possible.

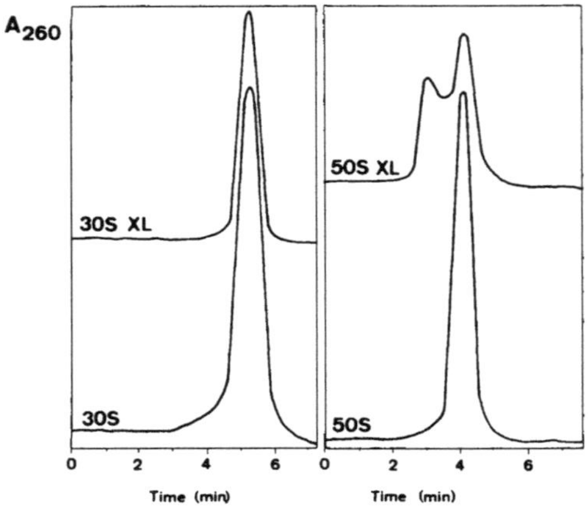


*Fig. 7:*

*The sucrose gradient centrifugation profile of 30S and 50S ribosomal subunits crosslinked by 200 mM cystamine.*

# IDENTIFICATION OF CROSSLINKED PROTEINS

The presence of crosslinked proteins after reaction with bifunctional reagents was checked by diagonal PAGE using 3 x 3 cm micro gels according to (26). The proteins were separated by diagonal gel electrophoresis in both dimensions under the same conditions (urea gel), except that after the first dimension the gel was soaked in cleavable reagents to regenerate monomeric proteins. The treated first dimension gel was polymerized onto the second dimension gel and the electrophoresis then was performed exactly as in the first dimension. The gels were stained with Coomassie Brilliant Blue. The cleaved crosslinked proteins have a greater electrophoretic mobility in the second dimension and appear below the diagonal as free non crosslinked proteins.

The use of the micro 2D-gels allows the detection of crosslinked proteins in the nanogram level (in case of Coomassie stained gels) and in picogram levels (using the silver stain technique).

The comparison of 2D-electropherograms of modified and unmodified subunits allows identification of crosslinked proteins, which appear as new spots after crosslinking and dissappear after cleavage by reduction or oxidation.

As example fig.8 shows the identification of the 30S crosslink S13-S19 from B. stearothermophilus. Further, the immunological approach allows identification of these proteins involved in the crosslink. Fig.9 shows as example an immunoblot of a LiCl fraction containing the 50S crosslink L23-L29 from B.stearothermophilus which was separated by HPLC techniques as described in (16). One nitrocellulose sheet was stained with antisera directed against the L23 protein, another nitrocellulose sheet was incubated with anti L29 serum. Only two spots are visible, one in the position of L23 or

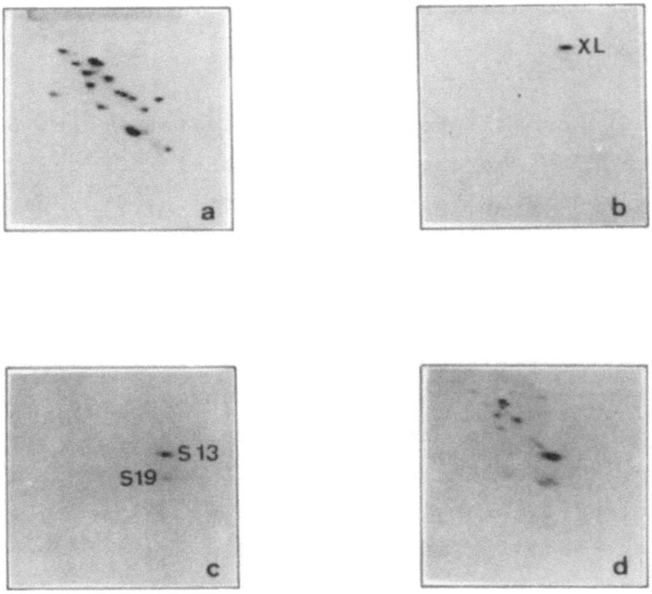


*Fig. 8:*

*Identification of proteins involved in the crosslink.*

*a) Two-dimensional electropherogram of 30S ribosomal total proteins from B. stearothermophilus.*

*b) 2D-electropherogram of the purified diepoxybutane crosslinked complex S13-S19 before cleavage.*

*c) 2D-electropherogram of the S13-S19 crosslink cleaved with sodium periodate.*

*d) 2D-electropherogram of S13-S19 crosslink mixed with 30S total proteins as internal standard for easy identification.*

L29, respectively and the second spot corresponds to the crosslinked protein pair L23-L29.
The immunological data confirmed the results derived from the 2D-electropherograms.

In addition microsequencing of the purified protein pair before and after cleavage allows a clear prediction of the proteins involved in the crosslink, except if the crosslinked residues are modified at the N-terminal amino-acids of the peptide chain or are monovalently modified.

## ISOLATION OF PROTEIN CROSSLINKS

Several methods were tested for isolation of protein-protein crosslinks.It became obvious that the crosslinked protein pairs are more heterogenous than unreacted mixtures, as the bifunctional reagents also react monovalently with many groups of the proteins and change their properties. The separation of such a complex mixture is very difficult and only sensitive separation methods can be applied. Different HPLC techniques were used for separating of crosslinked protein and peptide pairs,eg.,size exclusion, reversed phase and ion exchange HPLC.

In some cases also salt extractions combined with HPLC techniques allow successful and quick purifications.

Usually separation of crosslinked proteins from one dimensional SDS-PAGE is very difficult, as on an average crosslinked ribosomal proteins have similar molecular weights and it is difficult to purify the crosslinks without any prefractionation step.

More satisfactory is elution of crosslinked proteins from two-dimensional gels, which allows a higher resolution of the many proteins.

Limitation of this method is the relative low recovery of the

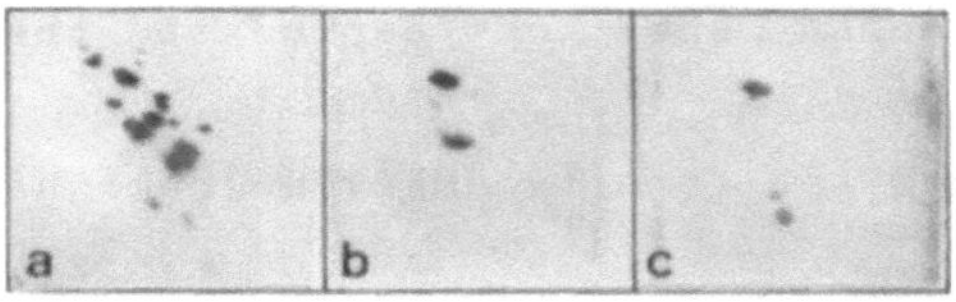


*Fig. 9:*

*Immunological analysis of the 50S crosslink L13-L19 of B. stearothermophilus.*

*a) 2D-electropherogram of total proteins of 50S subunits stained with amido black.*

*b) Nitrocellulose sheet after blotting of crosslinked proteins and incubation with anti L23 serum.*

*c) Nitrocellulose sheet incubated with anti L29 serum and stained by the aid of an alkaline phosphatase conjugate of anti-goat-IgG.*

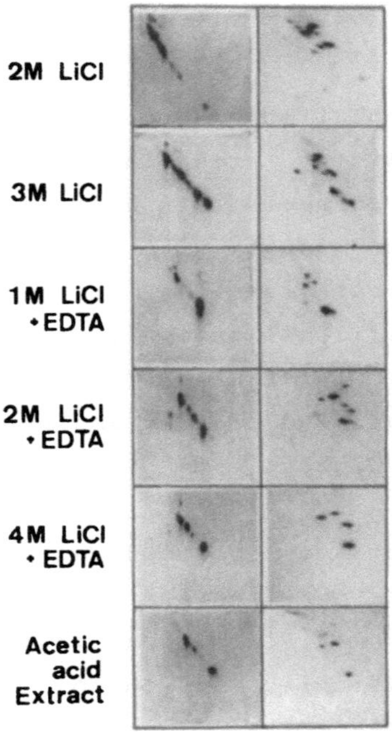


*Fig. 10:*

*Prefractionation of diepoxybutane crosslinked 30S subunits of B. stearothermophilus. Each fraction was analysed by diagonal (left) and 2D-PAGE (right). Only one crosslink is formed in high yields in the 2M LiCl + EDTA extraction.*

eluted proteins from the gel with yields about 50 %. This method was applied for isolation of the S13-S19 crosslink from E.coli (Pohl,T.& Kamp,R.M, paper in preparation).

**Prefractionation of ribosomal protein-protein crosslinks**

The isolation profile of the crosslinked proteins can be subdivided into groups using LiCl extraction as described in detail (16). Proteins were extracted stepwise with different LiCl and EDTA concentrations. Fig.10 shows LiCl extracts of 50S subunits crosslinked by diepoxybutane. Only fraction number 4 extracted from the ribosomal core fraction with 2M-EDTA contains high amounts of the crosslink. Prefractionation of the 50S ribosomal proteins by this procedure allows about a tenfold enrichment of the 50S crosslink.

**Isolation of crosslinked proteins by HPLC**

Protein-protein pairs prefractionated by LiCl extraction were purified on HPLC.

In one instance size-exclusion chromatography allowed purification of a crosslinked protein pair derived from the 30S LiCl extract of B.stearothermophilus to homogenity.

The purification was done on a TSK 2000 SW column (7x300mm) using 0.1 M ammonium acetate pH 4.1 as eluent.Recovery on size exclusion columns is more than 90% and allows quick isolation of pure crosslinks as identified by 2D-PAGE.

Usually crosslinks were purified by combining two different HPLC techniques. Fig.11a shows purification of 30 S crosslinks after modification with cystamine. The crosslinked proteins were first isolated on the gel filtration column TSK 2000 SW

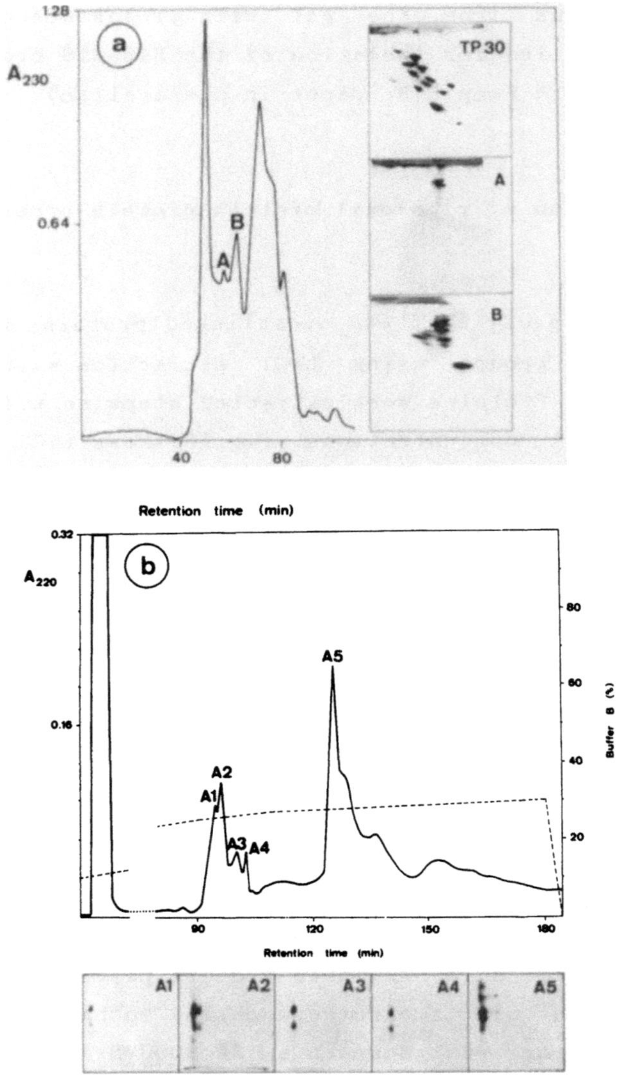


*Fig. 11:*

a) *Separation of cystamine crosslinked 30S ribosomal proteins on size-exclusion HPLC column TSK 2000 SW. Peak A contains three crosslinks and peak B four crosslinks and some non crosslinked proteins.*

b) *Repurification of peak A containing crosslinks on reversed phase column Nucleosil C4.*
*Diagonal electropherograms A1 - A5 show purified crosslinked complexes.*

without any prefractionation. In the second step, fractions containing the crosslinks were repurified using aqueous 0.1% TFA and isopropanol as buffers, on the reversed phase column Nucleosil C4 (300 Å pore size, 5μ particle size; 250x4 mm column size). Some fractions contain pure protein crosslinks as identified by diagonal gel electrophoresis (fig.11b).
Fractions containing not completely purified crosslinks can be applied for further purification on another type of reversed phase column, e.g.,a diphenyl column with variation in the elution,e.g., buffers or gradient.

## ISOLATION OF CROSSLINKED PEPTIDES

The purified crosslinks were used for futher studies at the peptide level.
For the presentation of the strategies for determination of the crosslink at the peptide or amino acid level , crosslink S13–S19 from E.coli is chosen as representative.
The crosslinked protein pair was fragmentated by dilute acetic acid hydrolysis (2% acetic acid, 15h, 105°C) to obtain peptides specifically cleaved after aspartic acid.
The peptides were separated on the small pore size reversed phase column Lichrospher C18 (Merck,F.R.G.) (fig.12). Each peak was rechromatographed on the same column for check of purity.
If one homogeneous peak was obtained after reinjection onto the column, the same fraction was cleaved in 10 μl of 10 mM sodium periodate for 10 min and directly applied onto the column.
Only one peptide peak disappears in the chromatogram after periodate cleavage and two new peaks become visible.
This peptide pair corresponds to the crosslinked peptides of ribosomal proteins S13–S19. The sequencing of these peptides allows the determination of the amino acids involved in the

crosslink.

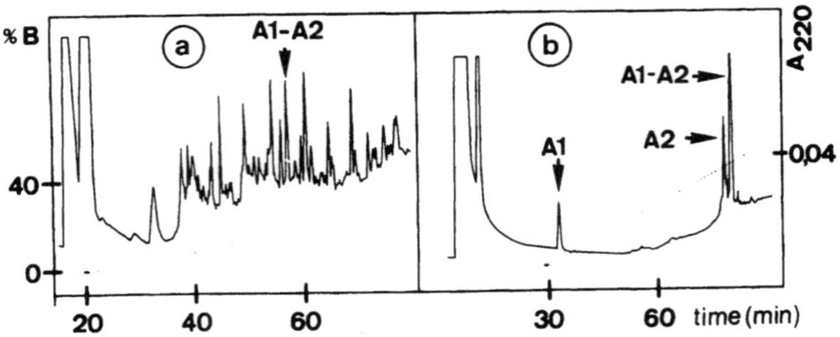


*Fig. 12:*

a) *Separation of peptides after partial hydrolysis (2% acetic acid; 15 h; 110°C) on Vydac C4 column. Crosslinked protein pair A1-A2 is indicated by arrow.*

b) *Separation of the cleaved peptides of crosslink A1-A2 on LiChrospher column. Purified peptides are indicated by arrows.*

**CONCLUSION**

The identification of the protein neighbourhoods in ribosomes of different organisms allows a better understanding of the evolution of bacteria.

The described cleavable reagents facilitate the identification of crosslinked proteins and peptides in the low nanogramm range using micro gel systems. Also the reproducible production of crosslinks and the formation of specific crosslink bonds are important criteria in the choice of suitable reagents.

The purification of proteins involved in the crosslink is possible by sensitive methods such as HPLC or alternatively by a combination of these techniques with prefractionation steps using salt extraction of ribosomes.

Crosslinking studies of ribosomal proteins permit the

construction of more detailed structural models of the ribosomal subunits on the amino acid level.

The described reagents can futher be used for chemical modification of proteins in order to study changes of their physical or biochemical properties.

## ACNOWLEDGMENTS

The author thanks Prof.B.Wittmann-Liebold for support and comments during the work and the preparation of the manuscript and Jürgen Brockmöller for providing some of figures.

## LITERATURE

1. Traut, R.R., Lambert, J.M., Boicav, G., Kenny, J.W., in Ribosomes: Structure,Function and Genetics [Chamblis G., Craven, G.R., Davies, J., Davies,. K., Kahan, L., & Nomura, M. eds.] pp.89–110, University Park, Baltimore (1980)

2. Peretz, H., Towbin, H. & Elson, D., Eur.J. Biochem. **63**, 83–92, (1976)

3. Lutter, L., Bode, U., Kurland, C.G. & Stöffler, G., Mol. Gen. Genet. **29**, 167–176, (1974)

4. Lutter, L.C.,Ortanderl, F. & Fasold, H., FEBS Lett. **48**, 144–159, (1974)

5. Lutter, L.C., Kurland, C.G. & Stöffler, G., FEBS Lett. **54**, 144–159, (1975)

6. Lutter, L.C., Zeichardt,H., Kurland, C.G. & Stöffler, G., Mol. Gen. Genetic. **119**, 357–366, (1972)

7. Expert-Bezancon, A., Barritault, D., Clegg, J.C.S., Milet, M. Khouvine, Y. & Hayes, D.H., FEBS Lett., **59**, 64–69, (1975)

8. Sommer, A. & Traut, R.R., J. Mol. Biol., **106**, 995–1015, (1976)

9. Kenny, J.W. & Traut. R.R., J. Mol. Biol., **127**, 243–263, (1979)

10. Traut, R.R., Ferris, R.J., Gavino, G.R., Lambert, J.M., Butler, P.D. & Kenny, J.W., in Cell. Regulation and Malignant Growth (Ebashi, S., ed.) pp. 101–113, Japan. Sci. Press, Tokyo/Springer-Verlag, Berlin, (1985)

11. Uchiumi, T., Kikuchi, M., Terao, K. & Ogata, K. J., Biol. Chem., **260**, 5675–5682, (1985)

12. Allen, G., Capasso, R., & Guarlerzi,C., J. Biol. Chem. **254**, 9800–9806, (1979)

13. Stöffler, G., Bald, R., Kastner, B., Lührmann, R., Stöffler-Meilicke, M., Tischendorf, G., & Tesche, B., in Ribosomes: Structure, Function and Genetics (Chambliss, G., Craven, G.R., Davies, J., Davies, K., Kahan, L. and Nomura, M. eds.) pp. 171–205, University Park Press Baltimore,(1980)

14. Moore, P.B. in Ribosomes: Structure, Function and Genetics (Chambliss, G., Craven, G.R., Davies, K., Kahan, L. and Nomura, M. eds.) pp. 111–133, University Park

Press,Baltimore, (1980)

15. Novotny, V., May, R.P., & Nierhaus, K.H., in: Structure, Function and Genetics of Ribosomes (Hardesty, B. & Kramer, G. eds.) pp. 101–111, Springer Verlag, New York, Berlin,Heidelberg, (1986)

16. Brockmöller, J. & Kamp, R.M., Biol. Chem. Hoppe-Seyler, **367**, 925–935, (1986)

17. Kamp, R.M., Brockmoller., J. & Wittmann-Liebold, B., Chromatographia **22**, 7–12, (1986)

18. Kamp, R.M., Bosserhoff, A., Kamp,D. & Wittmann-Liebold, B., J. Chrom. **317**, 181–192, (1984)

19. Kamp, R.M., Yao, Z-J. & Wittmann-Liebold, B., Hoppe Seyler's Physiol. Chem. **364**, 141–155, (1983)

20. Expert-Bezancon, A., Barritault, D., Milet, M., Guerin, M-F. & Hayes, D.H., J. Mol. Biol. **117**, 603–629, (1977)

21. Packman, L.C. & Perham, R.N., Biochem. **21**, 5171–5175 (1982)

22. Lutter, L.C., Ortanderl, F. & Fasold, H., FEBS Lett. **48**, 288–292, (1974)

23. Bäumert, H.G., Sköld, S.-E. & Kurland, C.G., Eur. J. Biochem. **89**, 353–359, (1978)

24. Cover, J.A., Lambert, J.M., Norman, C.M. & Traut, R., Biochem. **20**, 2843–2852, (1981)

25. Uchiumi, T., Terao, K. & Ogata, K., J. Biochem. **88**,

1033–1044, (1980)

26. Brockmöller, J. & Kamp, R.M., Biol. Chem. Hoppe-Seyler's, **366**, 901–906, (1985)

# TOWARDS THE IDENTIFICATION OF THE NUCLEOTIDE-BINDING SITE OF TUBULIN

Klaus Linse and Eva-Maria Mandelkow

Max-Planck-Unit for Structural Molecular Biology
Ohnhorststrasse 18, D-2000 Hamburg 52, F.R.G.

## ABSTRACT

By comparing tubulin with several nucleotide binding proteins we have predicted four possible regions of interaction between the protein and its cofactor GTP (Mandelkow et al., 1985a). In this report we describe a strategy to identify the amino acid residues that participate in the binding of GTP. It involves direct photoaffinity labelling, limited proteolysis, identification of peptides by antibody end labelling, isolation and sequencing of peptides.

## INTRODUCTION

The alpha-beta-heterodimer of tubulin binds one non-exchangeable and one exchangeable GTP (GTPn and GTPe), one of which (GTPe) is hydrolyzed during microtubule assembly. Proteins that are regulated by GTP are currently the subject of extensive studies, particularly since some of them are oncogene products. Sequence homologies have been detected among GTP-binding proteins and other nucleotide binding proteins (Sternberg and Taylor, 1984; Leberman and Egner, 1984). The homologies with tubulin are summarized in Fig. 1 (see Mandelkow et al., 1985a). They are classified as regions I-IV, all of which probably represent loops connecting the C-terminal ends of beta-strands with the N-terminal starts of alpha-helices. The aim of this study is to provide experimental evidence for the theoretical predictions and to learn which amino acids are crucial for the control of microtubule assembly by GTP.

Modern Methods in Protein Chemistry, Vol. 3

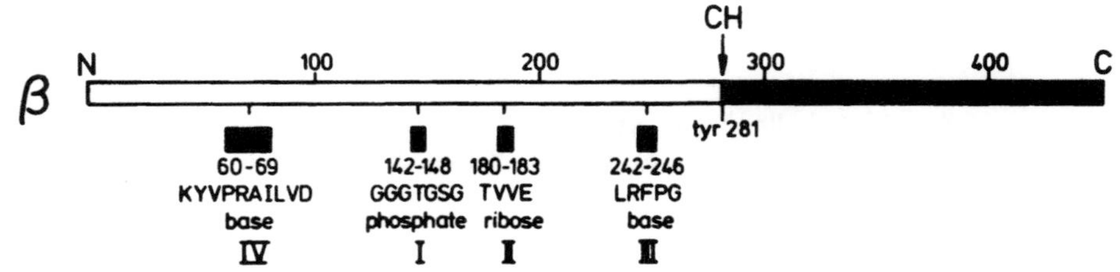


Fig. 1: Diagram of the beta-tubulin chain, domain structure, and homologies with nucleotide binding proteins. CH = chymotryptic cleavage site at tyr281 (Mandelkow et al., 1985a). I-III = regions of homology with ATP or dinucleotide binding proteins (e.g. dehydrogenases, pp60-src, see Sternberg and Taylor, 1984), corresponding to probable binding sites of phosphate, ribose, and base. IV = homology with GTP-binding proteins (e.g. elongation factor Tu, p21-ras, see Leberman and Egner, 1984).

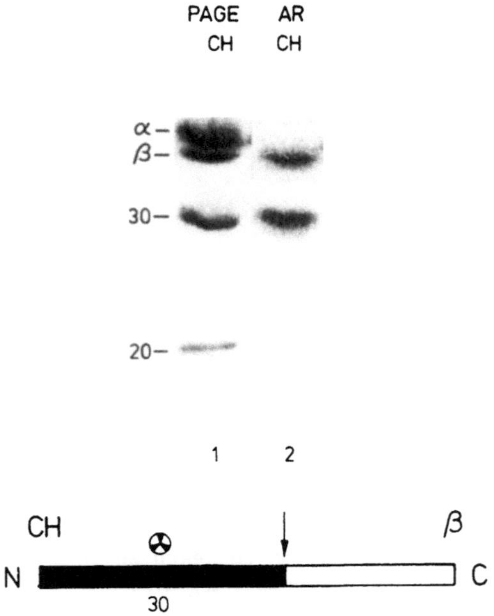


Fig. 2: Limited proteolysis of tubulin by chymotrypsin after direct photo-crosslinking with radioactive GTP. Lane 1: SDS-PAGE showing that beta-tubulin is cleaved into a 30 kDal (N-terminal) and a 20 kDal (C-terminal) fragment. Lane 2: Autoradiogram showing the label of the exchangeable GTP on the beta but not on the alpha chain, and on the 30 kDal fragment of beta-tubulin (see diagram below; * indicates the GTP label).

## MATERIALS AND METHODS

Microtubule protein from porcine brain and phosphocellulose-purified tubulin were prepared as described (Mandelkow et al., 1985a). The reassembly buffer was 0.1 M PIPES pH 6.9, with 1 mM each of $MgSO_4$, EGTA, GTP, and DTT. The protein was stored in liquid nitrogen until use.

Direct photoaffinity labelling of tubulin by (alpha-32P)GTP (1 mCi/ml, 3000 Ci/mmol) was done at protein concentrations around 50-100 uM and concentrations of labeled GTP around 0.1 uM, using UV irradiation at 254 nm (HBO 100W/2 Super Pressure Hg Lamp, Spindler and Hoyer).

SDS-PAGE (linear or gradient gels, with or without urea) was performed following Laemmli (1970), using 0.5 mm microslab gels.

Immunoblotting on nitrocellulose sheets was done following Bittner et al. (1980), with modifications (Kirchner and Mandelkow, 1985).

Autoradiography of GTP-labeled peptides was done from destained wet gels, dried gels, and nitrocellulose sheets, using Kodak X-Omat AR-5 film (exposure time 1-8 days).

Limited proteolysis of tubulin was done with the following enzymes: alpha-chymotrypsin (Sigma), Staph. aur. V8 protease (Miles), subtilisin type VIII (Boehringer), and clostripain (Sigma). The reactions were quenched with 2 mM PMSF or 40 ug/ml alpha-macroglobulin.

Purification of peptides by electrophoresis and elution by diffusion were performed as described (Mandelkow et al., 1985a).

Sequencing was done using a gas phase sequencer (Applied Biosystems; Organogen, Heidelberg).

## RESULTS

The strategy for the identification of the nucleotide-binding region involved several successive approximations.

(a) Subunit of microtubules: The core of microtubules contains alpha- and beta-tubulin. When total tubulin is labeled with GTP and analyzed by SDS-PAGE and autoradiography the nucleotide appears only on beta-tubulin which therefore carries the site for the exchangeable GTP (Fig. 2; compare top bands in lanes 1 and 2). This is consistent with the results of Geahlen and Haley (1977), Nath et al. (1985, 1986), and Hesse et al. (1985), but in contrast to those of Maccioni and Seeds (1983).

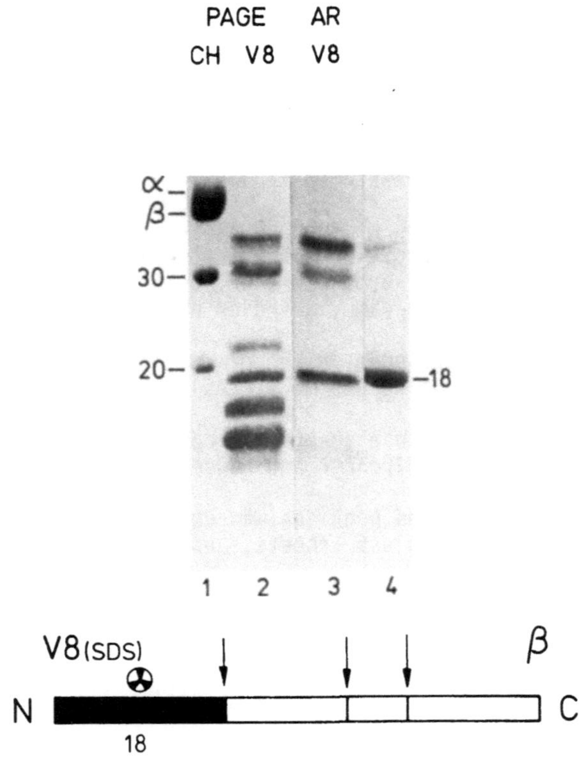


Fig. 3: Limited proteolysis of tubulin by Staphylococcus aureus V8 protease after photo-crosslinking with GTP and denaturation in 1% SDS. Lane 1: SDS-PAGE of chymotryptic fragments used as markers. Lane 2: SDS-PAGE of V8 protease digest. Lane 3: Autoradiogram of lane 2, showing that the GTP label is incorporated into three major fragments, the smallest one at 18 kDal. Lane 4: SDS-PAGE of the 18 kDal fragment after preparative gel electrophoresis and elution by diffusion. Sequence analysis shows that this fragment contains the N-terminus (see diagram below, left arrow; the other two arrows indicate cleavage sites for the larger radioactive fragments seen in lane 3).

(b) Domain of beta-tubulin: Both tubulin subunits consist of two domains that can be separated by limited proteolysis using trypsin (specific for alpha-tubulin; cleavage at arg339) or chymotrypsin (specific for beta-tubulin; cleavage at tyr 281; see Mandelkow et al., 1985a). When GTP-crosslinked tubulin is cleaved with chymotrypsin the label is found on the N-terminal domain of beta tubulin (Fig. 2; compare lower bands). This coincides with the domain where the nucleotide binding is predicted since it contains all four regions of homology (cf. Fig. 1).

We now proceed to determine which part of the N-terminal domain contains the base binding site. In particular, we want to distinguish the two predicted regions of homology III and IV.

(c) GTP-binding peptides within the N-terminal domain: GTP-crosslinked tubulin was briefly cleaved by V8 protease in 1% SDS. The lowest well-defined fragment showing radioactivity is at 18 kDal (Fig. 3, lane 4). This band was excised from the gel, eluted by diffusion, and six N-terminal residues were determined by gas-phase sequencing, showing the N-terminus of beta-tubulin (met-arg-glu-ile-val-his). Thus the base binding occurs within the N-terminal 18 kDal region, and the V8 protease does not remove the N-terminus from beta-tubulin.

(d) Identification of peptides by antibody end-labeling: A further restriction in the nucleotide binding site was achieved by a combination of limited proteolysis and antibody end-labeling of the GTP-crosslinked protein. In order to use the method one needs a label for one of the ends of the polypeptide chain. As shown earlier (Mandelkow et al., 1985a), the region of highest antigenicity of both tubulins is near their C-termini. Several polyclonal and monoclonal antibodies were investigated. For our purposes, the most suitable one was the monoclonal antibody against beta-tubulin obtained from Amersham. Using limited proteolysis by V8 protease and subtilisin its binding site was located between 3 and 5 kDal of the C-terminus (data not shown).

The next step consisted of finding large proteolytic fragments that bind the C-terminal antibody label but not GTP. One such peptide is a 43 kDal fragment produced by limited proteolysis with clostripain (Fig. 4, lane 1). It contains the antibody site (lane 2) but not the GTP label (lane 3). This means that the base-binding site must be located within the N-terminal 7-10 kDal.

(e) Outlook: The results described so far are summarized in Fig. 5. They allow us to discriminate between the regions III and IV, both of which are implicated in the binding of the base as judged by their homologies with other proteins (see Discussion). Only region IV is directly compatible with the experimental results. The next stage consists of isolating nucleotide-binding peptides by HPLC and determining the amino acid to which the guaninosine base is linked covalently by UV irradiation; these experiments are in progress.

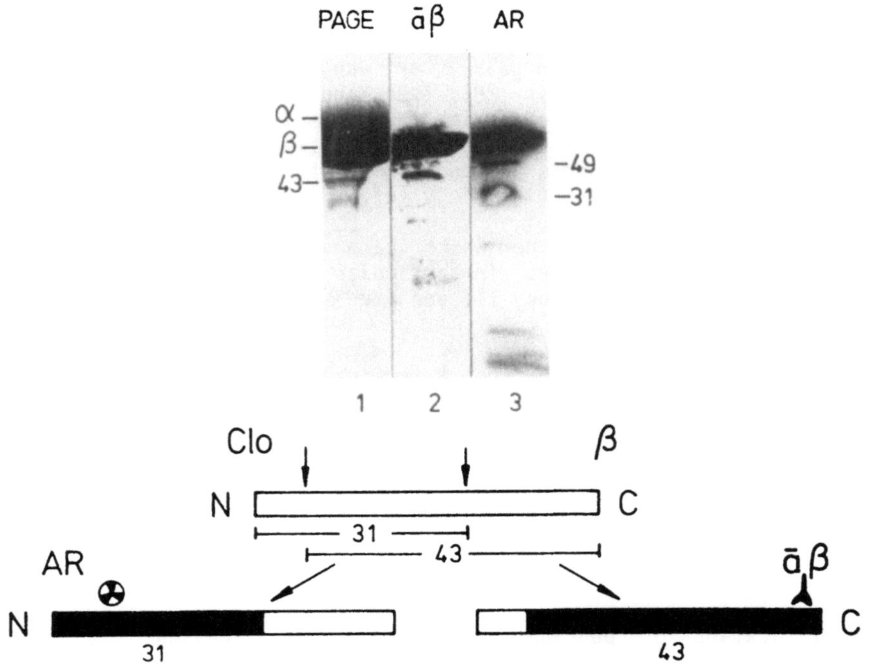


Fig. 4: Limited proteolysis of tubulin with clostripain after photo-crosslinking with GTP, and identification of peptides by antibody end labeling. Lane 1: SDS-PAGE showing alpha- and beta-tubulin and major fragments at 43 and 31 kDal. Lane 2: Immunoblot of the gel, showing that the 43 kDal fragment contains the epitope of the monoclonal antibody and therefore is a C-terminal fragment of beta-tubulin. Note that the 31 kDal fragment is not labeled, but a faint band at 49 kDal can be seen. Lane 3: Autoradiogram of the gel, showing the GTP label on several fragments, especially at 31 and 49 kDal, but not on the C-terminal fragment of 43 kDal. This means that the GTP label is located within 10 kDal of the N-terminus of beta-tubulin.

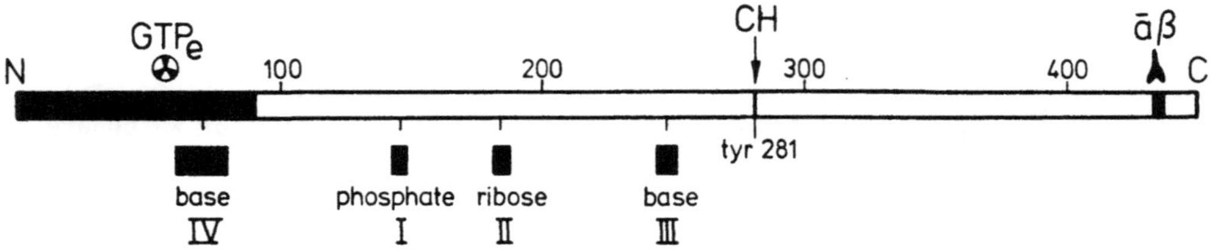


Fig. 5: Diagram summarizing the results. The epitope of the antibody is about 3-4 kDal away from the C-terminus, the photo-crosslinked GTP is found within the N-terminal 10 kDal of beta-tubulin. This supports region IV (see Fig. 1) as a possible base binding site.

**DISCUSSION**

We have determined the approximate location of the guanosine-binding region of tubulin and compared it with predicted binding regions, derived from homologies with other nucleotide binding proteins. The base was cross-linked to the protein by direct photoaffinity labelling and the bound nucleotide was identified by the radioactivity of the alpha-phosphate.

The peptides used in this work were generated by limited proteolysis of chymotrypsin, Staph. aureus V8 protease and clostripain. One base-binding peptide was determined directly by partial sequence-analysis. Other peptides were identified using C-terminal end-labelling by a beta-specific monoclonal antibody. In order to use this label, the epitope was first located and found to be between 3 and 5 kDal of the C-terminus.

(a) The mechanism of direct photoaffinity labelling.

Direct photocrosslinking of nucleotides has been used by several groups in order to identify the binding sites (e.g. ATP on myosin, Maruta and Korn, 1981; GTP on beta-tubulin, Nath et al. 1985; NAD on diphtheria toxin, Carroll et al., 1985). This reaction usually involves decarboxylation and covalent binding of acid residues (asp, glu) to the base. In some cases this causes the disruption of the bond between the base and the ribose (Carroll et al., 1985). We have not yet identified the residue involved in the photo-crosslinking. However, the GTP appears to remain intact since the crosslinked base can be detected via the alpha-32P phosphate by autoradiography.

(b) Identification of GTP-binding peptides

The combination of end-labelling by antibodies and specific enzymatic cleavage fo the protein provides a convenient tool for identifying peptides of interest (Matsudaira, 1985). In order to use the method, we had to confirm first that one of the antibodies tested bound near a terminus. It turned out that the epitope of the beta-specific monoclonal antibody (Amersham) was located between 3 and 5 kDal of the C-terminus. This is consistent with predictions of epitopes based on hydrophilicity whose maximum is near the C-terminus (Mandelkow et al., 1985a).

The location of the GTP-binding region was carried out in several consecutive steps. The first was the demonstration that the exchangeable GTP binds to beta-tubulin but not to alpha-tubulin, in agreement with earlier reports (Nath et al., 1985; Geahlen and Haley, 1977, Hesse et al., 1985). The next question was to find out which of the two major domains of beta-tubulin binds the exchangeable GTP. As shown previously, the two domains of beta-tubulin can be separated by chymotrypsin (Mandelkow et al., 1985a). The large fragment of 30 kDal binds the GTP. This is expected from the predictions (see below). However, since the homologies with other nucleotide binding proteins suggested two possible regions of base-binding, separated by about 170 amino acid residues (III and IV in Fig. 1), a further restriction of the binding site was necessary.

In order to distinguish between the two possibilities, it was necessary to find a small peptide with GTP-binding activity. One example was an 18 kDal fragment produced by V8 protease digestion. Partial sequencing showed that it contained the N-terminus. This is consistent with region IV of Fig. 1 as a site of GTP-binding.

A further restriction is achieved by end-labelling analysis using monoclonal antibodies. The idea is to identify peptides produced by limited proteolysis on the basis of an end-labelled terminus. We chose a commercial antibody against beta-tubulin (Amersham) whose epitope was shown to bind between 3 and 5 kDal away from the C-terminus. Thus, for our purposes, the antibody can be used as C-terminal end-label.

Using a variety of proteases, we searched for the largest polypeptide that contains the C-terminus but not the base-binding site. A 43 kD fragment produced by chlostripain fulfilled this requirement. This implies that the base-binding must occur within 10 kDal or less of the N-terminus.

### (c) Comparison with homologous proteins.

Recently, two classes of nucleotide-binding proteins have been investigated with respect to their homologies. The first class contains several dehydrogenases and other nucleotide- and dinucleotide-binding proteins, including proteins related to pp60-src (Sternberg and Taylor, 1984). The second class contains the elongation factors, G-proteins and proteins related to p21-ras (Leberman and Egner 1984; Halliday, 1984).

Beta-tubulin shares features of both classes. The homologies with the ATP and dinucleotide-binding proteins are: (I) a glycine-rich stretch between residue 142 and 148 (GGGTGSG), (II) residues 180 - 183 (TVVE), (III) residues 242 - 246 (LRFPG, see Fig. 1). (The sequences quoted are those of beta-tubulin; the numbering of residues follows that obtained after optimal alignment of alpha- and beta-tubulin, see Krauhs et al., 1981). Region I probably represents the turn between a beta-strand and an alpha-helix as part of a Rossman fold (Rossman et al., 1975) which is involved in the binding of the phosphates. Region II is thought to represent the end of another beta-strand that could contribute to the binding of the ribose, whereas region III, the end of another beta-strand, may be near the base. The sequences of both alpha- and beta-tubulin are particularly comparable with the FAD-binding domain of glutathione reductase (Schulz et al., 1982). Note that the presumptive binding regions of phosphate, ribose and base occur in the direction from the N- to the C-terminus along the polypeptide chain.

The homologies with the GTP-binding proteins also include a glycine-rich region (residues 18-23 in EF-Tu, 10-15 in p21; Halliday, 1984; Leberman and Egner, 1984). Another region of homology thought to be involved in base-binding of EF-Tu or p21 occurs in the vicinity of residues 60-70 of both tubulins (IV in Fig. 1). Thus the homologies with different classes of proteins support two different base-binding regions, of which only region IV is consistent with the direct photo-crosslinking experiments presented here.

This result is notable in two respects. All the proteins mentioned here have their presumptive base-binding regions downstream from the phosphate-binding region by about 80-120 residues. This order appears to be reversed in tubulin since region IV is upstream from I. This apparent discrepancy need not to be real; it may simply reflect our present poor understanding of GTP-binding pockets which must involve more than one stretch of the polypeptide chain. In fact, the conserved residues may not even be the ones directly responsible for hydrogen bonding to the nucleotide, a large part of which may be provided by the main chain rather than the residues (compare, for example, the details of the FAD-binding pocket in glutathione reductase, Schulz et al., 1982).

It is thus likely that the base-binding regions of the ATP- and GTP-binding proteins derived from homologies represent only one part of the respective polypeptide chain approaching the base; conversely, the homologous region III of tubulin may still be near the guanosine base without being directly photo-crosslinked. It is also possible that different parts of the protein take over the base-binding function in different conformational states, e.g. before and after the hydrolysis of the alpha-phosphate. Note that region IV contains two acid groups in beta-tubulin that could catalyse the photo-crosslinking reaction (Asp 69 or Glu 71), while there is none in region III.

The only crystallographic study of a GTP-binding protein thus far is that of EF-Tu (Jurnak, 1985; La Cour et al. 1985). These authors suggest that the sequence NKCD (135 - 138 in EF-Tu) binds the base and propose the same function for similar sequences in other GTP-binding proteins such as p21 or transducin. If one aligns beta-tubulin and EF-Tu on the basis of the homologies between residues 60 - 70 of beta-tubulin the sequence NKCD (135-138) in EF-Tu would correspond to DLEP (69-72) in beta-tubulin. The relationship between these sequences is not immediately obvious; however, both of them could provide hydrogen bonds to the base.

The study presented here was undertaken in order to learn more about the vicinity of the GTP-binding site of tubulin by comparison with other nucleotide-binding proteins. It seems very likely that the N-terminal GTP-binding domain adopts the general conformation of alpha/beta-proteins (central beta-pleated sheet sandwiched between alpha-helices, see Schulz and Schirmer, 1979); this would be compatible with the X-ray pattern of microtubules (Mandelkow, 1986). Thus it has been possible to propose an approximate folding for part of the tubulin chain (Mandelkow et al., 1985b). Further studies are underway to identify the location of the base-binding residue directly by separation of peptides using HPLC and sequencing.

ACKNOWLEDGEMENTS: We thank Dr. Karin Kirchner for help with some of the digestion experiments and Dr. Eckhard Mandelkow for discussions on protein structure. This project was supported by the Deutsche Forschungsgemeinschaft (DFG) and the Bundesministerium für Forschung und Technologie (BMFT).

## REFERENCES


Bittner,M., Kupferer,T. and Morris,C.F. (1980). Anal. Biochem. 102, 459-471.

Carroll,S., McCloskey,J., Crain,P., Oppenheimer,N., Marschner,T. and Collier,R.(1985). Proc. Natl. Acad. Sci. U.S.A. 82, 7237-7241.

Geahlen, R.L. and Haley, B.E. (1977). Proc. Natl. Acad. Sci. U.S.A. 74, 4375-4377.

Halliday, K.R. (1984). J. Cyc. Nucl. and Prot. Phosphor. Res. 9, 435-448.

Hesse, J., Maruta, H. and Isenberg, G. (1985). FEBS Lett. 179, 91-95.

Jurnak, F. (1985). Science 230, 32-36.

Kirchner, K. and Mandelkow, E. (1985). EMBO J. 4, 2397-2402.

Krauhs,E.,Little,M.,Kempf,T.,Hofer-Warbinek,R.,Ade,W. and Ponstingl,H.(1981). Proc. Natl. Acad. Sci. U.S.A. 78, 4156-4160.

La Cour, T.F.M., Nyborg, J., Thirup, S. and Clark, B.F.C. (1985). EMBO J. 4, 2385-2388.

Laemmli, U.K. (1970). Nature 227, 680-685.

Leberman, R. and Egner, U. (1984). EMBO J. 3, 339-341.

Maccioni, R.B. and Seeds, N.W. (1983). Biochem. 22, 1572-1579.

Mandelkow, E.-M., Herrmann, M. and Rühl, U. (1985a). J. Mol. Biol. 185, 311-327.

Mandelkow, E.-M., Kirchner, K. and Mandelkow, E. (1985b). In "Microtubules and Microtubule Inhibitors 1985", pp. 31-47, M. De Brabander and J. De Mey, eds., Elsevier, Amsterdam.

Mandelkow, E. (1986). Methods in Enzymology 134, 633-657. R.B. Vallee, ed., Academic Press, New York.

Maruta, H. and Korn, E.D. (1981). J. Biol. Chem. 256, 499-502.

Matsudaira, P.T. and Burgess, D.R. (1978). Anal. Biochem. 87, 386-396.

Matsudaira, P., Jakes, R., Cameron, L. and Atherton, E. (1985). Proc. Natl. Acad. Sci. U.S.A. 82, 6788-6792.

Nath, J.P., Eagle, G.R. and Himes, R. H. (1985). Biochem. 24, 1555-1560.

Nath, J.P. and Himes, R. H. (1986). Biochem. Biophys. Res. Comm. 135, 1135-1143.

Rossmann, M.G., Liljas, A., Brånden, C.-I. and Banaszak, L.J. (1975). In "The Enzymes", 3rd ed., vol. 11, pp. 61-102, P. D. Boyer, ed., Academic Press, New York.

Schulz, G.E. and Schirmer, R.H. (1979). Principles of Protein Structure. Springer Verlag, Heidelberg.

Schulz, G.E., Schirmer, R.H. and Pai, E.F. (1982). J. Mol. Biol. 160, 287-308.

Sternberg, M.J.E. and Taylor, W.R. (1984). FEBS Letters 175, 387-392.

Towbin,H., Staehelin,T and Gordon,J. (1979). Proc. Natl. Acad. Sci. U.S.A. 76, 4350-4354.

# CRYSTALLOGRAPHIC AND IMAGE RECONSTRUCTION STUDIES ON RIBOSOMES

A. Yonath
Dept. of Structural Chemistry, Weizmann Inst. of Science, Rehovot, Israel
Max-Planck-Arbeitsgruppen für Strukturelle Molekularbiologie, Hamburg
H.G. Wittmann
Max-Planck-Institut für Molekulare Genetik, D-1000 Berlin (Dahlem), FRG

## Introduction

The translation of genetic information into a polypeptide chain involves specific interactions of ribosomes with mRNAs, aminoacyl-tRNAs, several ions and small molecules as well as a number of proteins including initiation, elongation and termination factors. This intricate and accurate process of biosynthesis of protein molecules occurs in a similar manner on the ribosomes of all organisms. Ribosomes consist of two different subunits which associate upon initiation of protein biosynthesis. Each subunit is a defined assembly of proteins and ribonucleic acid chains with a characteristic sedimentation coefficient (e.g. 30S and 50S for small and large ribosomal subunits from bacteria). During the last two decades a vast amount of information has been accumulated about the function and the chemical, biological and genetic properties of ribosomes (1-4). This has shed light on the entire process of protein biosynthesis, though the understanding of the detailed mechanism is still severely limited by the lack of a molecular model. To this end, we undertook crystallographic studies.

As object for crystallographic studies, ribosomal particles are of enormous size, with no internal symmetry. Furthermore, they are instable and flexible. In spite of this, as a result of an intensive systematic exploration of crystallization conditions and development of an innovative

Modern Methods in Protein Chemistry, Vol. 3

experimental technique for fine control of the volume of the crystallization drop (5), procedures for *in vitro* growth of crystals of intact ribosomal particles have been developed. These procedures proved to be suitable for the reproducible production of ordered three-dimensional crystals and two-dimensional sheets of whole ribosomes (6, 7) as well as of the large ribosomal subunits (8-17). A list which represents the current status is given in Table I.

There is a strong correlation between crystallizability and biological activity. Inactive ribosomal particles cannot be crystallized. Moreover, in all cases, the crystalline material retains its biological activity for long periods, in spite of the natural tendency of ribosomes to disintegrate and in contrast to the short life time of isolated ribosomes in solution. Preservation of activity in the crystalline state accords well with the hypothesis that when external conditions (e.g. hibernation) demand prolonged storage of potentially active ribosomes in living organisms, temporary periodic organization occurs *in vivo* (18, 19). Currently, three-dimensional crystals and two-dimensional sheets grow from virtually all preparations of active large (50S) ribosomal subunits of *Bacillus stearothermophilus* and of *Halobacterium marismortui*. However, due to the intricate nature of the particles, the exact conditions for the growth of well ordered crystals must still be slighty varied for each ribosomal preparation. Moreover, for each crystal form the quality of the crystals depends, in a manner not yet fully characterized, on the procedure used in preparation of the ribosomal subunits and on the bacterial strain.

## Crystallographic Studies

Synchrotron radiation provides the most intense, intrinsically parallel X-ray beam. Thus it permits diffraction patterns to be obtained with short exposures and allows the definition of a very fine focussed beam. Synchrotron radiation is essential for crystallographic data collection from crystals of ribosomal particles. This is due to the large unit cell dimensions (Table I), the shape of the crystals, their fragility and their short life times. Using synchrotron radiation, the unit cell parameters of four

**Table 1.**
Packing parameters of two- and three-dimensional crystals of ribosomal particles

| Source | Crystal Form | Cell Dimensions, (Å) determined by | |
|---|---|---|---|
| | | electron microscopy | X-ray crystallography |
| 70S E.coli | A* | 340x340x590; 120°; $P6_3$ | |
| 70S *Bacillus stearothermophilus* | 2D,AS* | 200x440; 93° | |
| 50S *Halobacterium marismortui* | 1, P* | 310x350; 105° | |
| | 2, P | 148x186; 95° | 147x181; 97° |
| | 3, P | 170x180; 75° | 182x584x186; 109°; $P2_1$ |
| 50S *Bacillus stearothermophilus* | 1, A | 130x254; 95° | |
| | 2, A | 156x288; 97° | |
| | 3, A | 260x288; 105° | |
| | 4, A | 405x405x256; 120° | |
| | 5, A | 213x235x315; 120° | |
| | 6**, A | 330x670x850; 90° | 360x680x920 orthorhombic |
| | 2D, A | 145x311; 82°; P2 | |
| | 2D,AS* | 148x360; 109°; P1 | |

* Crystals are grown by vapor diffusion from alcohols or their mixtures (A), polyethylene glycol (P), or by mixing with ammonium sulphate (AS).

** Same form and parameters for crystals of the large ribosomal subunits of -L11 mutant of the same source.

crystal systems have been determined and interpretable crystallographic data from two of them have been collected.

The best three-dimensional crystals grown so far are of the 50S ribosomal subunits from two bacteria: Bacillus stearothermophilus and Halobacterium marismortui. Bacterial 50S ribosomal subunits are of molecular weight of about 1,600,000 daltons. Their approximate size of 150x170x180Å was determined by electron microscopy as well as a variety of biophysical methods (for reviews see 1, 3). They are composed of 2 RNA chains, which comprise two thirds of the total weight, and 30-35 different proteins.

(1) 50S ribosomal subunits from Halobacterium marismortui

In contrast to ribosomal particles from eubacteria, the halophilic ribosomal particles are stable and active at high salt concentrations. Thus they provide a system for crystallization from salts or non-volatile compounds (15-17). Initially, crystals of these particles have been grown by the standard vapor diffusion technique (21). At this stage fairly disordered micro crystals appeared very quickly, occasionally within a few hours. It was observed that as for other biological systems, there is an inverse correlation between the rate of crystal growth and the internal order of the crystals.

The size of the crystals was increased together with enhancement of their internal order by slowing down the crystallization process. This was achieved in two steps, based on seeding and on variations of the relative concentrations of the ions which are in the natural growth media of these bacteria (K+ and Mg++). Larger and somewhat better ordered crystals were obtained as a result of a drastic reduction of the concentration of KCl in the crystallization mixture as well as in the reservoir (Fig. 1 and in 15, 16). A combination of this procedure with sophisticated seeding led to the growth of well ordered crystals (17). Advantage has been taken of the delicate equilibrium of mono- and divalent ions needed for the growth of the halobacteria as well as of the major role played by the Mg++ concentration in crystallization of ribosomal particles. It was found earlier that three-dimensional crystals of 50S ribosomal subunits from Bacillus stearothermophilus grow in relatively low Mg++ concentration, whereas the production of two-dimensional sheets requires a high Mg++ concentration, at which growth of three-dimensional crystals is prohibited (14). Similarly, for spontaneous crystal growth of 50S subunits from Halobacterium marismortui, the lower the Mg++ concentration is, the thicker the crystals are. With these points in mind, a variation of the standard seeding procedure has been developed. Thin crystals of the 50S subunits from Halobacterium marismortui, grown spontaneously under the lowest possible Mg++ concentration, are transferred to mixtures in which the Mg++ concentration is so low that the transferred crystals dissolve, but after several days new microcrystals can be observed. These are well ordered and 10-20 fold thicker than the original seeds (Fig.1).

Monoclinic (P2 ) crystals of the 50S subunits from Halobacterium marismortui grow as fragile thin plates with a maximum size of 0.6 x 0.6 x 0.2 mm. They diffract to a resolution of 6 Å (Fig. 2), and have relatively small, compactly packed unit cells of a = 182 ± 5 Å, b = 584 ± 10 Å, c = 186 ± 5 Å, and beta = 109°, in contrast to the "open" structure of the large crystals of Bacillus stearothermophilus (Table I and Fig. 3).
Although between -2°C and 4°C up to 15 photographs can be taken from an individual crystal, the high resolution terms appear only on the first 2-3 X-ray diffraction patterns. Hence over 260 crystals were irradiated in order to obtain a complete data set, and X-ray diffraction data were obtained from crystals which were aligned only visually, with a similar algorithm to that applied for rhino virus (20).

Two of the unit cell edges (a=182 Å and c=186 Å) are approximately the size of the 50S ribosomal subunits from eubacteria. The third unit cell edge (b=584 Å) is parallel to a twofold screw axis and corresponds to the thin dimension of the crystals. Judging from the approximately known size of the 50S particles (1-4) and from the dimensions of the unit cell, it is clear that each asymmetric unit contains two particles.

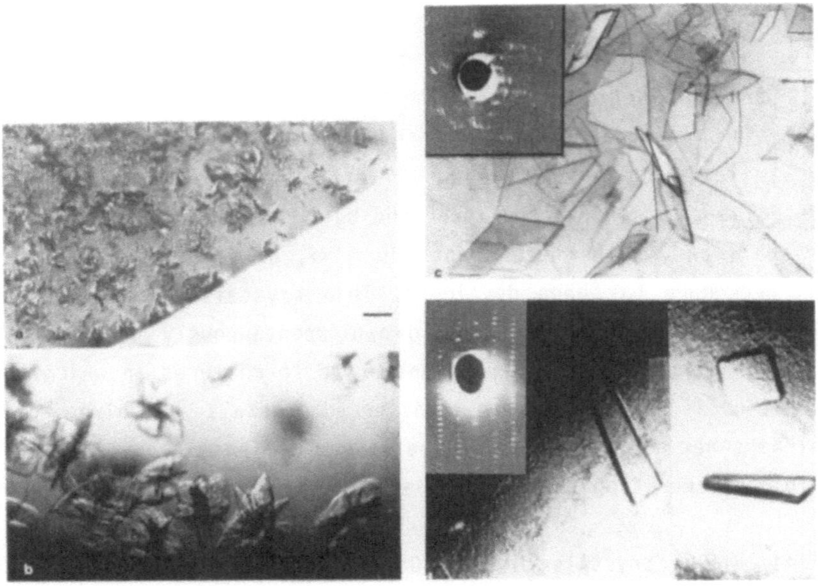

**Fig.1** Growth of large, ordered three-dimensional crystals of the 50S ribosomal subunits from Halobacterium marismortui by vapor diffusion at 19°C (Bar length = 0.2mm).
a) Microcrystals obtained within 1-2 days from 7-8% polyethylene glycol in the presence of 2.5 M KCl, 0.5 M $NH_4Cl$, 0.15-0.20 M $MgCl_2$ and 10 mM spermidine in the crystallization mixture (pH = 5.0-5.2), and equilibrated against 3.0 M KCl, 9% polyethylene glycol, 0.5 M $NH_4Cl$ and 0.20 M $MgCl_2$.
b) Crystals obtained within 2-3 days in droplets containing lower KCl concentration than used in (a). The droplet of 4-5% polyethylene glycol in the presence of 1.2-1.7 M KCl, 0.5 M $NH_4Cl$, 0.10 M $MgCl_2$ and 10 mM spermidine.
c) Crystals obtained within 3-5 days from droplets similar to those used for (b), equilibrated with reservoirs of lower KCl concentrations. The droplet of 4-5% polyethylene glycol in the presence of 1.2 M KCl, 0.5 M $NH_4Cl$, 0.05-0.10 M $MgCl_2$ and 10 mM spermidine in the crystallization mixture (pH = 5.0-5.6), and equilibrated against 1.7 M KCl, 9% polyethylene glycol, 0.5 M $NH_4Cl$ and 0.10 M $MgCl_2$. An X-ray diffraction pattern taken perpendicular to the thin axis of the crystals, obtained under conditions similar to those described in Fig. 2, is inserted.

d) Crystals obtained by seeding of crystals from (c) in a crystallization drop containing 5% polyethyleneglycol, 1.2 M KCl, 0.5 $NH_4Cl$, and 0.03 M $MgCl_2$, at pH = 5.6, which was equilibrated with 7% PEG, 1.7 KCl, 0.5 M $NH_4Cl$ and 0.03M $MgCl_2$, pH = 5.6.
Seeds were small, well-shaped crystals, transferred in a stabilization solution of 7% polyethylene glycol in 1.7 M KCl, 0.5 M $NH_4Cl$, and 0.05 M $MgCl_2$ at pH = 5.6.
An X-ray diffraction pattern taken perpendicular to the thin axis of the crystals, obtained under conditions described in Fig. 2, is inserted.

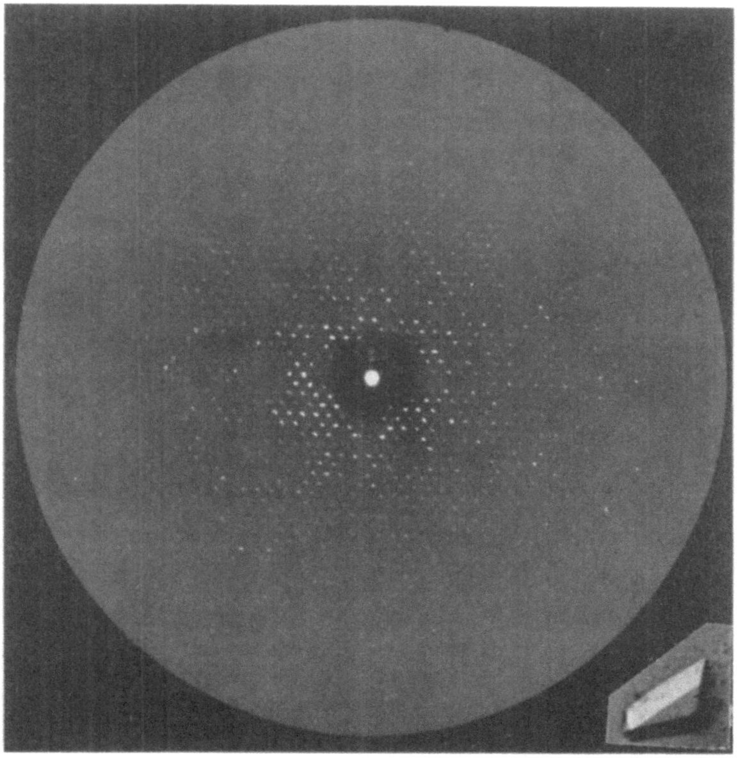

**Fig.2** 1° rotation pattern of the h 0 l zone obtained from the inserted crystal. The pattern was obtained at 4°C with synchrotron radiation (X11 station at EMBL/DESY operating at 5 GeV, initial current 34 mA). The X-ray beam of wave length 1.51Å was collimated by vertical and horizontal slits (0.3x0.3mm). Exposure time = 2 min, crystal to film distance = 235mm. The high mosaic spread of the crystals is the reason for an apparent "precession" pattern. Insert: A crystal grown under the condition of Fig. 1(d).

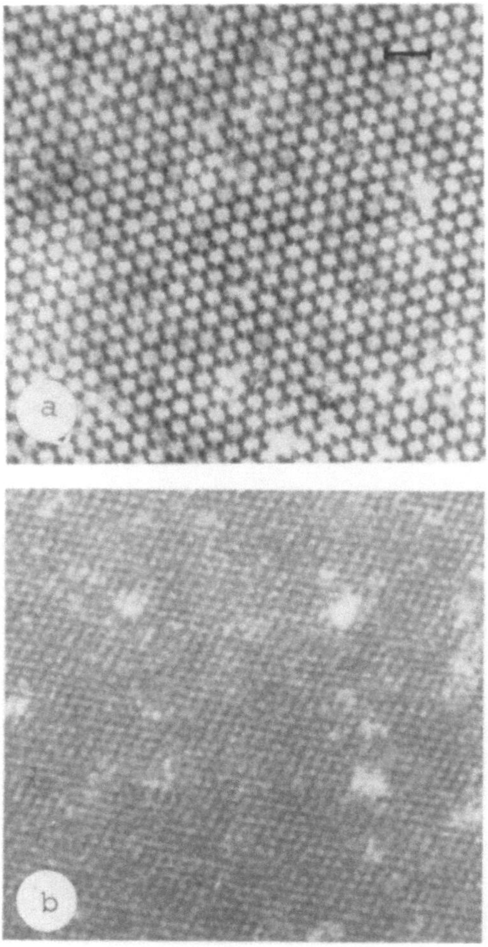


**Fig. 3** Electron micrographs of positively stained (2% uranyl acetate), thin sections of crystals of the 50S subunits from Bacillus stearothermophilus (a) and Halobacterium marismortui (b) similar to those shown in Figs. 5(b) and 1(d), respectively. The crystals have been fixed in 0.2% glyceraldehyde and embedded in resin ERL 4206.
Micrograph (a) shows the characteristic "open" packing of crystal forms # 4 and # 6 (Table I) in contrast to the compact packing shown in (b).

(2) 50S ribosomal subunits from Bacillus stearothermophilus

Because ribosomes from eubacteria fall apart at high salt concentrations, volatile organic solvents have to be used as precipitants. A modified version of the standard hanging drop technique, using glass plates, was developed (Fig. 4 and in 5). The crystallization droplets, which contain no precipitant or an extremely small quantity of it, were equilibrated with a reservoir containing the precipitant. Attempt to increase the size of these crystals by a drastic slow down of the crystallization process failed, since ribosomal particles from this source are very unstable, and may deteriorate before they are able to aggregate and form proper nucleation centers (21).

Growing crystals from volatile organic solvents imposes many technical difficulties in manipulating, data collection and heavy-atom derivatization. In fact, any handling of the crystals, such as removing or reorienting the crystals, or replacing the growth medium with a different solution, is extremely difficult. Thus, for this system seeding was virtually impossible. However, reducing the size of the exposed surface of the crystallizing droplet led to the production of large crystals. This was achieved by growing crystals directly in X-ray capillaries (Fig. 4).

Currently crystals from the 50S subunits of wild-type and mutated (see below) Bacillus stearothermophilus which may reach length of 2.0 mm and a cross-section of 0.4 mm (Fig. 5) are obtained at 4°C from mixtures of methanol and ethylene glycol (10 - 13). Since most of the crystals grow with one of their faces adhering to the walls of the capillaries, it is possible to irradiate them without removing the original growth solution. Experiments prove that crystals kept in growth solution last significantly longer in the X-ray beam and rarely develop as many cracks as crystals which have been slightly dried. Consequently, the X-ray diffraction patterns extend to higher resolution, and data of better quality may be obtained. Although most of the crystals grow with their long axes parallel to the capillary axis, a fair number grow in different directions. Thus, it was possible to determine the unit cell constants (Table I) and to obtain

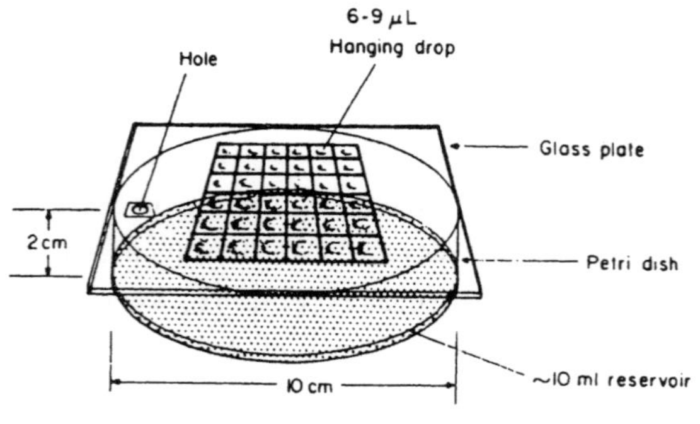


a

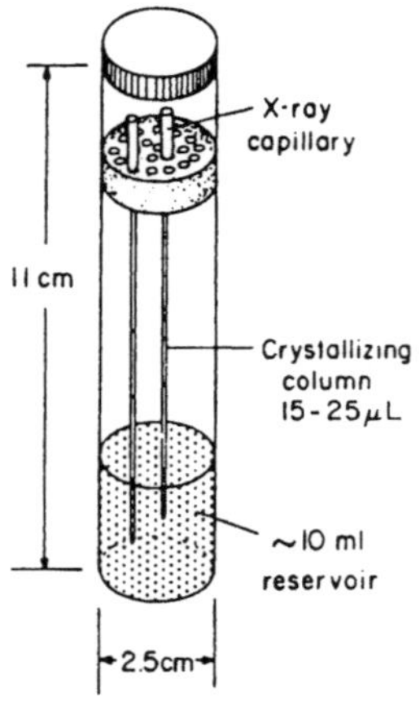


b

**Fig. 4.** a. The crystallization system used for vapor diffusion in capillaries. b. The crystallization system used for vapor diffusion in hanging drops on glass plates.

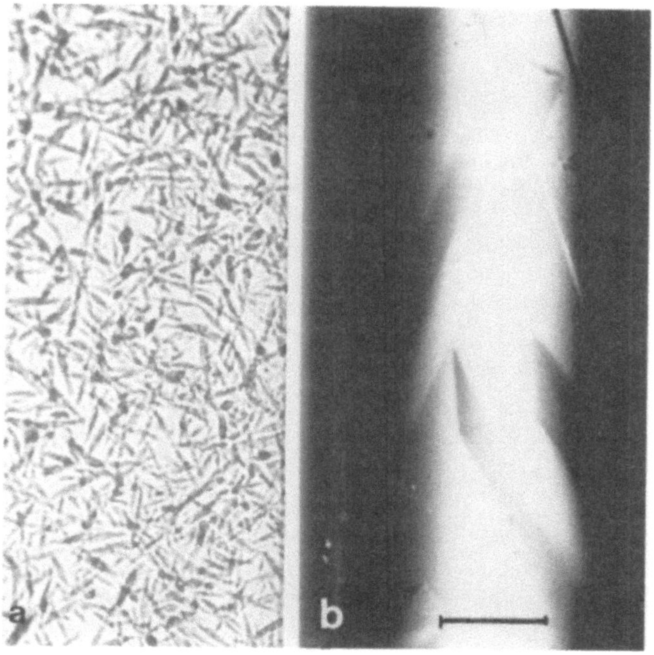


Fig. 5 Left: Crystals of 50 subunits from B. stearothermophilus grown in glass petri dishes by vapor diffusion at 4°C. Crystallization mixture of 7 microliters of 50S subunits (10-20 mg/ml) in H-I buffer (2), 10 mM spermine, 1% methanol, 10 mM Hepes or glycine buffer, pH=8.4, was equilibrated with 10% methanol, 1.0 M NaCl, pH=8.4.
Right: Crystals of the same subunits grown in 0.5 mm X-ray capillaries by vapor diffusion at 4°C. Crystallization mixture of 20 microliters 50S subunits (same condition as for the left sample), was equilibrated with a reservoir of 12% methanol, 12% ethyleneglycol, 0.5 M NaCl, pH=8.4.
Bar length = 0.2 mm.

diffraction patterns from all of the zones without manipulating the crystals (Fig. 6).

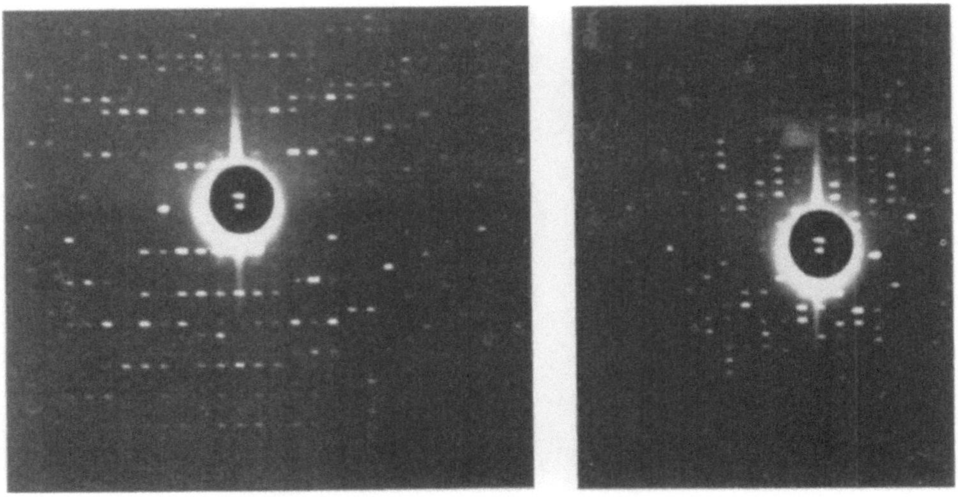

**Fig. 6.** X-ray diffraction patterns from crystals shown in Fig.5(b), obtained at -4°C with synchrotron radiation (A1 station at CHESS/CORNELL University, operating at 5GeV, current 30-40 mA) with 0.3 mm collimated X-ray beam of wave length of 1.55 Å, on a HUBER precession camera equipped with a He path. Exposure time = 3 min, crystal to film distance = 200 mm.

The crystals, of space group P222 or $P2_12_12$ are loosely packed (Fig. 3) in a unit cell of 360 x 680 x 920 Å. Though we propose that the ribosomes are packed in a primitive lattice, at low resolution the <u>h</u>0<u>l</u> zone does show extinctions, characteristic of a centered lattice (<u>h</u> + <u>l</u> = 2n + 1). The current crystals diffract to 13 Å resolution at best (typically 15-18 Å). They often last a few hours in the synchrotron beam permitting up to 35 photographs to be taken from an individual crystal (under optimal conditions, still photographs require about 5-8 minutes exposure and a 1.5° rotation about 10-15 minutes).

Oriented arcs and distinct spots, extending to 3.5 Å, with spacings similar to those measured from diffuse diffraction patterns of ribosome gels and extracted rRNA (22-24) have been detected on several diffraction patterns of single crystals as well as on those of samples containing large numbers of microcrystals. For aligned crystals the average arc length is ± 30°. Such patterns may arise from partial orientation of the nucleic-acid component within the particle.

The process of crystal growth is initiated by nucleation. Although many biological molecules and complexes have been crystallized, little is known about the mechanism of nucleation. Theoretical models have been developed for the nucleation of crystals of small molecules (25-26). However, most of the data currently available concerning the process of nucleation of crystals of biological systems are based on rather indirect evidence, such as monitoring aggregation under crystallization conditions by scattering techniques (27). Because ribosomal particles are large enough to be detected by electron microscopy, crystals of ribosomal particles provide an excellent system for direct investigation of nucleation. The crystallization process was interrupted before the formation of mature crystals, and the crystallization medium was examined by electron microscopy. It was found that the first step in crystal growth is unspecified aggregation and that nucleation starts by a rearrangement within the aggregates (21).

## Phase Determination

### (1) Heavy atom derivatives

The most common method in protein crystallography to derive phases is multiple isomorphous replacement (MIR). For an object as large as the 50S ribosomal subunit, it is necessary to use extremely dense and compact compounds as heavy atom derivatives. Heavy-atom clusters are most suitable for this purpose. Examples of candidates for this purpose are tetrakis(acetoxymercuri)methane which was the key heavy atom derivative in the structure determination of nucleosomes and the membrane reaction center (28, 29), and an undecagold cluster, in which the gold core has a diameter of 8.5 Å (30).

Preliminary experiments show that crystals of Halobacterium marismortui soaked in 1-2 mM gold cluster for 6-12 hours are isomorphous to the native crystals, and their X-ray patterns show intensity differences.

Because the surface of the ribosomal subunit is a composite of a variety of potential binding sites for such clusters, in parallel to soaking experiments attempts to bind the heavy-atom derivatives covalently to a few specific sites on the ribosomal particles are in progress. This may be achieved either by direct interaction of a heavy-atom cluster with chemically active groups such as -SH or ends of rRNA (31) on the intact particles prior to crystallization or by covalent attachment of a cluster to tailor-made carriers which bind to one or a few specific sites on ribosomes. The gold cluster has been prepared as a monofunctional labeling reagent (30). This compound interacted, in turn, with exposed -SH groups on the entire 50S ribosomal subunit. It has been shown that for Bacillus stearothermophilus only two ribosomal proteins, L11 and L13, participated in the reaction and for Halobacterium marismortui predominantly one protein was involved.

Appropriate carriers may be antibiotics (32), DNA oligomers complementary to exposed single-stranded rRNA regions (33) and Fab molecules specific to ribosomal proteins. Since most of the interactions of these materials are characterized biochemically, the crystallographic location of the heavy-atom compounds will not only be used for phase determination but will also reveal the location of specific sites on the ribosome.

Alternatively, such clusters may be attached to chosen sites on isolated ribosomal components which will subsequently be incorporated into particles in which they are missing. A mutant of Bacillus stearothermophilus which lacks protein L11 was obtained by growing cells in the presence of thiostrepton at 60°C (34). The 50S mutated ribosomal subunits crystallize in two and three dimensions under the same conditions as, and are isomorphous to, those obtained from the 50S ribosomal subunits of the wild type (12). This shows that L11, the missing protein, is not involved in crystal forces in the native crystals. Furthermore, protein L11 has only one sulfhydryl group, and binding of N-ethylmaleimide to it does not reduce the activity and crystallizability of the reconstituted modified particles.

(2) Neutron diffraction, molecular replacement and electron microscopy

Phase information may also be obtained by neutron diffraction, as performed for the nucleosomes (35). In addition, the application of real- and reciprocal space rotation searches (36) should be feasible by taking advantage of the non-crystallographic symmetry within the asymmetric units. In later stages, high resolution information obtained from crystallographic studies of isolated individual ribosomal components may be used for iterative phase determination by the molecular replacement methods, assuming that the conformations of crystallized isolated components are sufficiently similar to their conformations within the particle.

The large size of ribosomal particles, which is an obstacle for crystallographic studies, permits direct investigation by electron microscopy. Electron microscopy of crystalline ribosomal particles can be used to locate and orient the particles within the crystals. A model obtained by three-dimensional image reconstruction of two-dimensional sheets may be used for gradual phasing of low resolution crystallographic data. These phases, in turn, may help locate heavy-atom sites for the isomorphous replacement technique. To this end, we have initiated three-dimensional image reconstruction studies.

## Three-Dimensional Image Reconstruction

Three-dimensional image reconstruction permits determination of three-dimensional structures from periodically ordered arrays. It is useful for determination of structures of large biological macromolecules and assemblies at moderate resolution. In spite of some inherent limitation three-dimensional image reconstruction has grown in popularity. It involves averaging, by Fourier transformation, of images obtained by electron microscopy of tilt series of periodically organized identical objects.

Most of the two-dimensional sheets from prokaryotic ribosomal subunits have been grown _in vitro_ from low molecular weight alcohols (AL) by vapor diffusion in hanging drops (14, 37). Recently, mixtures of salts and alcohols have been used for the growth of two-dimensional sheets in depression slides or on electron microscopy grids (ST sheets), (7, 14, 38). Well ordered two-dimensional sheets from the 50S ribosomal subunits from _Bacillus stearothermophilus_ have been subjected to three-dimensional image reconstruction studies at 30 Å resolution (40).

Both (AL) and (ST) two-dimensional sheets consist of small unit cells: 145 $\pm$ 3 x 311 $\pm$ 14 Å, gamma = 89 $\pm$ 2.6° for the alcohol grown arrays, and 148 $\pm$ 10 x 370 $\pm$ 20 Å, gamma = 109 $\pm$ 3° for those grown from salts. These cell dimensions are close to those of forms 1 and 2 (41) of the three dimensional crystals from the same source (Table I). Both (AL) and (ST) sheets are well ordered, and optical diffraction patterns of electron micrographs of negatively stained specimens with gold-thioglucose extend to 30 Å and 28 Å, respectively (Fig. 7).

Because only a small fraction of the particles in the crystallization medium comprises the two-dimensional sheets, these cannot be separated from the rest of the crystallization mixture. Thus, evidence for the integrity of the ribosomal particles in the sheets may be obtained indirectly by testing the biological activity and the migration profile of the entire crystallization medium.

Images were recorded at a minimum dose, of the specimens tilted at angles up to 55°, at intervals of 5°, with respect to the incident electron beam. Ordered area images were selected by optical diffraction and scanned with a 25 micron raster, corresponding to 9 Å at the specimen. For Fourier transforms, optical densities of arrays (512 x 512) were used, and reciprocal

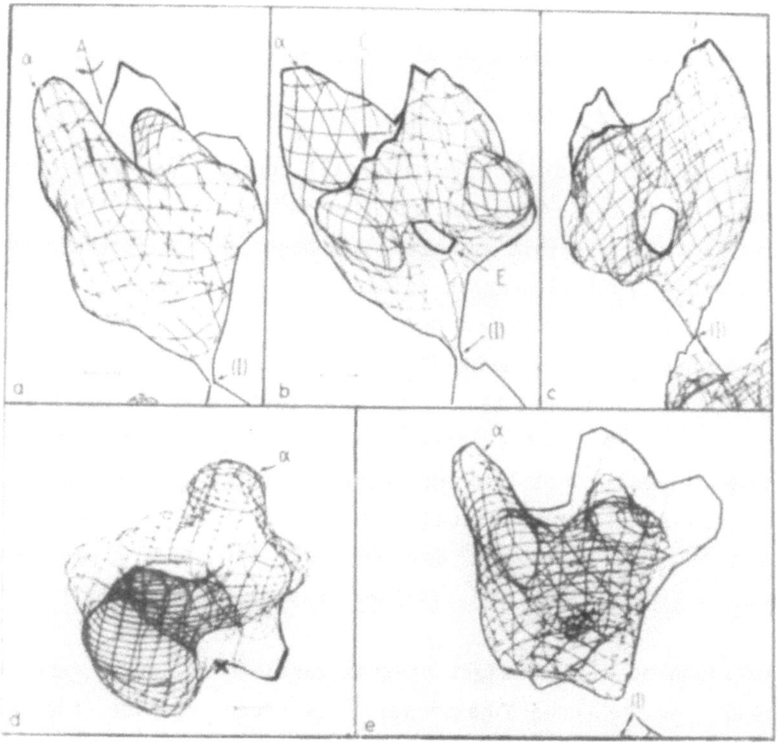


**Fig. 7** Computer graphic display of the outline of the reconstructed model of the 50S ribosomal subunit at 30Å resolution. The front is contoured with heavy lines, the back with lighter ones.
(a) A side view of the model. The entire particle and part of a second one are shown. The arrow (I) is pointing at the crystal contact between the two particles. (A) marks the approximate axis around which the model was turned to obtain the view shown in (b).
(b) The model shown in (a) rotated about the (A) axis. (C) points at the cleft between the projecting arms, at the site it turns into the tunnel. The exit point of the tunnel is marked (E).
(c) The model shown in (a) and (b), viewed into the branch of the tunnel from the exit point.
(d) A view into the tunnel from the cleft.
(e) The model, viewed in a projection which resembles models derived from electron microscopy of single particles.

lattices were fitted to the calculated diffraction patterns. Amplitudes and phases were extracted from peaks of at least two-standard deviations above background.

For both (ST) and (AL) crystals each unit cell contains two particles with dimensions similar to those obtained by other methods (1-4). The alcohol-grown sheets contain a twofold symmetry. Although it appears that the (ST) sheets also contain crystallographic symmetry, it was found that this holds only for the low resolution data. Three-dimensional image reconstructions were therefore carried out assuming $P2_1$ symmetry for the (AL) crystals, and P1 symmetry for the salt-produced ones. In a test case, in which twofold symmetry was introduced for the latter system, it was found that some of the fine details easily resolved in regular reconstructions became less interpretable. In all reconstructions, the two particles in the unit cell show the same essential features. However, when no symmetry was applied, generally one particle in the unit cell was resolved better than the other.

Based on the known molecular weight of this particle ($1.6 \times 10^6$), and on the value obtained from the three-dimensional image reconstruction thickness of the sheets (160-170Å), the calculated density of a 50S particle is 1.3-1.4 gr/cm , and the Vm for a hypothetical crystal is 2.6-2.7Å /dalton. Both are in good agreement with values tabulated and calculated for other large nucleoprotein structures (39-42).

Interparticle contacts within the sheets are clearly revealed in the resulting three-dimensional map. The best defined contact area is of 10-15 Å diameter, compatible with the regular intermolecular contacts found in crystals of proteins of nucleic acids. As details as small as these are beyond the resolution of our map, we cannot yet identify the nature of these contacts.

Several striking features have been revealed in the resulting model (Fig. 7). The concave surface of the particle consists of several protrusions of 25-30Å diameter, the approximate size of globular proteins of molecular weights typical of many ribosomal proteins. Several projecting arms, two of which are longer than the others, are arranged radially around the edge (upper side of the particle shown in Fig. 7), near the presumed interface with the 30S subunit. A narrow elongated cleft is formed between the projecting arms and turns into a tunnel of a diameter of up to 25Å and 100-120Å length. Another long arm is located on the other side of the particle (bottom of the particle shown in Fig. 7).

Thirteen reconstructed particles were studied. The tunnel is present in all reconstructions of the (ST) sheets independent of the staining material. Furthermore, indications of such a tunnel are detected in some filtered images of particular tilts (Fig. 8). In every reconstructed particle there is a region of low density which branches off the tunnel to form a Y (or V) shape and terminates on the opposite side of the particle (Fig. 7). As yet the exact nature of this region cannot be determined. It may be a loosely packed protein region, but in some reconstructed models the density of this region is so low that it appears as a branch of the main tunnel.

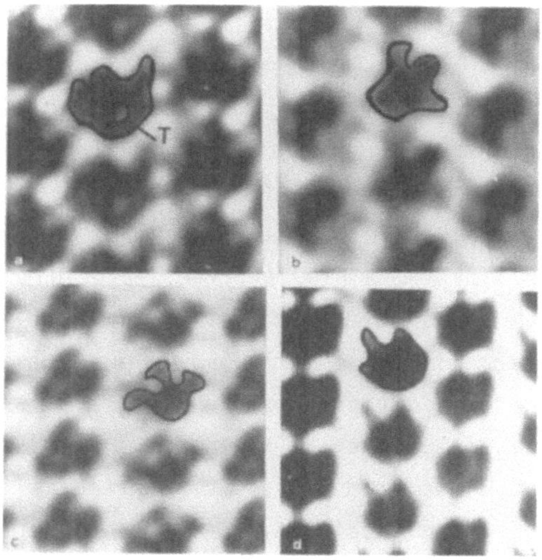


**Fig. 8** Filtered images of electron micrographs in which the depicted view resembles that derived from electron microscopy of single particles.
(a) (AL) sheets stained with uranyl-acetate. Tilt angle of 25°.(b) (ST) sheets stained with gold-thioglucose, tilted by 35°, 20° and 30°. In (b) and (c) "conventional views" can be seen (tilts of 35°and 20°). (d) The tunnel T can be detected (tilt of 30°).

Three-dimensional image reconstruction was applied to two-dimensional sheets stained with uranyl acetate, too. Although the arrays used were large and well ordered, the diffraction patterns of their micrographs extended only to 32Å, and were of lower quality than those from gold-thioglucose. This might stem from the fact that uranyl acetate is not a pure negative stain and may interact with selected parts of the particle (presumably the rRNA). This interaction may depend on the accessibility to the stain and therefore be somewhat irregular. Consequently, the reconstructed model shows less details. For the (ST) sheets, the essential features - the concave shape, the tunnel and the projecting arms - are resolved.

Comparison of this model with that obtained from the same crystals (ST) stained with gold-thioglucose shows two regions (on one of the long arms and on the body of the particle) where uranyl acetate, acting as a positive stain, is incorporated into the particle. This may indicate that in these regions the rRNA is concentrated and/or easily exposed to the stain.

Several models for the 50S ribosomal subunits from E. coli have been suggested previously based on electron microscopial visualization, reconstruction or averaging of single particles. Our model can be positioned so that its projected view resembles the usual image seen when single particles are investigated by electron microscopy (Fig. 7). In addition, there are a few filtered images of two-dimensional sheets tilted by certain angles, which show the same shape and include the characteristic features that have been visualized by electron microscopy of single particles (Fig. 8). At the same time there are, however, major discrepancies in the nature of the gross structural features between our model and the others, which probably stem from the basic differencies between visualization of isolated particles in projection and the inherently more objective character of structure analysis by diffraction methods. There are a few preferred orientations in which the isolated particles tend to lie on grids. As a result of the contact with the flat grids, their projected views are likely to be somewhat distorted, and their surfaces may appear flat rather than curved. In contrast, particles within the crystalline sheets are held together by interparticle contacts. These contacts construct a network which may stabilize the conformation of the particles and decrease, or even eliminate, the influence of the flatness of the grids. Thus, the particles within ordered arrays provide a suitable starting point for reconstruction studies.

In order to understand why some of the features revealed in our studies have not yet been detected in previous investigations, we have reconstructed a model selecting from the entire set of diffraction data only the low resolution (55Å) terms. In spite of the low resolution the tunnel is clearly resolved. However, at this resolution the particle is almost spherical and shows only two thick and short arms instead of the elongated ones

resolved in the high resolution studies. It should be mentioned that at the same resolution (55Å), a similar feature, a narrow elongated region of low density was also detected on ribosomes from lizard by three-dimensional image reconstruction (43).

The functional significance of the tunnel is still to be determined. However, originating at the presumed site for actual protein biosynthesis and terminating on the other end of the particle, and being of a diameter large enough to accomodate even the largest amino acids, this tunnel appears to provide the path taken by the nascent polypeptide chain. Furthermore, this tunnel is of a length which could accomodate and protect from proteolytic enzymes, a peptide of about 40 amino acids in an extended conformation (44-46). It remains to be seen whether the tunnel terminates at a location compatible with that assigned by immune electron microscopy as the exit site for the growing polypeptide chain (47).

Recently ordered two-dimensional sheets of the whole ribosome from _Bacillus stearothermophilus_ have been obtained, using a similar method to that used for the growth of the (ST) sheets (Fig. 9). Stained with uranyl acetate these sheets diffract to 35Å, and seem to be suitable for three-dimensional image reconstruction.

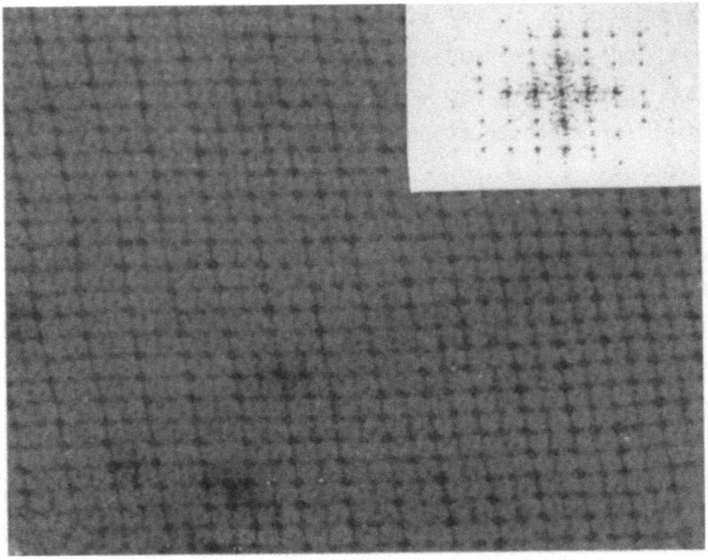

**Fig. 9** Image of salt grown two-dimensional sheet of 70S from Bacillus stearothermophilus, stained with uranyl acetate. Magnification 28.000.

Acknowledgement

This work was supported by BMFT (05 180 MP BO), Minerva and NIH (GM 34360-01A1) research grants.

References

(1) Chambliss, G., G.R. Craven, J. Davies, L. Kahan and M. Nomura. eds. 1980. Ribosomes: Structure, Function and Genetics. Univ. Park Press, Baltimore.

(2) Wittmann, H.G.. 1982. Annual Rev. Biochem. 51, 155-187.

(3) Wittmann, H.G.. 1983. Annual Rev. Biochem. 52, 35-65.

(4) Hardesty, B. and G. Kramer, eds. 1986. In: Structure, Function and Genetics of Ribosomes. Springer Verlag, Heidelberg and N.Y..

(5) Yonath, A., J. Müssig and H.G. Wittmann. 1982. J. Cell Biochem., 19, 145-155.

(6) Wittmann, H.G., J. Müssig, H.S. Gewitz, J. Piefke, H.J. Rheinberger and A. Yonath. 1982. FEBS Letters 146, 217-220.

(7) Piefke, J., T. Arad, I. Makowski, H.S. Gewitz, B. Hennemann, A. Yonath and H.G. Wittmann. 1986. FEBS Letters, 209, 104-106.

(8) Yonath, A., J. Müssig, B. Tesche, S. Lorenz, V.A. Erdmann and H.G. Wittmann. 1980. Biochem. Inter. 1, 428-435.

(9) Yonath, A., J. Piefke, J. Müssig, H.S. Gewitz and H.G. Wittmann. 1983. FEBS Letters, 163, 69-78.

(10) Yonath, A., H.D. Bartunik, K.S. Bartels and H.G. Wittmann. 1984. J. Mol. Biol. 177, 201-206.

(11) Yonath, A., M.A. Saper, I. Makowski, J. Müssig, J. Piefke, H.D. Bartunik, K.S. Bartels and H.G. Wittmann. 1986. J. Mol. Biol. 187, 633-636.

(12) Yonath, A., M.A. Saper, F. Frolow, I. Makowski and H.G. Wittmann. 1986. J. Mol. Biol., 192, 161-162.

(13) Yonath, A., M.A. Saper and H.G. Wittmann. 1986. In: Structure, Function and Genetics of Ribosomes (B. Hardesty and G. Kramer, eds.), Springer Verlag, Heidelberg and N.Y., pp. 112-127.

(14) Arad, T., K.R. Leonard, H.G. Wittmann and A. Yonath. 1984. EMBO J.3, 127-131.

(15) Shevack, A., H.S. Gewitz, B. Hennemann, A. Yonath and H.G. Wittmann. 1985. FEBS Letters, 184, 68-71.

(16) Shoham, M., J. Müssig, A. Shevack, T. Arad, H.G. Wittmann and A. Yonath. 1986. FEBS Letters, 208, 321-324.

(17) Makowski, I., F. Frolow, M.A. Saper, H.G. Wittmann and A. Yonath. J. Mol. Biol., in press.

(18) Barbieri, M.. 1979. J. Supramol. Struct. 10, 349-357.

(19) Liljas, A.. 1982. Progr. Biophys. Mol. Biol. 40, 161-198.

(20) Rossmann, M.G. and W. Erickson. 1983. J. Appl. Cryst. 16, 629-636.

(21) Yonath, A., G. Khavitch, B. Tesche, J. Müssig, S. Lorenz, V.A. Erdmann and H.G. Wittmann. 1982. Biochem. Inter. 5, 629-636.

(22) Klug, A., K.C. Holmes and J.T. Finch. 1961. J. Mol. Biol. 3, 87-100.

(23) Zubay, G. and M.H.F. Wilkins. 1960. J. Mol. Biol. 2, 105-112.

(24) Langridge, R. and K.C. Holmes. 1962. J. Mol. Biol. 5, 611-617.

(25) Gibbs, J.W.. 1928. In: Collected Works of J.W. Gibbs.Longmas Green, N.Y., pp. 55-70.

(26) Zettelmeyer, A.C.. 1969. In: Nucleation. Marcel Dekker, New York.

(27) Kam, Z., H.B. Shore and G. Fehder. 1978. J. Mol. Biol., 123, 539-557.

(28) Deisenhofer, J., O. Epp, K. Mikki, R. Huber and H. Michel. 1984. J. Mol. Biol. 180, 385-398.

(29) Richmond, T., J.T. Finch, B. Rushton, D. Rhodes and A. Klug. 1984. Nature 311, 533-537.

(30) Yang, M. and P.A. Frey. 1984. Biochem. 23, 3857-3862.

(31) Odom, O.W.Jr., D.R. Robbins, J. Lynch, D. Dottavio-Martin, G. Kramer and B. Hardesty. 1980. Biochemistry 19, 5941-5947.

(32) Nierhaus, K.H. and H.G. Wittmann. 1980. Naturwissenschaften 67, 234-288.

(33) Hill, W.E., B.E. Trappich and B. Tassanakajohn. 1986. In: Structure, Function and Genetics of Ribosomes (B. Hardesty and G. Kramer, eds.), Springer Verlag, Heidelberg and N.Y., 233-252.

(34) Schnier, J., H.S. Gewitz, B. Leighton. 1986. In preparation.

(35) Bently, G.A., A. Lewit-Bently, J.T. Finch, A.D. Podjarny and M. Roth. 1984. J. Mol. Biol. 176, 55-77.

(36) Rossmann, M.G..1972. In: Molecular Replacement Method. A Collection of Papers on the Use of Non-Crystallographic Symmetry. 1977. Gorden & Breach, New York.

(37) Clark, M.W., K.R. Leonard and J.A. Lake. 1982. Science 216, 999-112.

(38) Yonath, A., K.R. Leonard and H.G. Wittmann. 1987.Science, in press.

(39) Matthews, B.W.. 1968. J. Mol. Biol. 33, 491-497.

(40) Hogle, J.M.. 1982. J. Mol. Biol. 160, 663-668.

(41) Liljas, L., T. Unge, A. Jones, K. Fridborg, S. Lovgren, U. Skoglund and B. Strandberg. 1982. J. Mol. Biol. 159, 93-108.

(42) Finch, J.T., L.C. Lutter, D. Rhodes, R.S. Brown, B. Rushton, M. Levitt M. and A. Klug. 1977. Nature 269, 29-32.

(43) Milligam, R.A. and P.N.T. Unwin. 1986. Nature 319, 693-696.

(44) Malkin, L.I. and A. Rich. 1967. J. Mol. Biol. 26,329-346.

(45) Blobel, G. and D.D. Sabatini. 1970. J. Cell. Biol. 45, 130-145.

(46) Smith, W.P., P.C. Tai and B.D. Davis. 1978. Proc. Natl. Acad. Sci. USA 75, 5922-5925.

(47) Bernabeau, C. and J.A. Lake. 1982. Proc. Nat. Acad. Sci., USA 79, 3111-3115.

# Cell Culture Techniques for Testing of Biologically Active Peptides and Drugs: Clonogenic Assays Using Agar-Containing Glass Capillaries

H. R. Maurer

Institut für Pharmazie der Freien Universität Berlin
D-1000 Berlin 33, Germany

## Introduction

Nowadays excellent methods are available to isolate and characterize peptides and proteins by biochemical and biophysical parameters on a molecular level. However the determination and evaluation of the biological activities of the isolated peptides and proteins still raise serious problems. Many biological assays are subject to wide variations and potential artifacts. In addition, some assays are time-consuming and tedious, thus delaying the progress of the biochemical work. This is in particular true for most in vivo assays. Moreover increasing costs of animal experimentation and stringent animal protection laws have fostered the development of sophisticated in vitro assays. Indeed various improvements of cell and tissue culture methods do support these trends and seem to justify further work to refine these techniques (1,2). Although in vivo assays using animals are undoubtedly closer to any clinical situation than in vitro assays using cell cultures, some arguments favour the application of in vitro assays:

- Less sample quantities to be tested;
- less time spent, less animals, less costs;
- thus more substances can be assayed.

Originally tissue culture started with the culture of whole fragments of explanted organs. In the past a great deal of cell biology was learnt from relevant organ culture studies. It should be emphasized that organ cultures will preserve cell and matrix interaction, retain histological

Modern Methods in Protein Chemistry, Vol. 3

and biochemical differentiation for longer and will normally remain in a non-growing steady state for several days and even weeks. However the effects of any agents and drugs to be tested on organ cultures are usually difficult to quantitate and evaluate, since variations in geometry and constitution greatly prejudice the reproducibility of organ culture tests.

Cell cultures on the other hand usually constitute an artificial system with randomly dispersed, single cells at the start of the culture. Such cells may remain single in suspension or may form a monolayer on a support with the cells attached to each other. However the original three-dimensional network of precisely positioned cells is generally not attained. The cells have thus lost the biological and biochemical properties associated with the tissue architecture and may adopt "new" growth and differentiation characteristics. They may be propagated and may give rise to defined cell populations or cell lines that retain specific uniform properties for a number of passages. Such cell lines can usually be preserved by freezing, thus enabling repeated studies. They can be purified by growth in selective media, physical cell separation and cell cloning. These advantages are prerequisites for reproducible studies, e.g. on drug effects.

However it should be kept in mind that cell lines rarely reflect the phenotypic expression of the tissue from which they are derived. Apart from differentiated tumor cells, culture conditons may select for the proliferating cell compartment of the tissue or induce partially differentiated cells to revert to a precursor state (1). Indeed the processes of proliferation and differentiation may reveal distinct qualitative and quantitative differences between in vitro and in vivo upon careful analysis.

Nevertheless the advantages of cell cultures mentioned above allow, at least, a follow-up of drug effects on the growth properties of cells. A great number of experimental studies seem to support the view that such cultures may provide meaningful pharmakological and toxicological data.

- It may well be that spheroid cultures will, in the future, avoid some of the disadvantages of cell cultures while retaining some properties of organ cultures (3).

## Cell cultures for in vitro screening of growth effects by agents

A variety of assays have been developed to measure the stimulatory and/or inhibitory effects of agents on the proliferative capacity of cells in culture. The methods can be divided into

a) Cellular assays: Cell number, properties of morphology, karyology, cloning capacity are determined;
b) Biochemical assays: Uptake or incorporation of nucleic acid precursors ($^3$H-thymidine or $^3$H-uridine) are determined.

However, among the latter method, the widely used $^3$H-thymidine uptake procedure is subject to several potential artifacts rendering results obtained with this method meaningless (4). In order to avoid misinterpretations a counter-check with other methods to measure cell proliferation is greatly recommended. On the other hand cellular assays are more time-consuming and require appropriate experience in cell morphology.

## Clonogenic assays with hematopoietic cells

So-called hematopoietic stem cells from mouse and human bone marrow can be induced, in vitro, to form colonies, provided the appropriate hormones (colony stimulating factors = CSFs) and culture conditions have been selected. Such hematopoietic colony cultures are of interest for in vitro cytotoxicity testing of a great number of drugs. It has been increasingly recognized that serious side effects of many drugs involve injuries of hematopoietic organs. The Rote Liste, in which $>$ 85 % of all drugs registered in the Federal Republic of Germany are compiled, lists even 34 groups of drugs for which side-effects on the hematopoietic organs have been reported. Therefore hematopoietic stem cells seem to provide distinguished targets to assay for presumed drug effects, be they stimulatory or inhibitory including cytotoxic.

Hematopoietic colony assays in agar-containing glass capillaries

Based on the pioneering studies by the groups of Sachs and Metcalf, methods have been developed during the last decades to clone, in agar or methylcellulose, the stem cells of virtually all hematopoietic cells (erythrocytes, lymphocytes, granulocytes, macrophages, megakaryocytes) (5). The in vitro colony assays of most research groups utilize Petri dishes into which the stem cells are seeded in an agar-medium on top of another agar-layer serving as a feeder layer. Since this double-layer technique requires considerable expenditure, particularily when applied to test many biochemically isolated fractions, the group of Maurer has adopted and greatly refined, since 1976, a micro-procedure (6): Instead of Petri dishes glass capillaries are utilized that offer a number of advantages:

- Less materials (~10 %) : Cells, media, test samples; thus lower costs and more samples can be tested.
- No agar-double layer: Easier and quicker test set-up.
- Lower agar-concentration (o.18 versus 0.3 % in Petri dishes). Better cell staining after culture.
- Linear alignment of colonies: Thus easier and quicker colony counting (with a dissecting microscope); optical evaluation by simple light scattering densitometry (7).
- Reduced risk of contamination by bacteria and fungi.
- Higher cloning (plating) efficiencies (up to 5 fold (8,9), see below).

The glass capillaries have an i. d. of 1.38 mm, are 127 mm long and are filled with 30 - 70 µl of cell suspension (10). E. g. $7.5 \cdot 10^3$ mouse bone marrow cells/75 µl (per capillary) will yield 20 - 30 granulocytic colonies, 75 L1210 leukemic cells/30 µl 50 colonies. Fig. 1 shows the steps of the procedure for T-lymphocyte (11) and for granulocyte (10) colony assays. The dimensions of the capillaries, open at both ends, are selected such that homogeneous colony is maintained along the entire gel. The coefficient of variation of colony numbers does not exceed $\pm$ 5 % from capillary to capillary (12). Filling of the capillaries is easy: The capillaries are put onto the tip of an automatic pipette, by

Figure 1

## Standard-Colony-Assays in Glass Capillaries

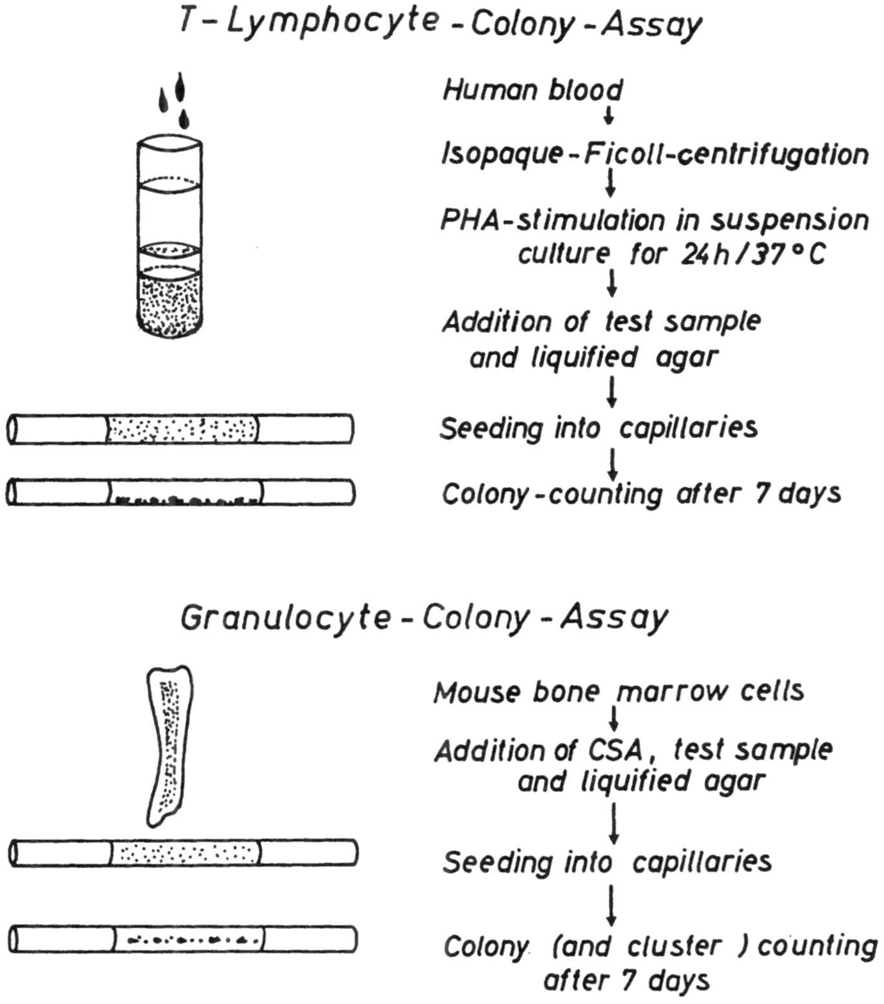

which the cell suspension in liquified agar is sucked in. Optical evaluation by light scattering densitometry is suited for special tasks (12,13). For this the capillaries are drawn through the light path of a spectrophotometer by means of a densitometer attachment (Fig. 2). The light scattered by the colonies inside the capillaries is received via 2 mirrors (7). The agreement of visual colony counting by the microscope and light scattering densitometry is excellent (Fig. 3). This arrangement allows the follow-up of daily growth of individual colonies of one capillary (12) (Fig. 4). For kinetic questions the effect of a granulopoietic inhibitor could be evaluated starting at day 3 of culture (13) (Fig. 5).

The cloning efficiencies obtained with the agar capillaries were in all cases tested higher than those found with the standard Petri dish (35 mm) or microtiter plate technique (96 wells). This was true for both mouse granulocytic (GM-CFC) and L1210 leukemic colonies (9) (Figs. 6 and 7).

Applications of the agar capillary technique using hematopoietic cells

By means of the clonogenic assays using agar-containing capillaries granulocyte and lymphocyte chalone activities have been extensively determined (14 - 16). The advantage of the micro techniques became evident in the course of the many biochemical purification processes: The expenditure of materials to detect chalone activities in the many fractions was greatly reduced as compared with the common Petri dish method.

Recently Fetsch and Maurer (17) prepared from calf spleen, after ethanol extraction and acetone precipitation, ultrafiltrates of which the 500 - 1000 Da fraction yielded a relatively high granulocyte chalone activity. Following FPLC (Mono Q anion exchanger, Fig. 8) a fraction was isolated that revealed a remarkable feature: An inhibitory activity specific for granulopoiesis with a maximum at 1 : $10^3$ -$10^4$ fold dilution (Fig. 9). This corresponded to a concentration of 5 . $10^{-9}$ - $10^{-10}$ M, presumably of a peptide, since further analysis using acidic hydrolysis and HPLC indicated the presence of Asp, Glu among others. The bell-

Figure 2

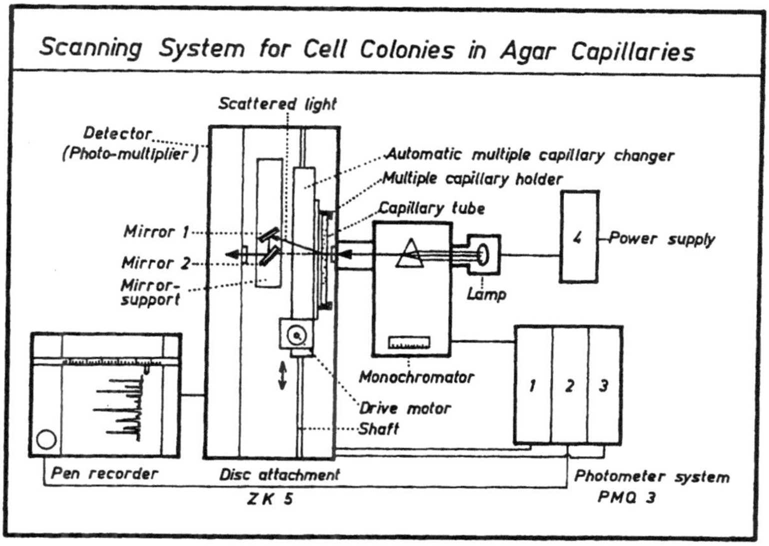


Figure 3

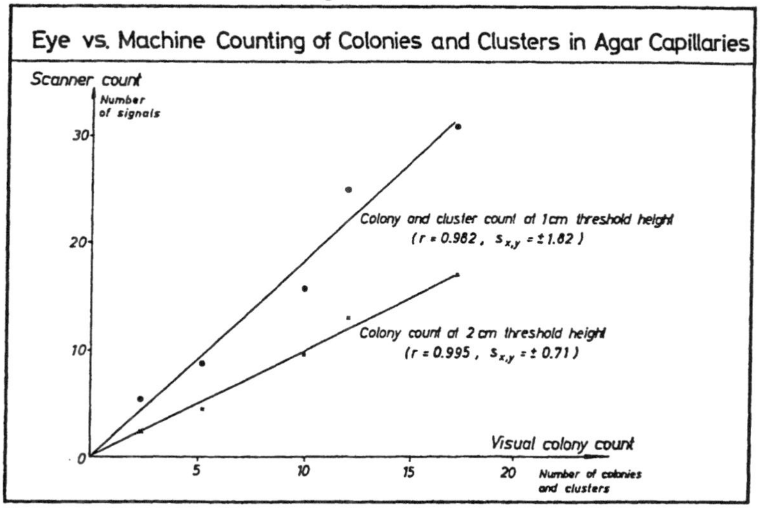

Figure 4

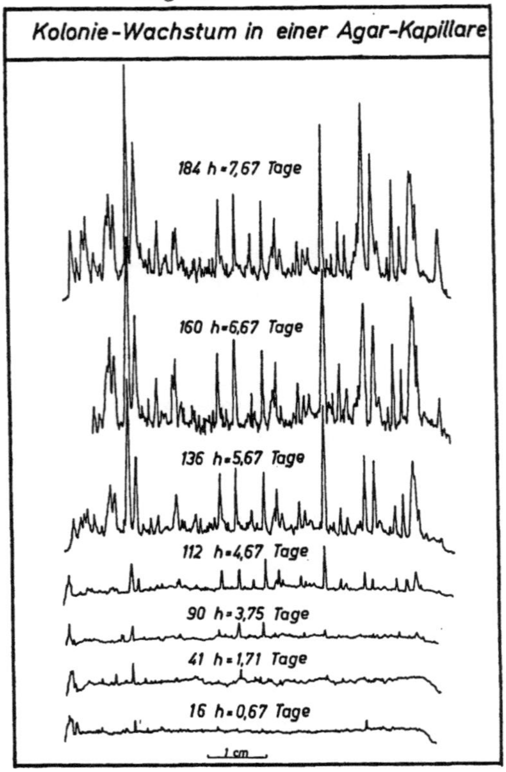


Figure 5

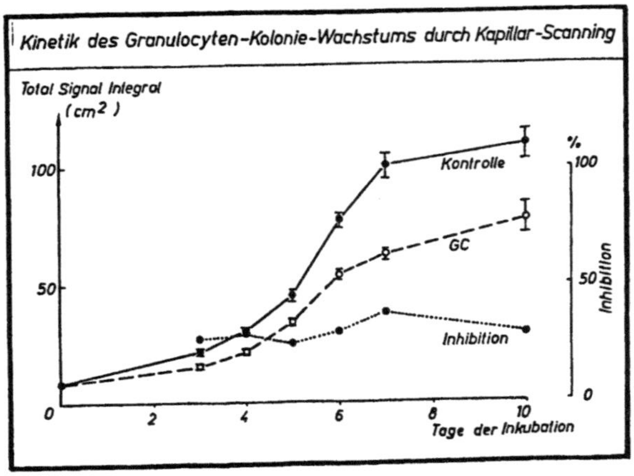

Figure 6

*Mouse GM-CFC : Cloning Efficiencies (C.E.) in Different Clonogenic Assay Systems*

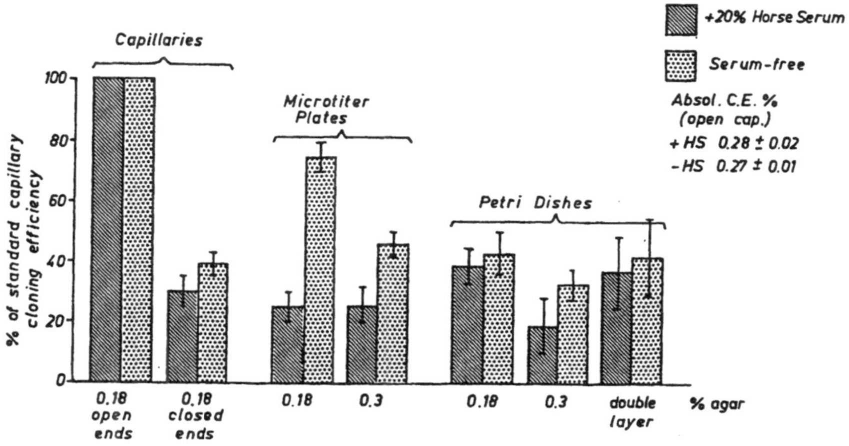


Figure 7

*L 1210 Cells: Cloning Efficiencies (C.E.) in Different Clonogenic Assay Systems*

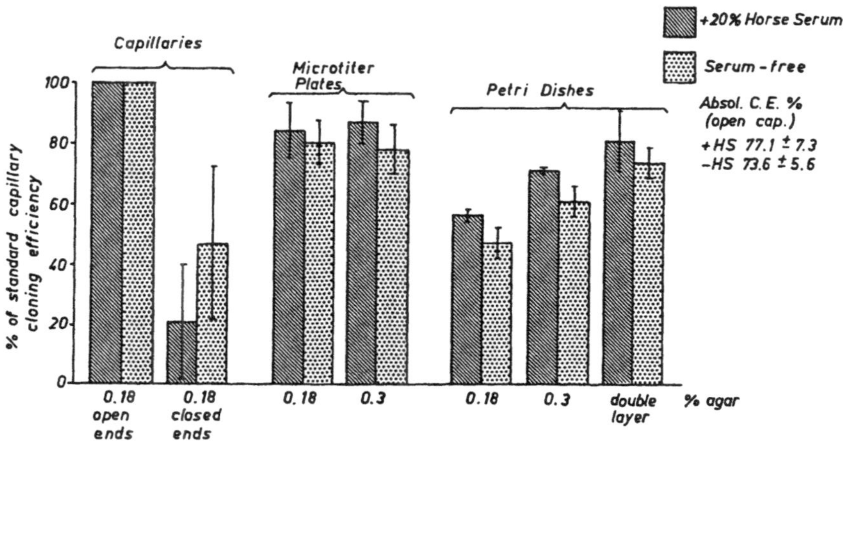

Figure 8 a

**Calf Spleen Granulopoiesis Inhibitor:**
**EtOH Extr.→ Acetone Prec.→ Ultrafiltration→ FPLC of 0.5-1 kDa Fraction**

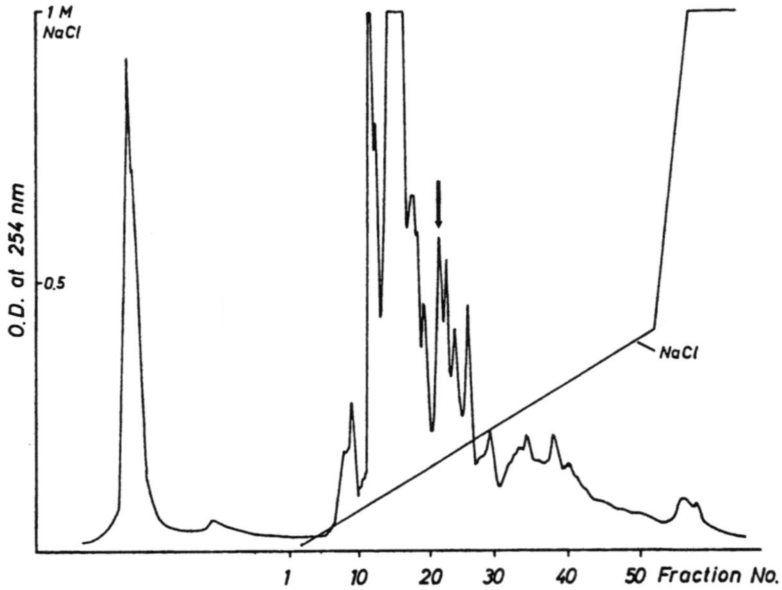


Figure 8 b

**Calf Spleen Granulopoiesis Inhibitor: Activity of FPLC Fractions of 0.5-1 kDa Ultrafiltrate**

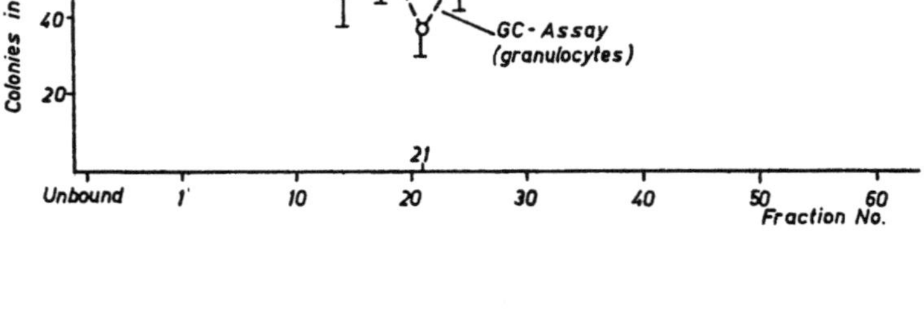

Figure 9 a

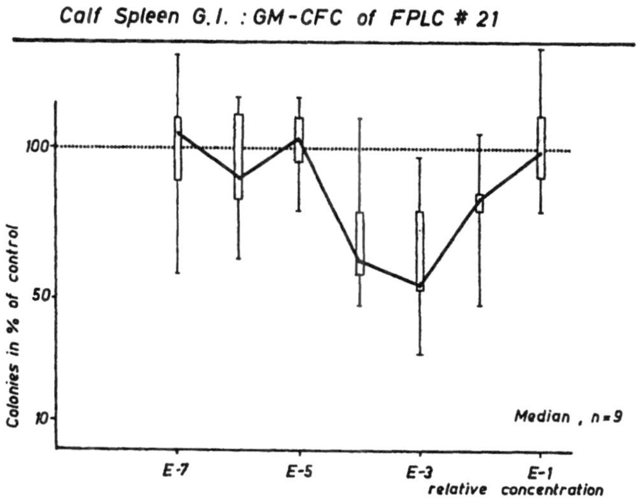


Figure 9 b

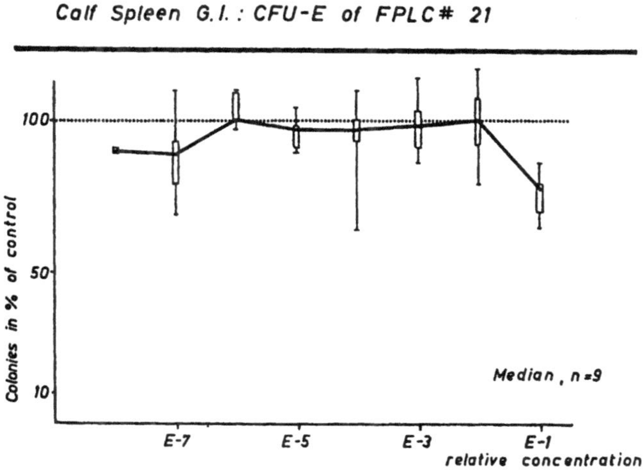

shaped dose-response curve was striking and precluded any cytotoxicity as a potential artifact. Indeed the distinction between a true physiological proliferation inhibitor (chalone) and a mere cytotoxic factor is a difficult problem. Clear-cut evidence requires techniques that allow extended dose-response curves. The micro capillary technique favours these prerequisites.

Using the granulocyte and the T-lymphocyte assays Maurer et al. (18,19) have studied the influence of cytostatic and immunosuppressive methylhydrazones and found that the clonogenic assays were much more sensitive than the widely used $^{3}H$-thymidine uptake method.

Braun and Maurer (20-22) recently provided an instructive example for how to combine in vitro and in vivo assays in order to obtain meaningful data on the cytotoxicity of distinct drugs. The epoxide-containing valeriana drugs valtrate and dihydrovaltrate had shown alkylating and cytostatic properties. Cytotoxic activities were also found in the micro clonogenic assays, valtrate being more toxic than the mutagenic and alkylating agent epichlorohydrine. The valtrate effects were not reversible upon washing of the cells treated. This indicated a direct cytotoxic effect on proliferating cells (20).

But is this also true in vivo? The drugs did neither affect the sensitive bone marrow, even after high p. o. or i. p. doses, nor the liver after p. o. doses. However, following i. p. administration the liver was injured (21). This led to the hypothesis that the drugs indeed attack the cells they approach first: In the gastrointestinal tract the mucosal cells, after i. p. also the liver cells; they are, not absorbed by the blood. These combined in vitro - in vivo data may give some clues as to the site of action and cytotoxicity of widely used tranquilizing drugs.

Moreover using the capillary clonogenic assays the presumed cytoprotective activity of somatostatin (by means of membrane stabilization) could be easily demonstrated (23): The dekapentapeptide appears to protect granulopoietic cells from the cytotoxic attack of the alkylating cytostatic drug bischloroethylnitrosourea (BCNU) only when the cells have been exposed to the peptide well before BCNU treatment (Fig. 10).

## Clonogenic assays with tumor cells

Tumor cells can also be cloned in agar to form colonies. This finding supports the increasingly recognized theory of the clonal evolution of tumors (24). The theory assumes the existence of so-called tumor stem cells. Consequently an in vitro assay should utilize the clonogenic capacity of tumor cells to determine their sensitivity towards cytostatic drugs. Thus several research groups have developed appropriate in vitro colony assays and have demonstrated the suitability of such tests for chemosensitivity screening. Particularily the groups of Salmon (Tucson, USA) and v. Hoff (San Antonio, USA) have shown (24 - 28), that such assays can be used not only to predict the individual sensitivity of a cancer patient, but also to screen for potential, yet hitherto unknown cytostatic drugs.

Maurer and Osman (29) have refined and optimized the conditions for cloning various fresh and established tumor cells in agar-containing capillaries. They also found (30) that their micro method is about twice to 8 fold more sensitive than other methods to determine cell proliferation (cell counting and $^{3}$H-thymidine uptake). The most important steps for individual chemosensitivity screening are depicted in Fig. 11.

An index was developed that is free of any dimension and independent of pharmakokinetic differences of the drugs in order to facilitate the evaluation of the individual in vitro chemosensitivity of human tumors (31). Using several primary human tumors the advantages of the new approach were demonstrated (Fig. 12).

Figure 10

Cytoprotection by Cyclic Somatostatin of BCNU-Inhibited Granulocyte Colony Growth

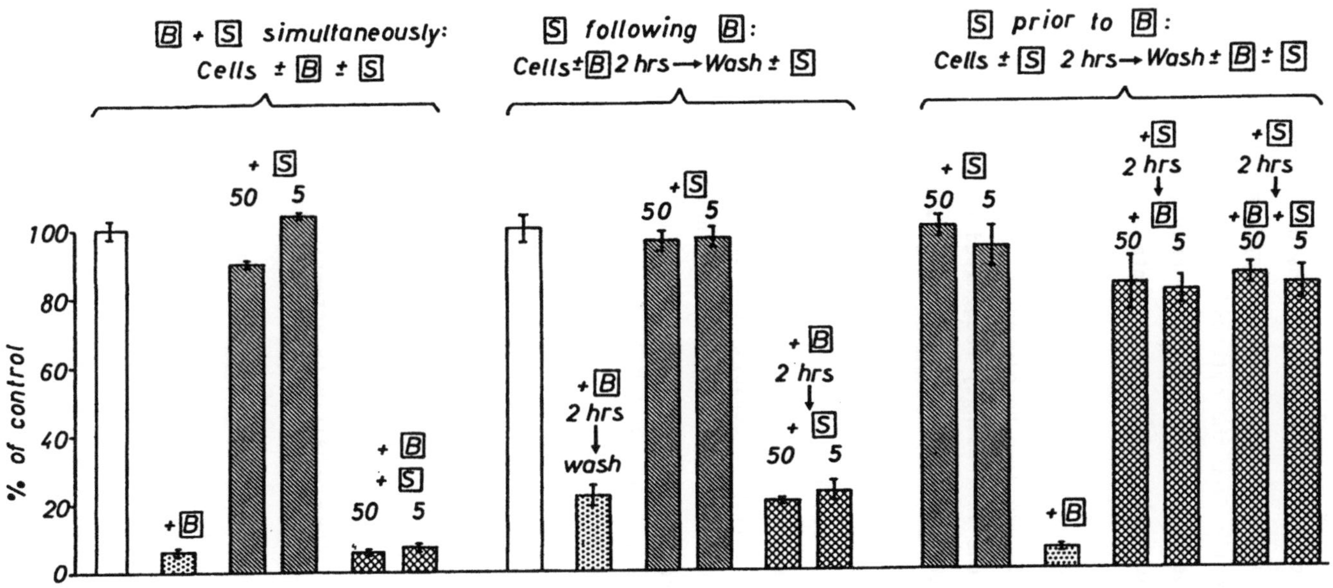

Figure 11

## Tumor Stem Cell Assay as On obiogram
(Cytostatic drug sensitivity testing in vitro)

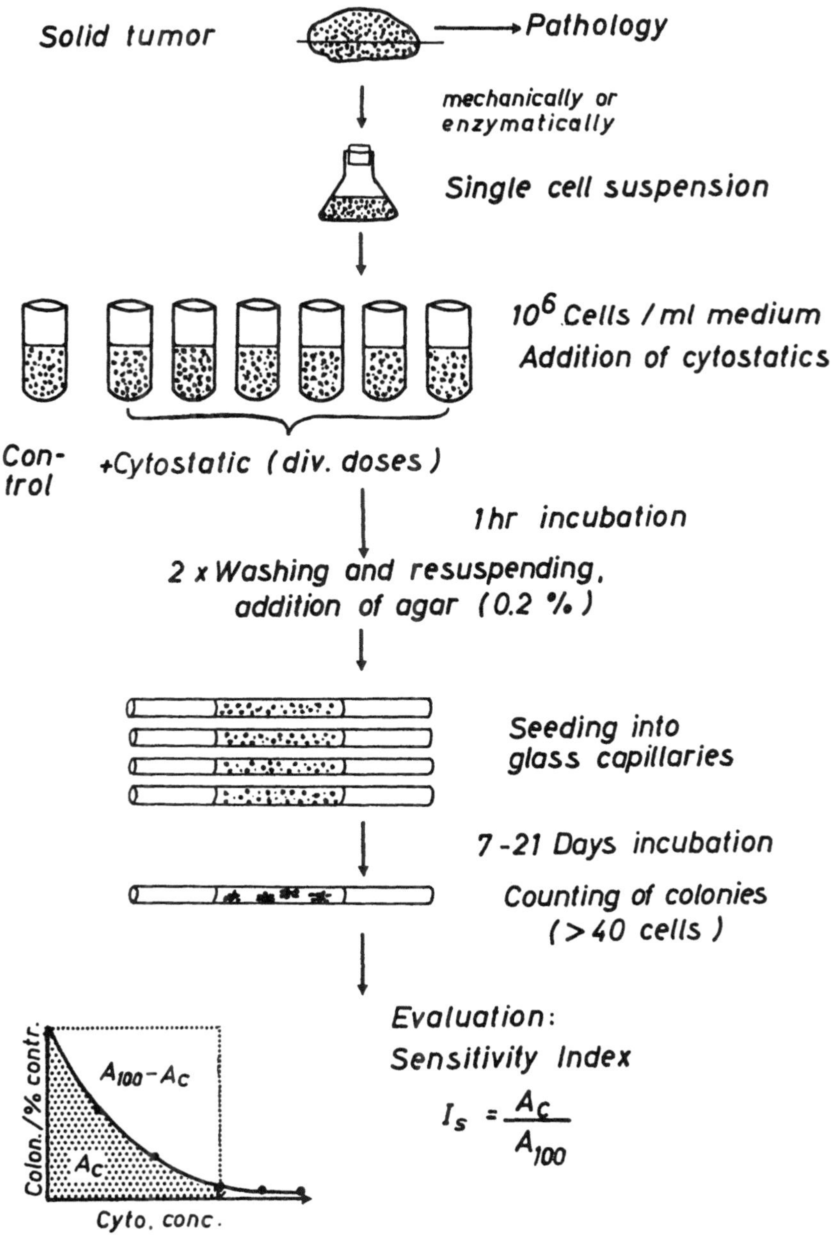

Using the micro clonogenic assay with tumor cells Osman and Maurer (32) found that certain tumor cells may produce polyamines which are metabolized to polyaminoaldehydes in the presence of polyaminooxidases. The latter enzymes are contained in fetal calf serum. This finding is of significance for optimizing culture conditions. Similarily the micro methods proved useful to evaluate the effect of the growth stimulating peptide gly-his-lys in the presence and absence of various horse serum concentrations (33) (Fig. 13).

To elucidate the potential mode of action of cytostatic methylhydrazones, L1210 and Ehrlich-ascites tumor cells were cloned in agar capillaries (19). Exposure of the cells to the drugs for 3 hours and subsequent washing, prior to seeding, yielded data suggesting that the plasma membrane affinity and the lipophilic character of the drugs are determining their bioavailability.

A comparative investigation was performed on guinea pig hepatoma cells in order to test the sensitivity against 7 cytostatic drugs in vitro and in vivo following intratumoral chemotherapy (34). The correlation of the in vivo/in vitro resistence approached 100 %, while the in vivo/in vitro sensitivity was found to be 80 %.

## Potential problems of the agar capillary assays

Potential problems may be encountered when the agar capillary assays are intended to substitute all applications so far known for the clonogenic Petri dish techniques. Therefore some possible drawbacks may be listed:

- It is impossible to substitute agar for methylcellulose, since highly viscous solutions of the latter cannot be drawn into the capillaries.
- The closed incubation form of the capillary does not permit administration of samples after the start of incubation.
- Since less colonies are usually counted per capillary (20 - 70) it is essential to determine the necessary number of cells to be seeded in order to obtain statistically reliable data.

Figure 12

*Bestimmung des Sensitivitäts-Index eines Portio-Carcinoms*

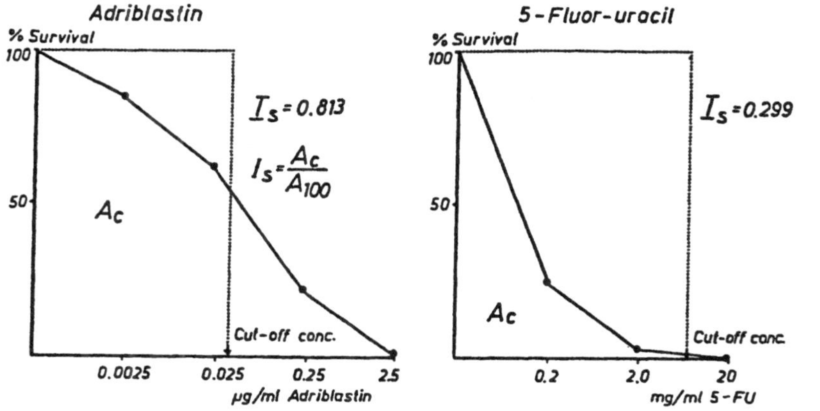


Figure 13

*Effect of Glycylhistidyllysine on Colony Growth of M1- and W25-Cells at different Horse Serum Concentrations*

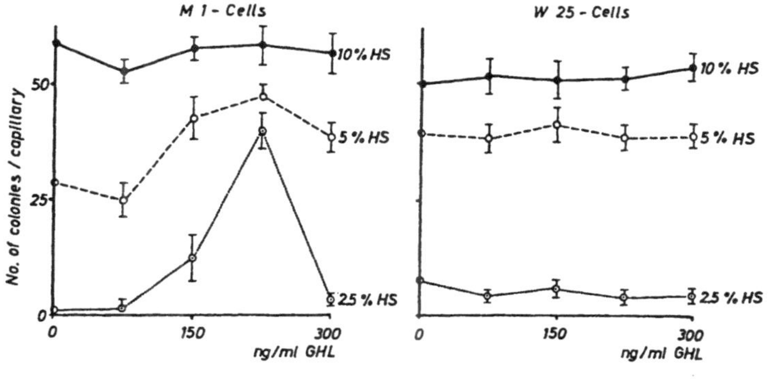

- Monolayer colonies of fibroblastoid and epithelial cells can be grown in capillaries, but are difficult to identify.
- In certain cases (e. g. erythropoiesis) it has not yet been possible to grow colonies in capillaries.

While one or the other obstacle may be relevant under particular conditions,it is certainly worthwhile to balance the pros and contras in each case. Nevertheless major advantages of the capillary assays have recently been acknowledged for large scale clinical studies (8).

Culture assays for functional capacities of differential cells

The capacity of cells to divide is an essential property and prerequisite for any cell culture assay. The above described clonogenic assays are based on this property. Consequently, for agents and drugs to be tested for their ability to affect cell growth, such assays may be recommended: Growth factors, hormones, chalones, cytostatic drugs, cytotoxic compounds etc. Besides clonogenic assays other methods have been applied to study drug effects on growth kinetics (1,2): Cell counting, viability tests (dye uptake), assays using $^{3}H$-thymidine labelling and mitotic index, metabolic tests (with $^{3}H$-labelled amino acids), histochemical and other cytological techniques, particularily flow cytometry, and last not least time-lapse cinemicrography.

In addition to these methods to measure effects on cell growth kinetics, assays are available that allow to detect effects on the genome level: Chromosome banding techniques and analysis of sister chromatide exchange (SCE). Such methods may be used for genotoxicity studies. Finally assays have been developed to test for mutagenicity and carcinogenicity using cell cultures, the procedures and problems of which have been described elsewhere in detail (1).

While all these techniques are dependent on proliferating cells - and in fact measure drug effects on cell proliferation - , one should not ignore that in vivo most cells have stopped dividing and are exerting special functions. In consequence, assays are needed, are increasingly developed and refined that measure the effects of agents and drugs on the functional capacity of terminally differentiated cells in vitro.

Nerve cell cultures have proven very suitable to test and screen for antagonists of peptide neurotransmitters like tachykinins, among which substance P antagonists have potential analgetic activity (35). For this an electrophysiological equipment was set up to follow the effects by such peptides, on the spontaneous action potentials of spinal cord neurons in culture.

Many more examples for assays measuring functional capacities of cultured cells may be cited, e. g. metabolic activities of liver (36) or arterial smooth muscle cells (37). Even leukemic cells can be used to follow the capacity, of biological response modifiers and distinct cytostatic drugs, to induce cellular differentiation; this can be measured with tests for phagocytosis, esterases, lysozyme activity etc. (38).

Advantages and drawbacks of drug screening by cell cultures

To summaraze, cell cultures offer several advantages for screening of drugs and bioactive agents like peptides:

- Reduction of animal experimentation and costs;
- Better control of experimental conditions;
- Easier statistical evaluation of data;
- Easier comparability of data among laboratories;
- Use of human cells from healthy specimen;
- Storage of cells by cryoconservation to facilitate reproducibility;
- Elucidation of mode of actions of drugs.

On the other hand one should not forget serious drawbacks and potential pitfalls that might be encountered with cell culture screening:

- Exclusion of factors upon which drug effects are dependent in vivo:
  - mode of application (p. o., i. v. etc.);
  - pharmakokinetics;
  - pharmakodynamics:
    absorption, distribution, biotransformation
    elimination, interactions

- hormones, chalones and other inhibitors.
- Selection of cell lines during extended passages;
- Lack of interaction of heterotypic cells at high density.

While the pros and contras have to balanced in each case, a large body of in vitro data clearly suggests that one single in vitro assay will rarely yield meaningful results; instead a panel of different cell culture assays should be utilized and the data comparatively analyzed. This will finally reduce false misinterpretations, may also reduce animal experimentation, but will not replace the latter completely.

References


1. Freshney, R.I.: Culture of animal cells - a manual of basic technique Alan R. Liss, Inc., New York (1983)

2. Freshney, R. I., editor: Animal cell culture - a practical approach, IRL Press Oxford - Washington DC (1986)

3. Hülser, D.F., Bräuner, T., Brümmer, F.: Interzelluläre Kommunikation in dreidimensional wachsenden Multizell-Späroiden. In: Ersatzmethoden zum Tierversuch. BMFT 1986, S. 103 - 124

4. Maurer, H.R.: Potential pitfalls of $^{3}H$-thymidine techniques to measure cell proliferation. Cell Tiss. Kinet. 14: 111-120 (1981)

5. Metcalf, D.: The hemopoietic colony stimulating factors. Elsevier Amsterdam 1984

6. Maurer, H.R.: In vitro Colony Growth of Granulocytes, Macrophages, T- and B-Lymphocytes in Agar Capillaries. Acta haematol. 62, 322-325 (1979)

7. Maurer, H.R. and Henry, R.: Automated scanning of bone marrow cell colonies growing in agar-containing glass capillaries. Exp. Cell Res. 103, 271-277 (1976)

8. Von Hoff, D.D., Forseth, B. J., Huong, M., Buchok, J.B., and Lathan,B.: Improved Plating Efficiencies for Human Tumor Cloned in Capillary Tubes versus Petri Dishes . Cancer Res. 46, 4012 - 4017 (1986)

9. Echarti, C., and Maurer, H.R.: Manuscript in preparation (1987)

10. Maurer, H.R. and Henry, R.: Colony growth of mouse bone marrow cells in agar containing in glass capillaries, Blut 33, 11 - 22 (1976)

11. Maurer, H. R., Maschler, R., Dietrich, R. and Goebel, B.: In vitro culture of lymphocyte colonies in agar capillary tubes after PHA-stimulation. J. Immun. 18, 353 - 364 (1977)

12. Maurer, H. R. and Henry, R.: Drug evaluation on haemopoietic cells in vitro. Arzneimittel-Forsch. (Drug Res.) 28, 601 - 605 (1978)

13. Maurer, H. R., Henry, R.. and Maschler, R.: Chalone inhibition of granulocyte colony growth in agar: Kinetic quantitation by capillary tube scanning. Cell Tiss. Kinet. 11, 129 - 138 (1978)

14. Maschler, R. and Maurer, H.R.: Screening for granulopoiesis inhibitors (chalones) by different assays. Hoppe-Seyler's Z. physiol. Chem. 359, 825 - 834 (1978)

15. Kastner, M. and Maurer, H.R.: Partial purification and characterization of an endogenous granulomonopoiesis inhibitor from calf spleen. Hoppe-Seyler's Z. physiol. Chem. 365, 129 - 135 (1984)

16. Maschler, R. and Maurer, H.R.: Screening for specific calf thymus inhibitors (chalones) of T-lymphocyte proliferation. Hoppe-Seyler's Z. physiol. Chem. 360, 735 - 745 (1979).

17. Fetsch, J., and Maurer, H.R.: Manuscript in preparation (1987)

18. Maurer, H.R., Maschler, R., Braun, R., and Dittmar, W.: Einfluß von cytostatischen und immunosuppressiven Methylhydrazonen auf die Myelo- und Lymphopoese in vitro. J. Cancer Res. Clin. Oncol. 95, 129-138 (1979).

19. Dittmar, W. Klitschka, G., Braun, R., Ali-Osman, F., Meckert, C., and Maurer, H.R.: In Vitro and in Vivo Investigations for the Development of Cytostatic Methylhydrazones. J. Cancer Res. Clin. Oncol. 110, 110 - 114 (1985)

20. Tortarolo, M., Braun, R., Hübner, G.E. and Maurer, H.R.: In vitro effects of epoxidebearing valepotriates on mouse early hematopoietic progenitor cells and human T-lymphocytes. Arch. Toxicol. 51, 37-42 (1982)

21. Braun, R., Dittmar, W., Hübner, G. E. und Maurer, H.R.: In-vivo-Einfluß von Valtrat/Isovaltrat auf Knochenmarkzellen der Maus und auf die metabolische Aktivität der Leber. Planta Medica 1-4 (1984)

22. Braun, R. Dieckmann, H., Machut, M., Echarti, C., und Maurer, H.R.: Untersuchungen zum Einfluß von Baldrinalen auf hämatopoetische Zellen in vitro, auf die metabolische Aktivität der Leber in vivo sowie zum Gehalt in Fertigarzneimitteln. Planta medica 446 - 450 (1986)

23. Maurer, H.R., Meckert, C., Ali-Osman, F. and Musil, J.: Cytoprotection by somatostatin in normal and malignant clonogenic cells against the in vitro cytotoxicity of bischlorethyl-nitrosourea (BNCU). Arzneimittel-Forsch. 34, 939 - 943 (1984)

24. Nowell, P.C.: The clonal evolution of tumor cell populations. Science 194, 23 - 28 (1976)

25. Salmon, S.E.: Cloning of Human Tumor Stem Cells: Alan Liss, Inc. N. Y. (1980)

26. Salmon, S. E. and v. Hoff, D.D.: In vitro evaluation of anticancer drugs with the human tumor stem cell assay. Semin. Oncol. 8, 377 (1981)

27. Salmon, S.E., and Trent, J.M., editors: Human Tumor Cloning. Grune & Stratton Inc. Orlando (1984)

28. Von Hoff, D.D., et al. Prospective clinical trial of a human tumor cloning system. Cancer Res. 43, 1926 - 1931 (1983)

29. Maurer, H.R. and Ali-Osman, F.: Tumor stem cell cloning in agar-containing capillaries. Naturwissenschaften 67,381 (1981)

30. Ali-Osman, F. and Maurer, H.R.: Comparison of cytostatic sensitivities of L1210 cells and human stimulated lymphocytes in three cell proliferation assays. J. Cancer Res. Clin. Oncol. 98, 221 - 231 (1980)

31. Ali-Osman, F., Maurer, H.R. and Bier, J.: In vitro cytostatic drug sensitivity testing in the human stem cell assay: A modified method for the determination of the sensitivity index. Tumor Diag. Ther. 4, 1 - 6 (1983)

32. Ali-Osman, F. and Maurer, H.R.: Stimulation of clonal tumor cell growth in vitro by inhibiting the serum polyamine oxidase activity. J. Cancer Res. Clin. Oncol. 106, 17 - 20 (1983)

33. Ali-Osman, F.: Entwicklung eines klonalen Mikro-Tumorstammzell-Assays und dessen Anwendung zur Cytostatica-Sensitivitäts-Testung menschlicher und tierischer Tumoren. Dissertation, Freie Universität Berlin, 1982

34. Ali-Osman, F., Bier, J., Bier, H., Siegel, T., Maurer, H.R. and Ohanian, S.: Correlation of intralesional in vivo chemotherapy of line 10 hepatoma with in vitro drug sensitivity. Int. J. Cell Cloning 1, 118 - 127 (1983)

35. Reuss, K., Wienrich, M., Harting, J.: Untersuchungen an Nervenzellen können Tierversuche in der Schmerzforschung einsparen. In: Ersatzmethoden zum Tierversuch. BMFT 1986, S. 11-29

36. Hümpel, H.: Metabolisierungsstudien von Xenobiotika mit Leberzellen und Leberzellkulturen und deren Vergleich mit entsprechenden in vivo-Versuchen. In: Ersatzmethoden zum Tierversuch. BMFT 1986, S. 161-171

37. Betz, E., Hämmerle, U.: Die Wirkungen von antiatherogenen Stoffen in Zellkulturen und Arterienwänden. In: Ersatzmethoden zum Tierversuch. BMFT 1986, S. 253 - 266

38. Lotem, J., and Sachs, L.: Potential Pre-Screening for Therapeutic Agents that Induce Differentiation in Human Myeloid Leukemic Cells Int. J. Cancer 25, 561 - 564 (1980)

# ELECTROTRANSFORMATION OF CULTURE CELLS

E. **Neumann** und E. **Boldt**
Faculty of Chemistry, Department of Physical and Biophysical Chemistry, University of Bielefeld, P.O. Box 8640, D-4800 Bielefeld 1, F.R.Germany

## Introduction

Electric pulse methods become increasingly important for cell biology and biotechnology. Especially, the techniques of electroporation and electrofusion are of particular interest for cell transformation. Electrotransfection or electrotransformation, i.e. the direct transfer of DNA (or mRNA) into cells by electric impulses, appears to develop as the method of choice for the artificial gene transfer into chemically untreated cells, too.

Electric field pulse-mediated uptake of DNA into cells followed by the expression of the introduced gene was first demonstrated in 1982 (1,2). The operative term 'electroporation' (electric pore formation) was introduced in order to specify the transient electropermeabilization of cell membranes as the physical basis of electrotransformation and of electrofusion. Applications of the electropermeabilization techniques include bacteria and yeast as well as plant and mammalian cells.

## Membrane Electropermeabilization

Both electroporative gene transfer and electrofusion apparently are based on the same fundamental event: the reversible, non-destructive electropermeabilization of cell membranes by short-duration high electric field pulses, discovered in 1972 (3,4). It was demonstrated that, above a threshold field strength, a controlled release of internal material from suspended membraneous vesicles (concomitant with uptake of solvent molecules) occurs in response to a field pulse in the

Modern Methods in Protein Chemistry, Vol. 3

microsecond range; there was no irreversible membrane damage (no dielectric breakdown). Apparently reversible changes in the membrane structure, of long-lived nature compared to the much shorter pulse duration, had been indicated by electro-optic techniques (5).
In contrast to reversible electroporation, irreversible and lethal effects on microorganisms caused by a whole series of similar electric field pulses had already been reported in 1967 (6,7): It was found that there is an inverse relationship between cell size and a critical external electric field strength $E_c$, necessary for, e.g., lysis of bacterial protoplasts (8). These results indicated that the cell membranes are the field-sensitive targets and, furtheron, that a critical transmembrane voltage $V_c$ (of about 1 Volt, independent of cell size) is built up by the external field pulses. Sale and Hamilton (8) suggested that the irreversible loss of the semipermeable membrane properties and ultimate lysis and cell death might be caused by conformational changes in the membrane structure.
In summary, short electric field pulses may either cause reversible or irreversible effects on membrane systems, depending on parameters such as the strength, duration, shape, and the number of the pulses. These early results remain fundamental also for the application of electric field pulses in the method of electroporative gene transfer. In the following we will focus on practical, experimental features important for electroporative gene transfer.

**Electroporative Gene Transfer**

Pulse Characteristics. High electric field pulses are conveniently applied to biological cells or other membrane particles by discharge of electric energy (current flow) through the suspension enclosed in a discharge chamber; this is readily achieved by the capacitor discharge (CD) technique adapted from fast relaxation kinetics (9). This technique was originally applied in the experiments on reversible electropermeabilization (3,4). Obviously, because of its ease and fairly low-costs, the CD technique has been preferred in most of the recent papers on electroporative gene transfer.

With respect to the time characteristics (shape) mainly two different types of electric field pulses are used for membrane electropermeabilization: either exponentially decaying pulses obtained by capacitor discharge, or rectangular (square) pulses delivered by, e.g., a cable discharge (10-12) or by special high voltage pulse equipment (6,13). Because of maximum voltage and power requirements in the discharge circuit the technical realization of the equipment for given pulse shapes may vary considerably. These requirements are mainly defined by the geometry (electrode distance and surface, volume) of the discharge chamber, by the size and nature of the cells to be electroporated as well as by the electric conductance of the suspension.

The discharge of a storage capacitor of capacity C, charged to the desired initial voltage $V_0$ by a power supply, ensures a defined high electric field pulse. If the inductance of the discharge circuit is negligibly small, the time characteristics of the field pulse within the discharge chamber of electrode distance d is given by (9,3):

$$E(t) = E_0 \exp(-t/\tau), \quad (1)$$

with $E_0 = V_0/d$ and $\tau = R \cdot C$ .

The initial electric field strength $E_0$ as well as the time constant of the electrical discharge $\tau$ (i.e. the time of the field to decrease to $E_0/e$) are given to characterize the applied electric field pulse. The resistance R of the discharge chamber filled with the cell suspension together with the chosen condensor capacity C determines the pulse duration ($\tau$).

Power supplies for electrophoresis, directly connected to a discharge chamber as described by Potter et al. (14), generally are not well-suited for delivering defined exponential pulses. Strong discrepancies to the voltage knob settings usually occur (15) and direct measuring of the pulse characteristics (by oscilloscope) is necessary.

A broad variety of discharge chambers differing in geometry, volume, electrode material and other characteristics has been designed and used in practice for the application of nearly uniform (homogeneous) electric field pulses to cell suspensions. Closed and open chambers, mostly of cubic or cylindrical shape with the filling volume between parallel rectangular or disc electrodes, have been described (3,6,13,16,17). The handling of fixed or variable volume is possible from several ml down to a few $\mu$l. Concentric electrode arrangements providing a moderate inhomogeneous field distribution have also been chosen; commercially a special 'pipetting' discharge chamber based on this geometry has been designed. There are means for temperature control as well as for sterile and easy handling. Even a combination of controlled centrifugation with *in situ* electrical pulsing was reported. Among the most easy and popular designs are disposable plastic spectrophotometer cuvettes ('cuvette chamber', semimicro, 1 cm pathlength, 0.4 to 0.5 cm electrode distance, up to 1 ml filling volume) with platinum or stainless steel paddle electrodes inserted (18) or aluminium foil electrodes directly glued to the cuvette walls (14). Most of the mentioned discharge chamber designs are meanwhile available as commercial equipments. In fact, commercial instruments for electroporation already emerged in 1984 (by W.Rauhaus, Fa. Dialog, Düsseldorf), and the number of manufacturers and models is still growing. For 'historical' reasons the first version of the original 'ELEKTROPORATOR®' is reproduced in Fig. 1.

Early data. After the finding of electric hemolysis of red blood cells (16) Auer et al. (19) demonstrated the uptake of genetic material into human red blood cells permeabilized by an electric field pulse. Applying the capacitor discharge technique ($E_0$ = 11 kV/cm; $\tau$ = 100 $\mu$s) Auer et al. succeeded in trapping $^3$H-labeled SV 40 DNA in the cells. Electric hemolysis as the first consequence of the pulsing was followed by a defined resealing procedure (90 min incubation at 37 °C) in the presence of the DNA or RNA. Up to 35 % of the totally added linearized SV 40 DNA was finally detected within the cells. The entrapped DNA was DNAse resistant due to the package. The authors noticed that DNA trapping was much more effective with the linearized than with the supercoiled form of SV 40 DNA.

In 1982 electroporative gene transfer with subsequent actual gene expression was first reported (1,2). Plasmid DNA carrying the

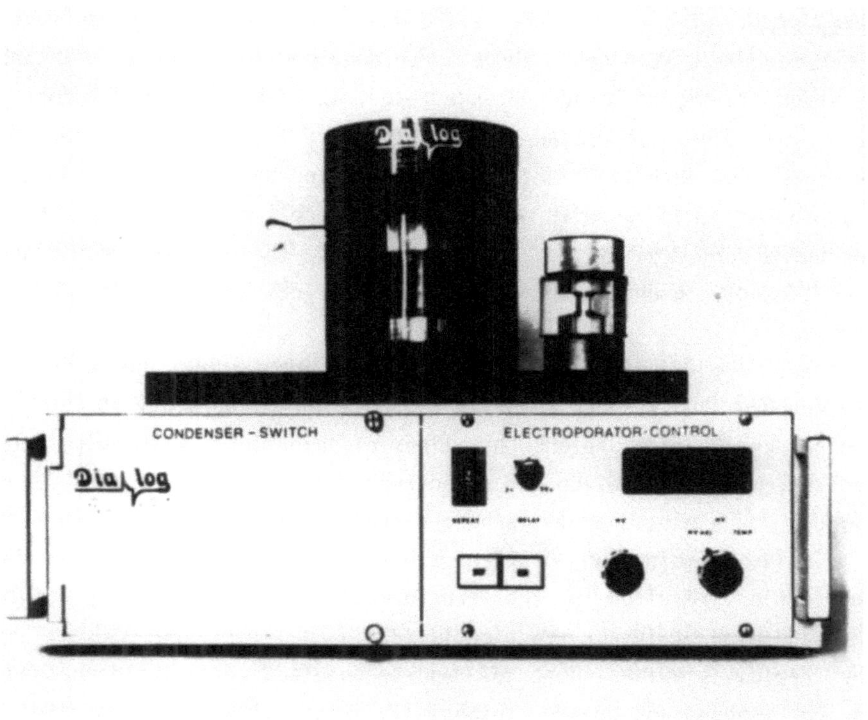


Fig. 1 The first commercial electroporation apparatus: **Elektroporator**® of the Fa. Dialog, Düsseldorf, developed by W. Rauhaus.

thymidine kinase (tk) gene was introduced into tk deficient mammalian culture cells (mutant L $tk^-$ mouse cells) by the application of short electric field pulses. Stable transformants surviving in the HAT selection medium proved the direct gene transfer and expression. Several main features of the procedure, already worked out in the early paper (2), became standard in current applications: (i) the highly conductive electroporation medium (HBS: HEPES buffered saline, with 140 mM NaCl); (ii) the fairly easy capacitor discharge (CD) technique for pulse application; (iii) the use of a few successive electric field pulses (e.g., three pulses at time intervals of several seconds); (iv) the necessity of a post-incubation of the cell-DNA mixture for about 10 min after pulsing, suggesting that the actual gene transfer is an after-field effect; (v) the existence of a fairly sharp optimum in electric field strength, above a threshold value, for a given pulse duration in the microsecond range (e.g., $E_0$ = (8 ± 0.5) kV/cm for $\tau$ = 5 $\mu$s); (vi) the limitation of transfer efficiency at higher field strengths by lethal effects; (vii) the increase of the number of transfected colonies with an increase in cell density and plasmid DNA concentration; (viii) the linearization of the plasmid resulting in significantly higher transformation yield compared to the circular form.

In fact, the first direct gene transfer into plant protoplasts by electroporation was achieved in 1983 in our laboratory (by Shillito, Bierth and Neumann) along the lines of experience with animal cells. Lateron, a combination of electroporation with PEG (polyethylene glycol) treatment greatly enhanced the transformation efficiency (20). A similar combination of PEG treatment and electric field pulse application resulted in the transformation of bacterial protoplasts (21) and even of intact yeast cells (22).

The rapidly growing number of reports on electroporative gene transfer in the subsequent years, especially since 1985, demonstrates the general applicability of this technique not only for mammalian, but also for plant, bacterial and yeast cells. The experimental parameters of the procedure have been investigated in more detail, including (i) the temperature (pulsing and post-incubation at 0°C recommended); (ii) the pulse shape (exponential vs. rectangular); (iii) the pulse duration ($\mu$s vs. ms range); (iv) the addition of divalent ions to the electroporation medium ($Ca^{2+}$, $Mg^{2+}$).

## Experimental Parameters

Microsecond pulses. The reversibility of membrane electropermeabilization has been linked to a sufficiently short duration of the external electric field pulses applied. Using the experience accumulated until 1981, the first electroporative gene transfer experiments were performed with electric field pulse durations in the $\mu$s-range in order to avoid irreversible destructive effects on the cell membranes. In fact, pulse times between 5 and 200 μs for exponential and square pulses have been chosen in a number of systematic studies on 'electrotransfection' and 'electrotransformation' (2,18,20,21,23-30). In some of the studies on electroporative gene transfer into plant protoplasts, equipment originally developed for cell electrofusion (square pulse generators and chambers) has been applied (28-30). The pulse times were also adapted from typical electrofusion conditions (1 to 100 μs), as well as the fairly high repetition rate in multipulse application. Because of power and field strength limitations, however, such equipment including dielectrophoresis set-up is not advised for systematic studies on electroporative gene transfer.

In the square pulse experiments with pulse times of 10 to 100 μs it is necessary to set a sufficiently low conductance of the electroporation medium. This limitation was circumvented by Hama-Inaba et al. (27) using a 'semi-rectangular' pulse shape generated by a time-controlled cut-off of a fairly slow capacitor discharge. In this manner pulse times of up to 50 μs can be realized even in high-conductance phosphate-buffered saline (PBS) medium.

A systematic comparison of exponential vs. rectangular pulse shapes with respect to electrotransfection efficiency is still lacking. In a paper dealing with this question (24) no clearcut conclusion could be drawn because, for technical reasons, different electroporation media for both types of experiments had to be used. Nevertheless, in these experiments with $tk^-$ chinese hamster ovary (CHO) cells, exponential pulses resulted in a higher yield and revealed more specifically the threshold of the electric field strength that is typical for the electropermeabilization phenomenon.

Millisecond pulses. The alternative pulse duration range used in practice is between 5 and 20 ms, initially obtained by chance (14); by directly discharging a standard electrophoresis power supply through a cuvette chamber using PBS as electroporation medium. Fromm et al. (15) showed that in this case, instead of the quoted 2000 V voltage setting, only some 120 to 150 Volts really are delivered to the cell suspension because of load mismatch within the discharge circuit. This creates an initial electric field strength of about 300 V/cm and yields a nearly exponential pulse with $\tau$ = 15 to 20 ms (32). Nevertheless, electrotransfection was successful and because of the widespread availability of the equipment the experimental procedure described by Potter et al. (14) was subsequently applied by many groups to a large variety of mammalian cell lines. Of course, no optimization with respect to electric field strength and pulse duration can be expected with only one 'standard' parameter setting. Recently, Chu et al. (33) reported on systematic reinvestigations of the electrotransfection of a variety of mammalian cell lines (ms-pulses by CD) and optimization with respect to various electroporation parameters.

Optimum electroporation conditions for transient expression and stable transformation of plant protoplasts were worked out by Fromm et al. (15,34) for ms-pulses delivered by CD. Fairly long electric field pulses were also applied to introduce TMV and CMV RNA and even whole TMV and CMV particles into tobacco protoplasts (35). The transformation of *intact* yeast cells with plasmid DNA was reported by Hashimoto et al. (22), who applied one or several ms-pulses in combination with PEG treatment and long incubation times. Thus, millisecond pulse durations have turned out to be also useful in electroporative gene transfer. These conditions became popular especially due to the reduced electric field strength necessary in combination with these longer pulse times. Charging voltages of only 200 to 400 V turn out to be sufficient (standard 0.4 cm electrode distance in cuvette chambers, pulse times between 2 and 20 ms) as reported for mammalian cells (32,33) and for plant protoplasts (15,34,35).

A systematic comparison of $\mu$s- vs. ms-pulses with respect to transfection efficiency using the same transfection system has been tried by Shigekawa (32). CHO cells where transfected with the pSVneo vector; pulses of 500 $\mu$s and pulses of 2 to 10 ms were compared. As already mentioned in the comparison between exponential and square pulse shapes, because of technical limitations no clearcut result was obtained. For the ms-pulses a fairly narrow peak was observed in the voltage dependence of transfection efficiency (1 kV/cm, at about 60 to 80 % cell survival). At higher fields the survival rate very quickly decreased, thus limiting the useful range of field strength. In contrast, a much broader dependence is seen for the shorter pulses (500 μs), but in this case the maximum transfection efficiency unfortunately could not be reached because of voltage limitation of the equipment.

Electric Field Strength. One of the fundamental characteristics of membrane electropermeabilization by short electric field pulses in the μs-range is the existence of a threshold electric field strength for the onset of reversible permeability changes (3). A corresponding critical transmembrane voltage of $V_c \approx 0.5$ to 1 V has been established in many cases. For practical purposes, the necessary critical external electric field strength $E_c$ can be estimated (4,8):

$$E_c \approx \frac{4\ V_c}{3\ D} \qquad (2)$$

$$E_c\ /\ \mathrm{kV\ cm^{-1}} \approx \frac{13.3 \text{ to } 6.7}{D\ /\ \mu\mathrm{m}} \qquad (3)$$

where D is the diameter of the suspended spherical cells. Such estimates are, however, only crude approximations. A variety of parameters such as, for instance, the cell membrane composition and the physiological state of the cells, will affect the value of $E_c$. Of course, the inverse relationship between $E_c$ and the dimensions of the cells is fundamental: electroporation of small bacteria requires much higher field strengths than that of big plant protoplasts (at the same pulse duration).

In this connection the relationship between pulse duration and critical electric field strength is practically important, but there is no quantitative general expression for the functional dependence between these parameters. Therefore, it became common practice to study the survival of cells after treatment with electric pulses of increasing field strength, as proposed in the original electric gene transfer experiments (1,2). The resulting 'survival curves' for given conditions (pulse duration and shape, medium, temperature) are the basis for selecting electroporation conditions for the actual gene transfer experiments. A typical example of a survival curve and of the field strength dependence of electrotransformation is shown in Fig.2. It has to be noted that there are noticeable differences between short-term survival (judged, e.g., within 2 h after pulsing by dye exclusion test) and long-term growth ability of electrically treated cells (26,27). Long-term stability is needed for successful electrotransfection.

Numerous reports have established that a relative optimum in electrotransfection efficiency exists at field strengths yielding survival rates between 80 and 30 % (1,2,15,22,23,26,27,33). These results can be qualitatively rationalized in the following way. If the field strength of the electric pulses is increased, reversible electroporation precedes the irreversible membrane permeability changes (break-down) leading to cell death. Successful electroporative gene transfer only occurs in the range of reversible electric field effects (reversible electroporation). Cell populations are usually non-synchronized and often have a fairly broad size distribution. A field pulse of given strength is only critical for the subpopulation of cells of proper size; this portion will become electrotransfected successfully. The lethal effects indicate the upper limit of the reversibility range of the electric field pulse effects.

Electric field strength limitations were critical in some experiments to electrotransform small objects like bacteria or yeast (21,23,32); the optimum field strength for maximum transformation efficiency was not revealed, partly because of technical reasons. We also may mention early attempts on electrotransformation of Bac. subtilis (Berg and Grunow, 1979, unpublished; Neumann and Süßmuth, 1982, unpublished) which were not successful with short µs-pulses in the field strength range available (up to about 30 to 40 kV/cm).

One recent example from our laboratory may illustrate the typical approach to elaborate suitable electroporation conditions. In attempts to electrotransform gram-positive amino acid producing Corynebacteria (Wolf, Pühler and Neumann, 1987, unpublished). At first, the survival of osmotically sensitized cells (spheroplasts) has been investigated at the pulse durations of 5 µs, 300 µs, and 5 ms; different media as well as different equipments had to be used. It was noticed that pulses of

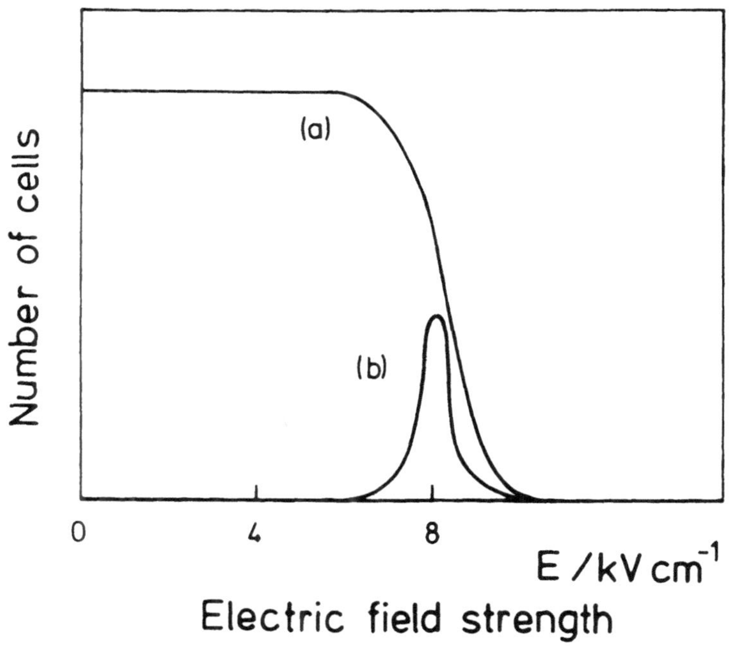


Fig. 2 Scheme (a) of survival curve and (b) of the number of electrotransformed cells.

even up to $E_0$ = 40 kV/cm with the 5 µs-pulses did not affect the viability of the bacterial spheroplasts. In contrast, with the longer pulse durations lethal effects were observed within the field strength ranges available (up to 25 kV/cm for the 300 µs-pulses and up to 6 kV/cm for the 5 ms-pulses). At field strengths of 20 kV/cm and 300 µs-pulses 60 % survival and stable transformants were obtained after regeneration and selection.

Electroporation Medium. Already in the very first electrotransfections of mammalian cell lines highly conductive media were used: HEPES buffered saline (HBS) (2) or PBS (1,14); they are now standard media (without divalent ions added). The cells are usually washed and resuspended just before electropermeabilization in one of these media. After mixing with the plasmid DNA the suspension is pulsed ('zapped') and, after an incubation period (see below), transferred to the growth medium.

The influence of divalent cations ($Mg^{2+}$,$Ca^{2+}$) on the transfection efficiency was investigated (1,2,15,30). While $Mg^{2+}$ ions increase the DNA adsorption on the cell surfaces, the actual overall electrotransfection efficiency is largest without addition of these ions (1,2). Therefore, divalent ions usually are not added to mammalian cells .

In the case of plant protoplasts, it has been found that, besides the necessary osmotic stabilization by sugars such as mannitol most oftenly 4 mM $CaCl_2$ is favourable (15). Hibi et al. (30) attempted to study systematically the effect of $Mg^{2+}$ and $Ca^{2+}$ ions on the electrotransfection of tobacco protoplasts by TMV RNA. Because of technical limitation they could only check the concentration range up to 100 $\mu$M; there was no effect of $Ca^{2+}$ ions, confirming Nishiguchi et al. (31) who found that there is no dependence on $Ca^{2+}$ ions up to 8 mM. On the other hand, $Mg^{2+}$ ions at 100 $\mu$M had a small promotive effect. The optimized electrotransformation protocol by Shillito et al. (20) for tobacco leaf protoplasts applies 6 mM $MgCl_2$. $Mg^{2+}$ ions have been reported to be essential for electric field mediated transformation of carrot protoplasts (28). In summary, with plant protoplasts usually some divalent ions up to about 5 mM are added to the electroporation medium.

The nature of the buffer used (e.g., PBS, HBS, or MES) has no marked influence on tobacco protoplasts if TMV RNA is electrically introduced. The yield is also independent of the pH value between pH 5 to 8 (35). Fromm et al. (15) found a dependence of the efficiency of transient expression (CAT activity) in electrotransformed carrot protoplasts on PBS concentration. However, this dependence most probably reflects the different pulse times. Therefore the different survival rates result from the increased electric resistance by dilution of the PBS buffer; see eq. (1). Chu et al. (33) studied the dependence of the CAT activity of mammalian cells after electroporative gene transfer on salt concentration in a HEPES buffer. The salt concentration was varied between 25 and 100 mM NaCl keeping the osmolarity constant by the addition of sucrose. The highest salt concentration (100 mM) gave the strongest CAT signal, and standard HBS buffer with 137 mM NaCl was even more suitable. It is fair to say that at present the role of different pulse time constants due to changing salt concentrations remains unclear.

Besides of proper choice of the electroporation medium for pulsing, special medium conditions may also be favorable for the membrane resealing process after pulsing, as has been claimed by Stopper et al. (25); this study beautifully confirms the early electroporation data of mouse L cells by Wong and Neumann (1) and Neumann et al. (2).

Incubation Times and Temperature. It was already noticed in the very first electroporative gene transfer experiments (2) that a post-incubation period for the cell-DNA mixture of up to 10 min after the electric field pulse application is necessary for effective gene transfer resulting in gene expression. This clearly indicates that the DNA uptake is an after-field effect which is triggered by the short-duration electric field pulse.

The longevity of the reversibly electropermeabilized membrane state is well-known (2) and has been studied systematically by following the uptake of normally impermeable molecules (drugs, dyes), added to the cells at different times after pulse application; see, e.g., (36,37). A strong temperature dependence of the resealing process has been found. It has been repeatedly reported that at low temperature the electropermeabilized state of the cell membranes can be maintained for minutes and even up to an hour (38). The slow resealing at low temperature fits well the observation made in electrotransfection experiments (14); the efficiency of gene expression can be increased by pulsing and post-incubation at low temperature. Thus, electroporative gene transfer experiments are usually performed at 0° to 4°C. According to a standard protocol oftenly used, the washed cells are resuspended in 'ice-cold' electroporation buffer and the plasmid DNA is added. Then this mixture is incubated for 5 to 10 min and one or several pulses are applied. After a post-incubation of about 10 min (all at the low temperature) the cells are transferred to growth medium at the necessary elevated temperature.

This protocol also reflects some of the supposed steps in the overall electrotransfection and electrotransformation procedures: (i) The necessary DNA/cell contact is established within the pre-incubation period by DNA adsorption on the cell membrane surface. Adsorption may be enhanced either by charge neutralization through the binding of divalent cations or by pretreatment with enzymes partially removing the extracellular matrix or cell walls. (ii) DNA transport across the cell membrane is initiated by one or several subsequent short electric field pulses. (iii) The actual permeation is a sequence of fairly slow (ms- to min-range) but still unknown 'transfer-steps' within the post-incubation period. (iv) The DNA transport is terminated by membrane resealing at elevated temperature. (v) If the DNA is inside the cell, obviously intracellular events lead to actual gene expression.

<u>Number of Electric Field Pulses.</u> The possible advantage of the application of several subsequent field pulses (e.g., 3 to 5 pulses at time intervals of several seconds) is not yet clear. Two simple effects might be envisaged in this respect. First, a mere summation of the favorable events might occur if, after the second pulse, other independent sites of DNA/cell contacts become electroporated and thus involved in DNA-transfer steps. Second, if the time between two pulses is small compared to the rotational diffusion time of the cells, a second pulse could enhance or accelerate, at the same electropermeabilized patch, the sequence of transfer steps following an event triggered by the first pulse. Most probably this has been the case in the study by Hama-Inaba et al. (27), where the time interval between two pulses was only 50 ms. Here, the efficiency in the double-pulse experiment was higher than that in an optimized single-pulse experiment.

<u>Linearization of plasmid DNA.</u> If tolerable, plasmid DNA should be linearized for electroporative gene transfer experiments. An enhancement of the gene expression yield between 3- and 20-fold with linearized plasmid DNA compared to the circular or supercoiled form

has been observed by several authors for different kinds of cells, for transient expression as well as for stable transformation (2,14,20,25,39,40). These observations point to a possible mechanism of electroporative DNA transfer and might be correlated with the size of the electropores or 'electrocracks' (41) necessary for transfer steps. It is expected that the permeation sites for the linear DNA molecules (fitting the diameter of a hydrated double-helical DNA 'rod') can be smaller than those necessary for a supercoiled, i.e., more 'globular', tertiary structure.

Addition of Carrier DNA. It is well-known from classical transfection, transformation, and microinjection procedures that the addition of 'carrier' DNA, i.e., of non-specific high-molecular DNA, leads to an increase in the level of gene expression of the specific plasmid transferred (42). A similar positive effect of the addition of carrier DNA to the cell-plasmid DNA mixture was also established in electrotransfection experiments (20,28,33). The nature of the carrier DNA was found to be important (33), in line with the data of other, non-electroporative cell transfection and transformation methods.
The mechanism of the effect of carrier DNA is still unknown. It is argued that it is linked to the activity of intracellular nucleases. In the presence of excess carrier DNA a sufficient amount of the specific foreign plasmid DNA obviously is preserved from cleavage.

## Conclusions

Recent progress in molecular genetics and in biotechnology demostrates that efficient and reliable gene transfer is an important prerequisite. It appears that direct transfer of DNA or mRNA by electroporation is not only simple and readily controlled but also generally applicable. The efficiency of the electric technique in many cases compares well with, and in several cases of special interest is clearly superior to, the chemical methods.
One of the main aims of further improvements of DNA transfer techniques is the development of conditions such that cells can be transformed

without excessive pretreatment, avoiding, e.g., protoplast formation. Electroporative gene transfer, to our opinion, is a likely candidate to realize these goals. In addition, easy and reliable handling of the electroporation method together with further technical development of equipment will make this method even more attractive to a broad range of cell biologists and biotechnologists.

## Acknowledgements

We gratefully acknowledge financial support by the DFG (grant NE 227/4 to E.N.).

## References

1. Wong, T.-K., Neumann, E., 1982. Electric field mediated gene transfer. Biochem. Biophys. Res. Commun. 107, 584-587.
2. Neumann, E., Schäfer-Ridder, M., Wang, Y., Hofschneider, P.H., 1982. Gene transfer into mouse lyoma cells by electroporation in high electric fields. EMBO J. 1, 841-845.
3. Neumann, E., Rosenheck, K., 1972. Permeability changes induced by electric impulses in vesicular membranes. J. Membr. Biol. 10, 279-290.
4. Neumann, E., Rosenheck, K., 1973. Potential differences across vesicular membranes. J. Membr. Biol. 14, 194-196.
5. Rosenheck, K., Lindner, P., Pecht, I., 1975. Effect of electric fields on light scattering and fluorescence of chromaffin granules. J. Membr. Biol. 20, 1-12.
6. Sale, A.J.H., Hamilton, W.A., 1967. Effects of high electric fields on microorganisms. I. Killing of bacteria and yeasts. Biochim. Biophys. Acta 148, 781-788.
7. Hamilton, W.A., Sale, A.J.H., 1967. Effects of high electric fields on microorganisms. II. Mechanism of action of the lethal effect. Biochim. Biophys. Acta 148, 789-800.
8. Sale, A.J.H., Hamilton, W.A., 1968. Effects of high electric fields on microorganisms. III. Lysis of erythrocytes and protoplasts. Biochim. Biophys. Acta 163, 37-43.
9. Eigen, M., DeMaeyer, L., 1963. Relaxation Methods. In: Techniques of Organic Chemistry, Volume 8(2) (S.L. Friess, E.S. Lewis, A. Weissberger, eds.). John Wiley, New York, pp. 895-1054.
10. Ilgenfritz, G., 1966. Chemische Relaxation in starken elektrischen Feldern, Dissertation, Universität Göttingen.

11. Grünhagen, H.H., 1973. Fast Spectrophotometric Detection System for Coupled Physical and Chemical Electric Field Effects in Solution. Biophysik 10, 347-354.
12. Lindner, P., Neumann, E., Rosenheck, K., 1977. Kinetics of permeability changes induced by electric impulses in chromaffin granules. J. Membr. Biol. 32, 231-254.
13. Kinosita, K., Jr., Tsong, T.Y., 1977. Hemolysis of human erythrocytes by a transient electric field. Proc. Natl. Acad. Sci. 74, 1923-1927.
14. Potter, H., Weir, L., Leder, P., 1984. Enhancer-dependent expression of human $\kappa$ immunoglobulin genes introduced into mouse pre-B lymphocytes by electroporation. Proc. Natl. Acad. Sci. 81, 7161-7165.
15. Fromm, M., Taylor, L.P., Walbot, V., 1985. Expression of genes transferred into monocot and dicot plant cells by electroporation. Proc. Natl. Acad. Sci. 82, 5824-5828.
16. Zimmermann, U., Pilwat, G., Riemann, F., 1974. Dielectric breakdown of cell membranes. Biophys. J. 14, 881-899.
17. Berg, H., Augsten, K., Bauer, E., Förster, W., Jacob, H.-E., Mühlig, P., Weber, H., Kurischko, A., 1984. Possibilities of cell fusion and transformation by electrostimulation. Bioelectrochem. Bioenerg. 12, 119-133.
18. Evans, G.A., Ingraham, H.A., Lewis, K., Cunningham, K., Seki, T., Moriuchi, T., Chang, H.C., Silver, J., Hyman, R., 1984. Expression of the Thy-1 glycoprotein gene by DNA-mediated gene transfer. Proc. Natl. Acad. Sci. 81, 5532-5536.
19. Auer, D., Brandner, G., Bodemer, W., 1976. Dielectric breakdown of the red blood cell membrane and uptake of SV 40 DNA and mammallian cell RNA. Naturwiss. 63, 391-391.
20. Shillito, R.D., Saul, M.W., Paszkowski, J., Müller, M., Potrykus, I., 1985. High efficiency direct gene transfer to plants. Bio/Technology 3, 1099-1103.
21. Shivarova, N., Förster, W., Jacob, H.-E., Grigorova, R., 1983. Microbiological implications of electric field effects. VII. Stimulation of plasmid transformation of Bacillus cereus protoplasts by electric field pulses. Z. Allg. Mikrobiol. 23, 595-599.
22. Hashimoto, H., Morikawa, H., Yamada, Y., Kimura, A., 1985. A novel method for transformation of intact yeast cells by electroinjection of plasmid DNA. Appl. Microbiol. Biotechnol. 21, 336-339.
23. Karube, I., Tamiya, E., Matsuoka, H., 1985. Transformation of Saccharomyces cerevisiae spheroplasts by high electric pulse, FEBS Lett. 182, 90-94.
24. Zerbib, D., Amalric, F., Teissie, J., 1985. Electric field-mediated transformation: Isolation and characterization of a $TK^+$ subclone. Biochem. Biophys. Res. Commun. 129, 611-618.
25. Stopper, H., Zimmermann, U., Wecker, E., 1985. High yield of DNA-transfer into mouse L-cells by electropermeabilizition. Biosciences 40, 929-932.
26. Riggs, C.D., Bates, G.W., 1986. Stable transformation of tobacco by electroporation: Evidence for plasmid concatenation. Proc. Natl. Acad. Sci. 83, 5602-5606.

27. Hama-Inaba, H., Shiomi, T., Sato, K., Ito, A., Kasai, M., 1986. Electric pulse-mediated gene transfer in mammalian cells grown in suspension culture. Cell Structur and Function 11, 191-198.
28. Langridge, W.H.R., Li, B.J., Szalay, A.A., 1985. Electric field mediated stable transformation of carrot protoplasts with naked DNA. Plant Cell Reports 4, 355-359.
29. Ou-Lee, T.-M., Turgeon, R., Wu, R., 1986. Expression of a foreign gene linked to either a plant-virus or a Drosophila promoter after electroporation of protoplasts of rice, wheat, and sorghum. Proc. Natl. Acad. Sci. 83, 6815-6819.
30. Hibi, T., Kano, H., Sugiura, M., Kazami, T., Kimura, S., 1986. High efficiency electro-transfection of tobacco mesophyll protoplasts with TMV RNA. J. General Virology 67, 2037-2042.
31. Nishiguchi, M., Langridge, W.H.R., Szalay, A.A., Zaitlin, M., 1986. Electroporation-mediated infection of tobacco leaf protoplasts with tobacco mosaic virus RNA and cucumber mosaic virus RNA. Plant Cell Reports 5, 57-60.
32. Shigekawa, K., 1987. Gene-Transfer in Prokaryotic and Eukaryotic Cells by Electroporation. In: Bio-Rad Labs, chapter 2, pp. 1-14.
33. Chu, G., Hayakawa, H., Berg, P., 1987. Electroporation for the efficient transfection of mammalian cells with DNA. Nucl. Acid. Res. 15, 1311-1326.
34. Fromm, M.E., Taylor, L.P., Walbot, V., 1986. Stable transformation of maize after gene transfer by electroporation. Nature 319, 791-793.
35. Okada, K., Nagata, T., Takebe, I., 1986. Introduction of functional RNA into plant protoplasts by electroporation. Plant Cell Physiol. 27, 619-626.
36. Jacob, H.-E., Förster, W., Berg, H., 1981. Microbiological implications of electric field effects. II. Inactivation of yeast cells and repair of their cell envelope. Z. Allg. Mikrobiol. 21, 225-233.
37. Teissie, J., Rols, M.P., 1986. Fusion of mammalian cells in culture is obtained by creating the contact between cells after their electropermeabilization. Biochem. Biophys. Res. Commun. 140, 258-266.
38. Knight, D.E., Baker, P.F., 1982. Calcium-dependence of catecholamine release from bovine adrenal medullary cells after exposure to intense electric fields. J. Membrane Biol. 68, 107-140.
39. Falkner, F.G., Neumann, E., Zachau, H.G., 1984. Tissue specificity of the initiation of immunoglobulin $\kappa$ gene transcription. Hoppe-Seyler's Z. Physiol. Chem. 365, 1331-1343.
40. Toneguzzo, F., Hayday, A.C., Keating, A., 1986. Electric field-mediated DNA transfer: Transient and stable gene expression in human and mouse lymphoid cells. Mol. Cell. Biol. 6, 703-706.
41. Sugar, I.P., Förster, W., Neumann, E., 1987. Model of cell electrofusion. Membrane elecroporation, pore coalescence and percolation. Biophysical Chemistry 26, 321-335.
42. Krens, F.A., Molendijk, L., Wullems, G.J., Schilperoort, R.A., 1982. In vitro transformation of plant protoplasts with Ti-plasmid DNA. Nature 296, 72-74.

AUTHOR INDEX

## Subject Index

Bei Fragen zur Produktsicherheit wenden Sie sich bitte an:
If you have any questions regarding product safety,
please contact:

Walter de Gruyter GmbH
Genthiner Straße 13
10785 Berlin
productsafety@degruyterbrill.com